Springer Series in Operations Research
and Financial Engineering

The Springer Series in Operations Research and Financial Engineering publishes monographs and textbooks on important topics in theory and practice of Operations Research, Management Science, and Financial Engineering. The Series is distinguished by high standards in content and exposition, and special attention to timely or emerging practice in industry, business, and government. Subject areas include: linear, integer, and non-linear programming including applications; dynamic programming and stochastic control; interior point methods; multi-objective optimization; supply chain management, including inventory control, logistics, planning and scheduling; game theory risk management and risk analysis, including actuarial science and insurance mathematics; queuing models, point processes, extreme value theory, and heavy-tailed phenomena; networked systems, including telecommunication, transportation, and many others; quantitative finance: portfolio modeling, options, and derivative securities; revenue management and quantitative marketing innovative statistical applications such as detection and inference in very large and/or high dimensional data streams; computational economics.

Antonius B. Dieker • Steven T. Hackman

QPLEX: A Computational Modeling and Analysis Methodology for Stochastic Systems

Springer

Antonius B. Dieker
Department of Industrial Engineering and
Operations Research
Columbia University
New York, NY, USA

Steven T. Hackman
H. Milton Stewart School of Industrial and
Systems Engineering
Georgia Institute of Technology
Atlanta, GA, USA

ISSN 1431-8598   ISSN 2197-1773   (electronic)
Springer Series in Operations Research and Financial Engineering
ISBN 978-3-031-74869-1   ISBN 978-3-031-74870-7   (eBook)
https://doi.org/10.1007/978-3-031-74870-7

Mathematics Subject Classification: 90-10, 90B15, 90B22, 62H22, 65C40, 60J22, 68U99

This Springer imprint is published by the registered company Springer Nature Switzerland AG
The registered company address is: Gewerbestrasse 11, 6330 Cham, Switzerland

If disposing of this product, please recycle the paper.

To our families

Preface

This book develops a computational methodology for modeling and analyzing a broad class of nonstationary stochastic systems with large state spaces. We call it *QPLEX*. It circumvents the curse of dimensionality inherent to working with large state spaces by imposing conditional independence and exploiting model dynamics. The ability to work directly with empirical distributions is baked into the methodology. We believe the QPLEX methodology can provide a foundation to support software for designing and controlling many practical stochastic systems.

Approach The QPLEX methodology does not rely on taking limits or solving stochastic differential equations, nor does it require characterizations of steady-state distributions. Instead, it iteratively generates distributions supported on a set with much lower cardinality than the state space. Functions of these output distributions yield approximate distributions of system performance over time. The same input always produces the same output distributions. Since QPLEX output is *not* noisy, it is tailor-made for sensitivity analysis and optimization, for instance, via gradient descent.

QPLEX output can be generated quickly and can result in excellent approximation quality. *Thousands* of transient distributions can be calculated in just a few seconds. For a wide variety of challenging model instances of a fundamental multiserver queueing system, the approximate distributions of the number of customers in the system and the virtual waiting time at each time epoch are essentially *indistinguishable* from the exact distributions.

We exploit fundamental work on *probabilistic graphical models* widely used in statistics and machine learning. The role played by data, however, is fundamentally different. Instead of automatically detecting patterns in data or making inferences from observed data, data are used as model input.

Model Specification The QPLEX methodology requires relatively few abstract model primitives to specify a model instance. These model primitives are flexible enough to specify a rich array of models. We draw examples mainly from the stochastic networks area since it provides a rich class of examples and we know it best. For instance, we will encounter classical multiserver queueing models, load balancing models, and queueing networks with blocking. These models can exhibit challenging characteristics, such as a relatively short operational horizon, time-varying arrival rates with periods where demand is far greater than resource capacity, time-varying general or data-driven service duration distributions, time-varying small-to-moderate numbers of servers, and complex routing of entities.

We introduce *distributional programs* as a notational device to specify models. The body of a distributional program is pseudocode that describes a stochastic simulation. The output of a distributional program is *deterministic* and represents the distribution of a sample produced by this simulation. It is often easier to read and understand distributional programs than lengthy algebraic expressions of the many variables in a typical model. Distributional programs separate specification and implementation yet serve as a bridge between the two.

Intended Audience Our presentation is intended to reach an audience with engineering, computer science, or mathematics background. Knowledge of probability and stochastic models taught at the advanced undergraduate level should suffice. We do not shy away from being precise using formal definitions, mathematical constructs, and proofs, but we have de-emphasized formalism where possible. Since we use computational tools that may not be familiar to the stochastic modeling community, we have made this book as self-contained as possible. We provide many examples to help the reader understand the concepts and calculations.

Organization This book begins with two prequel chapters. Chapter 1 sets forth the notation and terminology we use throughout this book. Chapter 2 develops the QPLEX methodology for a multiserver queueing model. This model has many practical applications in its own right and is a building block for subsequent models. It is simple enough to motivate, develop, and justify our efficient computational procedure from first principles without introducing the abstractions necessary for the general methodology.

The rest of the book consists of three parts. Each part begins with an introductory chapter that presents key terminology and an overview of the concepts developed in that part.

In Part I, we present model primitives, give model examples, and derive what we call the *QPLEX calculus* within two abstract setups. The main objective of Part I is to circumvent what we call the *curse of dimensionality for histograms* arising from letting go of the exponential distributions underlying the Markov paradigm.

In Part II, we present a methodology to reduce computational requirements by using graphs to represent conditional independence relationships among variables, which results in what we call the *graphical QPLEX calculus*. The main objective of Part II is to circumvent what we call the *curse of dimensionality for counters* arising from too many state variables.

In Part III, we present theoretical results that establish a foundation for the QPLEX methodology. We establish optimality properties of the QPLEX output and present sufficient conditions for exactness.

Acknowledgments We are grateful for the supportive environments at Georgia Tech and Columbia. We offer our heartfelt thanks to colleagues and friends at our institutions and elsewhere for their encouragement and interest. We are also grateful to the organizers of several conferences and seminar series who provided us with opportunities to present early versions of this work; we very much benefited from interactions with audience members. We also thank the students from the Fall 2023 Ph.D. course *Computational Stochastic Modeling* for their helpful comments and participation. Bob Foley, Dave Goldberg, Henry Lam, Dick Serfozo, and Craig Tovey each provided much-needed critical feedback and suggestions that have proven to be extremely helpful. Finally, we owe an immense debt of gratitude to Loren Platzman. He provided insightful editorial and technical suggestions on many drafts and presentations that proved invaluable in sharpening the exposition of our work. He also generously volunteered to create prototype software, which we used extensively throughout this project.

Disclaimer Since every sentence has been rewritten by each author seemingly infinitely often, each author confidently blames the other for any typos or mistakes that remain.

New York, NY, USA   Antonius B. Dieker
Atlanta, GA, USA   Steven T. Hackman
April 2024

Contents

Chapter 1
Preliminaries

This chapter introduces common notation and terminology used in this book. We focus on conventions we use in this book that are nonstandard in the operations research literature or for which there is no uniformly accepted standard. Sections 1.6–1.8 are not needed until Part II.

1.1 Sets and Functions

Sets The empty set is denoted by the symbol $\emptyset$. Fix two sets S and T. We write $S \subseteq T$ to mean that S is a subset of T (but could be equal to T) and write $S \subset T$ to mean that $S \subseteq T$ and $S \neq T$, i.e., that S is a proper subset of T. We write $S \backslash T$ for the set of elements of S that are not elements of T. A *partition* of a set S is a collection of sets $\{S_k\}_{k=1}^{K}$ such that $\bigcup_{k=1}^{K} S_k = S$ and $S_k \cap S_j = \emptyset$ for every distinct k and j.

For $M \geq 1$, the *Cartesian product* of the sets $X_1, \ldots, X_M$ is the set

$$\{(x_1, \ldots, x_M) : x_i \in X_i \text{ for every } i = 1, \ldots, M\}$$

and is denoted by $X_1 \times \cdots \times X_M$. (We also use "$\times$" for scalar multiplication.) To stress that $(x_1, \ldots, x_M)$ is an element of some Cartesian product, we often use boldface ("vector") notation and write x as shorthand for $(x_1, \ldots, x_M)$. Given some set $A \subseteq \{1, \ldots, M\}$ and some $x \in X_1 \times \cdots \times X_M$, x_A denotes the subvector of x associated with the indices in the set A. If $\{A_k\}_{k=1}^{K}$ is a partition of $\{1, \ldots, M\}$, then we may write x as $(x_{A_1}, \ldots, x_{A_K})$.

Functions and Maps Fix some function $f : X \to Y$. The sets X and Y are called the *domain* and *codomain* of f, respectively. We use the word *map* in place of function when the domain is a set of functions. For some set $\tilde{X} \subseteq X$, the *image*

© The Author(s) 2025
A. B. Dieker, S. T. Hackman, *QPLEX: A Computational Modeling and Analysis Methodology for Stochastic Systems*, Springer Series in Operations Research and Financial Engineering, https://doi.org/10.1007/978-3-031-74870-7_1

of $\tilde{X}$ under f is the set $\{f(x) : x \in \tilde{X}\}$ and is denoted by $f(\tilde{X})$. The set $f(X)$ is called the *range* of f. For $f(\tilde{X}) \subseteq \tilde{Y}$, we call the function $h : \tilde{X} \to \tilde{Y}$ defined by $h(x) = f(x)$ for $x \in \tilde{X}$ the $(\tilde{X}, \tilde{Y})$-*restriction* of f.

For functions $f : X \to \tilde{Y}$ and $g : Y \to Z$ such that $\tilde{Y} \subseteq Y$, the *composition* of g and f is the function $h : X \to Z$ defined by $h(x) = g(f(x))$ for $x \in X$ and is denoted by $g \circ f$. We write $(g \circ f)(x)$ for the image of x under $g \circ f$.

A function $f : X \to Y$ is called a *bijection* if there exists a function $g : Y \to X$ such that $(g \circ f)(x) = x$ for all $x \in X$ and $(f \circ g)(y) = y$ for all $y \in Y$. The function g is called the *inverse* of f. If f is a bijection, then the inverse g of f is also a bijection, and f is the inverse of g. Moreover, we have that $f(X) = Y$ and $g(Y) = X$.

Let e denote some logical expression involving one or more variables, such as $x + y < 4$. We write $1(e)$ to denote the function of these variable(s) that assigns one if e is true and zero if e is false, with its domain following from the context. We call such a function an *indicator function*.

Scopes Fix some $M \geq 1$, some sets $X_1, \ldots, X_M, Y$, and some function $f : X_1 \times \cdots \times X_M \to Y$. We say that the set $A \subseteq \{1, \ldots, M\}$ is a *scope* of f if $f(x^1) = f(x^2)$ for all elements x^1 and x^2 of $X_1 \times \cdots \times X_M$ with $x_A^1 = x_A^2$. Loosely speaking, the function f only uses the variables with indices in A, as knowledge of the other variables is irrelevant. For example, $\{2, 4\}$ is a scope of the function g on $\mathbb{R}^4$ defined via $g(x_1, x_2, x_3, x_4) = x_2 + x_4$. Note that $\{1, \ldots, M\}$ is trivially a scope of f and that $\emptyset$ is a scope of f if and only if f is a constant function.

The set of scopes of f is closed under intersection, namely, the intersection $A \cap B$ of two scopes A and B of f is itself a scope of f. To verify this claim, let x^1 and x^2 be two arbitrary elements of $X_1 \times \cdots \times X_M$ such that $x_{A \cap B}^1 = x_{A \cap B}^2$. We must show that $f(x^1) = f(x^2)$. Define $x \in X_1 \times \cdots \times X_M$ via

$$
x_i = \begin{cases} x_i^1 & \text{if } i \in A \backslash B \\ x_i^1 = x_i^2 & \text{if } i \in A \cap B \\ x_i^2 & \text{otherwise.} \end{cases}
$$

By construction, $x_A = x_A^1$ and so $f(x) = f(x^1)$, and similarly $x_B = x_B^2$ and so $f(x) = f(x^2)$, which shows that $f(x^1) = f(x^2)$, as required. Henceforth, when we refer to *the* scope of a function, it is understood that we mean the intersection of all scopes of the function. If the scope of f is A, then we sometimes abuse notation and write $f(x_A)$ instead of $f(x)$ in order to emphasize this property.

We often work with functions of variables represented by different letters, and writing the scope of such a function as a subset of indices can be a source of potential confusion. Consider, for example, the function on $\mathbb{R}^5$ defined as $f(z_1, z_2, y_1, y_2, y_3) = z_2 \times y_3$. For such a function, we write the scope as $\{z_2, y_3\}$. We stress that the z_2 and y_3 here should be viewed as tokens rather than values of variables, although we use the same symbols.

1.2 Probability Mass Functions

Definition and Terminology In this book, a *probability mass function* p is a nonnegative real-valued function on some countable set B with $\sum_b p(b) = 1$. We use the abbreviation *pmf* for probability mass function (plural *pmfs*). We speak of a *pmf on B* or a *pmf of Z* when we interpret p as a pmf of some B-valued random variable Z. This book focuses on discrete random variables, so we use "distribution" and "pmf" interchangeably. If the set B is a Cartesian product, a pmf on B can be thought of as the pmf of some random vector $\mathbf{Z}$. We then also say that p is a *pmf of $\mathbf{Z}$*. If $\mathbf{Z} = (Z_1, \ldots, Z_M)$ for some $M \geq 1$, then we also say that p is a *pmf of $Z_1, \ldots, Z_M$*.

If B is a set of integers and is clear from the context, we can identify a pmf with a vector. For instance, if $B = \{1, 2, 3, 4\}$, then $p = (0.5, 0, 0.4, 0.1)$ is shorthand for the pmf on B with $p(1) = 0.5$, $p(2) = 0$, $p(3) = 0.4$, and $p(4) = 0.1$. As we consider much more general sets B, we view pmfs as functions rather than vectors, so we do not use boldface notation for pmfs.

We often speak of pmfs on sets and pmfs of random variables (or vectors) interchangeably, even when we have not introduced any underlying random variables. For instance, suppose we introduce p as a pmf on some set B with elements written as (z_1, z_2). Even if B itself is not explicitly written as a Cartesian product of two sets, it should then be understood that p is a pmf of Z_1 and Z_2. More generally, the underlying random variables are written as the elements of B but in upper case.

Notational Conventions As we work with many random variables in this book, we introduce additional notation for working with multivariate pmfs that is slightly different from the customary notation in operations research to reduce notational clutter significantly. Conventional notation uses subscripts to identify marginal and conditional pmfs, but in this book, we use subscripts for other purposes, such as naming a pmf or emphasizing some functional dependence. We therefore employ notation used in some machine learning communities to distinguish between marginal, conditional, and unconditional pmfs. We simply refer to such functions as pmfs, where it follows from the context if they are marginal or conditional pmfs or neither. (The current section *does* use this terminology as we introduce our conventions.) These notational conventions can be viewed as appropriately combining conventional notation for probability measures and pmfs. We use the phrase "under the pmf p" or simply "under p" to mean "under the probability measure induced by the pmf p." Furthermore, we refer to the support of the probability measure induced by p simply as the support of a pmf p. If p is a pmf of some random variable Z or random vector $\mathbf{Z}$, then we also speak of the support of Z and $\mathbf{Z}$ (under p), respectively.

We illustrate our conventions for a pmf p of three random variables Z_1, Z_2, and Z_3, with the understanding that the conventions for any number of random variables (or random vectors) work similarly.

- The expressions $p(z_1)$, $p(z_1|z_2)$, and $p(z_1|z_2, z_3)$ stand for the marginal pmf of Z_1, the conditional pmf of Z_1 given $Z_2 = z_2$, and the conditional pmf of Z_1 given $Z_2 = z_2$ and $Z_3 = z_3$, viewed as functions of z_1, (z_1, z_2), and (z_1, z_2, z_3), respectively. As a result of this convention, we have, for instance, that

$$p(z_1) = \sum_{z_2, z_3} p(z_1, z_2, z_3) = \sum_{z_2} p(z_1, z_2)$$

and, whenever $p(z_1) > 0$, that

$$p(z_2|z_1) = \frac{p(z_1, z_2)}{p(z_1)}.$$

The lowercase letters can be thought of as "tokens" that indicate which marginal or conditional pmf the symbol p refers to rather than real numbers. For instance, $p(z_1)$ stands for a different function than $p(z_2)$ and should not be thought of as evaluating a pmf p of a single variable at two values z_1 and z_2.

We frequently construct a sequence of pmfs of the *same* random variables via an iterative scheme. To describe such iterative schemes with compact notation, we allow for the lowercase letters in the above convention to have a prime. For instance, without introducing new random variables Z_1', Z_2', and Z_3', we can define another pmf p' of Z_1, Z_2, and Z_3 (say) via

$$p'(z_1', z_2', z_3') = \sum_{z_1, z_2, z_3} p(z_1, z_2, z_3) \times \psi(z_1, z_2, z_3, z_1', z_2', z_3'),$$

where ψ needs to ensure that p' is a pmf.
- Instead of giving an abstract rule, the following representative examples illustrate how we evaluate (conditional) pmfs.

  - $p(Z_1 = 3)$ stands for the marginal pmf of Z_1 evaluated at 3.
  - $p(z_1, Z_2 = 2, z_3)$ stands for the joint pmf of Z_1, Z_2, Z_3, evaluated at $(z_1, 2, z_3)$, where z_1 and z_3 are typically fixed.
  - $p(z_1|Z_2 = 2)$ stands for the conditional pmf of Z_1 given $Z_2 = 2$.
  - $p(Z_1 = 1|z_2)$ stands for the conditional pmf of Z_1 given $Z_2 = z_2$ evaluated at 1, where z_2 is typically fixed.
  - $p(Z_1 = 1|Z_2 = 2, z_3)$ stands for the conditional pmf of Z_1 given $Z_2 = 2$ and $Z_3 = z_3$ evaluated at 1, where z_3 is typically fixed.

- To clarify which variables are fixed and which are free, we sometimes use "·" in place of one or more variables. For instance, $p(Z_1 = \cdot)$, $p(Z_1 = \cdot|z_2)$, and $p(Z_1 = \cdot|z_2, z_3)$ stand for the marginal pmf of Z_1, the conditional pmf of Z_1 given $Z_2 = z_2$, and the conditional pmf of Z_1 given $Z_2 = z_2$ and $Z_3 = z_3$, respectively. Here z_2 and z_3 are fixed.
- We frequently use our notation for a conditional pmf without having an underlying joint pmf on all variables used. For instance, we may use the notation

$p(z_1|z_2, z_3)$ without having an underlying joint pmf $p(z_1, z_2, z_3)$. In that case, we mean for $p(z_1|z_2, z_3)$ to be interpreted as a pmf of Z_1 that is parameterized by z_2 and z_3, i.e., it is a nonnegative function of z_1, z_2, z_3 for which the sum over z_1 yields 1. We use this convention most often for model primitives, in which case it is understood that this pmf only needs to be specified for values of z_2 and z_3 that are relevant in the context of the model.

Consistency of (Sets of) Pmfs Two multivariate pmfs are *consistent* if the intersection of their scopes is empty or if the marginal pmfs corresponding to this intersection are equal. For instance, two pmfs p and q of (Z_1, Z_2) and (Z_1, Z_3), respectively, are consistent if, for each possible value z_1 of Z_1,

$$\sum_{z_2} p(z_1, z_2) = \sum_{z_3} q(z_1, z_3).$$

Here, the scopes of p and q are $\{z_1, z_2\}$ and $\{z_1, z_3\}$, respectively, so the intersection of these scopes is $\{z_1\}$. Both the left-hand side and the right-hand side are marginal pmfs of Z_1. We call a collection of pmfs consistent if each pair of pmfs in this collection is consistent.

1.3 Histograms

In this book, a *histogram h* refers to any nonnegative integer-valued function on some countable set B with $\sum_{b \in B} h(b) < \infty$. We write $|h| = \sum_{b \in B} h(b)$ for the *size* of the histogram h. The diagram below depicts a histogram with the set B taken as $\{0, 1, 2, 3\}$ and $h(0) = 4$, $h(1) = 5$, etc. Other examples of sets B we use are $\{0, 1, \ldots\}$ and $\{1, 2, \ldots\}$. We do not make the set B explicit in our notation, as it is understood from the context. As such, like pmfs, a histogram can also be represented by a vector, such as $(4, 5, 3, 2)$ in the example below. Since a histogram is technically a function, we do not use boldface notation to represent it; instead, we use boldface notation to represent a vector of histograms.

0 1 2 3

Histograms are conceptually similar to pmfs and play a key role in this book. We will frequently encounter (vectors of) histogram-valued random variables. Their pmfs are functions on (vectors of) histograms and therefore real-valued functions of (vectors of) integer-valued functions. We often work with sets of such pmfs.

1.4   Random Variables and Stochastic Processes

Common Random Variables We use several common random variables throughout this book:

- $\text{ber}(\beta)$ denotes a Bernoulli random variable with success probability β.
- $\text{bin}(m, \beta)$ denotes a binomial random variable with parameters m and β. The
  random variable $\text{bin}(0, \beta)$ should be interpreted as zero with probability 1.
- $\text{pois}(\lambda)$ denotes a Poisson random variable with parameter λ.
- $\text{hypergeom}(N, K, n)$ denotes a hypergeometric random variable with population
  size N, number of success states K, and the number of draws n.

We use the generic symbol Pr to express probabilities for these common random
variables. As examples, we have, for $0 \leq x \leq m$, that

$$\Pr[\text{bin}(m, \beta) = x] = \binom{m}{x} \times \beta^x \times (1 - \beta)^{m-x},$$

and, for $\max(0, n + K - N) \leq k \leq \min(K, n)$, that

$$\Pr[\text{hypergeom}(N, K, n) = k] = \frac{\binom{K}{k} \times \binom{N-K}{n-k}}{\binom{N}{n}}.$$

Multivariate Hypergeometric Random Variables We also need the following
generalization of the hypergeometric random variable. Consider the following
experiment. An urn contains a total of $b_1 + \cdots + b_M$ balls with different labels
such that there are b_m balls with label m. Draw $s \leq b_1 + \cdots + b_M$ balls without
replacement, and let Φ_m denote the number of balls drawn that have label m. The
probability that $(\Phi_1, \ldots, \Phi_M)$ equals $(\phi_1, \ldots, \phi_M)$ with $\phi_1 + \cdots + \phi_M = s$ is
given by

$$\frac{\binom{b_1}{\phi_1} \times \cdots \times \binom{b_M}{\phi_M}}{\binom{b_1 + \cdots + b_M}{s}}.$$

If $s > b_1 + \cdots + b_M$, then we have that $(\Phi_1, \ldots, \Phi_M) = (b_1, \ldots, b_M)$ with
probability 1. Such a random vector $(\Phi_1, \ldots, \Phi_M)$ is called a multivariate hypergeometric random variable with positive integer parameters $b_1, \ldots, b_M$ and positive
integer parameter s. We denote a generic random variable with this distribution by
multivariateHypergeometric$(b_1, \ldots, b_M; s)$. For the special case $M = 2$, both Φ_1
and Φ_2 have a (classical) hypergeometric distribution.

Multinomial Random Variables This book heavily relies on histogram-valued
multinomial random variables, which generalize the commonly used vector-valued
multinomial random variables. Given a countable set B and a pmf p on B, a
realization of the histogram-valued multinomial random variable is a histogram on

B. We write mult(m, p) for such a histogram-valued random variable, where $m \geq 0$ denotes the number of trials. For each histogram $h = (h(b))_{b \in B}$ on B, we have that

$$\Pr[\text{mult}(m, p) = h] = \begin{cases} \frac{m!}{\prod_{b \in B} h(b)!} \prod_{b \in B} p(b)^{h(b)} & \text{if } |h| = m \\ 0 & \text{otherwise.} \end{cases}$$

The random variable mult$(0, p)$ should be understood as the zero histogram on B with probability 1.

If p is a pmf on some countable set B as before, then the mean of the multinomial variable mult(m, p) is

$$\sum_h h \times \Pr[\text{mult}(m, p) = h] = m \times p,$$

which is a nonnegative function on B. Here, the sum is taken over the (generally large) set of all histograms on B with size m. More generally, if $p_1, \ldots, p_K$ are pmfs on B and $m_1, \ldots, m_K \geq 0$ are nonnegative integers, then the mean of a sum of $K \geq 1$ independent multinomial random variables mult$(m_1, p_1), \ldots,$ mult(m_K, p_K) is

$$\sum_{k=1}^{K} m_k \times p_k$$

in view of

$$\sum_{h_1, \ldots, h_K} (h_1 + \cdots + h_K) \times \prod_{k=1}^{K} \Pr[\text{mult}(m_k, p_k) = h_k]$$

$$= \sum_{k=1}^{K} \sum_{h_k} h_k \times \Pr[\text{mult}(m_k, p_k) = h_k].$$

The mean is $\sum_{k=1}^{K} m_k \times p_k$ even if independence does not hold.

Markov Chains In this book, we work in discrete time and use the symbol t to denote a time epoch, which is understood to be a nonnegative integer. We work with several discrete-time stochastic processes, which are collections of random variables indexed by time. Markov chains are the most important stochastic processes for the development to follow.

A discrete-time stochastic process $\{Y^{(t)} : t \geq 0\}$ with some countable state space Ω_Y is a discrete-time *Markov chain* if each $Y^{(t)}$ takes values in Ω_Y and it possesses the *Markov property*. This means that, for each time t, the conditional probability that $Y^{(t+1)}$ equals $y^{(t+1)}$ given

$$Y^{(0)} = y^{(0)}, Y^{(1)} = y^{(1)}, \ldots, Y^{(t)} = y^{(t)},$$

viewed as a function of $y^{(0)}, \ldots, y^{(t+1)}$, has a scope that does not contain any of the past states $y^{(0)}, y^{(1)}, \ldots, y^{(t-1)}$.

In this book, we often consider a fixed period between times t and $t + 1$. We remove superscripts for random variables and their values to reduce clutter and distinguish those for time $t+1$ from those for time t by adding a prime. For instance, when t is fixed, we denote the random variables $Y^{(t)}$ and $Y^{(t+1)}$ by Y and Y', respectively. Similarly, we denote the *(time-t one-step) transition probabilities* by $p^{(t)}(y'|y)$ instead of $p^{(t)}(y^{(t+1)}|y^{(t)})$. Note that the superscript in $p^{(t)}$ is critical; it is customary to consider time-homogeneous Markov chains where these transition probabilities do not depend on t, but this assumption is specifically *not* made in this book.

We assume that the transition probabilities $p^{(t)}(y'|y)$ are specified for *all* y and y' in the state space Ω_Y for ease of exposition. Of course, to use a Markov chain in applications, the modeler only needs to specify these probabilities for relevant y, i.e., for y in the support of $Y^{(t)}$. Since a modeler typically does not know the support of $Y^{(t)}$ in advance, we assume the modeler specifies such transition probabilities for y in a set that is sure to contain this support, thereby avoiding the question of finding the support itself.

1.5   Distributional Programs

A *distributional program* is a notational device we created to indirectly specify a pmf. The body of a distributional program is pseudocode that describes a stochastic simulation. The output of a distributional program is the pmf of a sample produced by this simulation. Distributional programming is different from probabilistic programming used in Bayesian statistics because there is no underlying statistical inference task.

Here is a simple example.

```
1: function TRUNCATIONMOVE( )
2:     x ~ bin(2, 0.2)
3:     y ← min(x, 1)
4:     return y
```

Line 2 samples x from a binomial distribution, and Line 3 assigns a function of x to y. This program outputs a binomial pmf with its right tail truncated and the truncated probability mass moved to its right endpoint. Specifically, the output is the pmf p on $\{0, 1\}$ with $p(0) = 0.64$ and $p(1) = 0.36$.

Conceptually, the output of a distributional program might be obtained by running a Monte Carlo simulation an infinite number of times. In practice, it is obtained by enumerating over the (finite number of) sample values. Efficient calculation of the output of a distributional program is the subject of Chap. 14.

We adopt distributional programs in this book for two reasons. First, due to the number of variables needed to specify the calculations that arise even in the simplest applications of our methodology, as we experienced firsthand when writing this book, conventional algebraic representations of these distributions quickly become lengthy and impose heavy notational demands. Distributional programs are easier to read and understand. Second, we use distributional programs to clearly separate specification and implementation yet serve as a bridge between the two.

The remainder of this section is devoted to specifying the format and meaning of the pseudocode that specifies a distributional program.

Inputs The output pmf of a distributional program may be parameterized by inputs. In the distributional program below, the input is the integer n.

```
1: function TRUNCATIONMOVEREVISITED(n)
2:     x ~ bin(n, 0.2)
3:     y ← min(x, 1)
4:     return y
```

For $n = 3$, the output is the pmf p on $\{0, 1\}$ with $p(0) = 0.512$ and $p(1) = 0.488$.

Here is an example with two inputs, the pmfs p_1 and p_2.

```
1: function CONVOLUTION(p₁, p₂)
2:     x ~ p₁
3:     y ~ p₂
4:     z ← x + y
5:     return z
```

For the input pmfs p_1 with $p_1(3) = 0.4$, $p_1(4) = 0.6$ and p_2 with $p_2(1) = 0.7$, $p_2(2) = 0.3$, the output is the pmf on $\{4, 5, 6\}$ with $p(4) = 0.28$, $p(5) = 0.54$, $p(6) = 0.18$ (the convolution of p_1 and p_2).

Two distributional programs that generate the same output given the same input are called *equivalent* distributional programs. Throughout this book, we frequently replace a given distributional program with an equivalent one. We call this *refactoring*. We refactor distributional programs to promote efficiency or clarity.

Simple Sampling Statements A simple sampling statement takes the following form:

variable name $\sim$ *pmf expression*

The name on the left-hand side is the variable to which a new sample is assigned. The right-hand side is an expression that produces a pmf. This pmf is the conditional pmf of the new sample given all previous samples. In particular, if the right-hand side is a constant pmf, then the new sample will be statistically independent of all previous samples. A common random variable in a distributional program should be interpreted as the pmf of this random variable. For instance, bin(2, 0.2) stands for a binomial pmf with parameters 2 and 0.2.

Here is an example with two simple sampling statements.

1: **function** COMBINE( )
2:     $z \sim$ TRUNCATIONMOVEREVISITED(3)
3:     $x \sim bin(z + 1, 0.5)$
4:     **return** x

The output of this distributional program is the pmf p on $\{0, 1, 2\}$ with $p(0) = 0.378$, $p(1) = 0.5$, and $p(2) = 0.122$.

Multivariate Sampling Statements In each of the sampling statements in the examples so far, a single variable is sampled from a univariate pmf. Multiple variables can be sampled from a multivariate pmf, as illustrated in the distributional program below.

1: **function** MULTIVARIATESAMPLING( )
2:     $(x, y) \sim q$
3:     $z \leftarrow x^2$
4:     **return** z

If the input is the pmf q on $\{(2, 6), (3, 4), (2, 3)\}$ with $q(2, 6) = 0.5$, $q(3, 4) = 0.2$, $q(2, 3) = 0.3$, then the output is the pmf on $\{4, 9\}$ with $p(4) = 0.8$ and $p(9) = 0.2$.

If a variable sampled in a multivariate sampling statement is subsequently unused in the program, we may replace it with a general-purpose "throwaway" symbol "$-$." In the MULTIVARIATESAMPLING program above, the variable y in Line 2 is unused in the rest of the program. So, we may replace Line 2 above with the sampling statement $(x, -) \sim q$ instead, as shown below.

1: **function** MULTIVARIATESAMPLINGREVISITED( )
2:     $(x, -) \sim q$
3:     $z \leftarrow x^2$
4:     **return** z

We use the convention that the output of a distributional program without sampling statements is a point mass.

Return Statements A distributional program may specify a joint pmf. If so, the return must provide more than one value.

1: **function** MULTIPLERETURNVALUES( )
2:     $x \sim ber(0.4)$
3:     $y \sim bin(x, 0.75)$
4:     **return** (x, y)

The output is the pmf p on $\{(0, 0), (1, 0), (1, 1)\}$ with $p(0, 0) = 0.6$, $p(1, 0) = 0.1$, and $p(1, 1) = 0.3$.

Only returning some of the variables amounts to marginalizing a joint pmf. As an example, we modify the above program to return only y.

1: **function** MARGINALIZATION( )
2:     $x \sim ber(0.4)$

```
3:     y ~ bin(x, 0.75)
4:     return y
```

The output is the pmf p on $\{0, 1\}$ with $p(0) = 0.7$ and $p(1) = 0.3$.

A distributional program may have multiple return statements, as illustrated in the distributional program below.

```
1: function MULTIPLERETURNS( )
2:     x ~ bin(2, 0.8)
3:     if x ≤ 1 then
4:         return 0
5:     y ~ bin(3, 0.5)
6:     return y
```

The output is the pmf p on $\{0, 1, 2, 3\}$ with $p(0) = 0.44$, $p(1) = p(2) = 0.24$, and $p(3) = 0.08$.

Skip Statements A distributional program may contain skip statements. In a simulation, encountering a skip statement can be thought of as exiting without returning a value. The output of a distributional program with one or more skip statements is the pmf of the variable(s) returned by the program given the input but conditioned on not having reached any skip statement.

In the distributional program below, Line 4 is an exit point in the sense that Line 5 is never reached when the skip statement in Line 4 is reached.

```
1: function TRUNCATIONNORMALIZE( )
2:     x ~ bin(2, 0.2)
3:     if x > 1 then
4:         skip
5:     return x
```

The output is the pmf p on $\{0, 1\}$ with $p(0) = \frac{2}{3}$ and $p(1) = \frac{1}{3}$. Note how this output differs from the output of TRUNCATIONMOVE, the first example in this section.

Global Variables Throughout this book, any variable in a distributional program that is not part of its input nor assigned in the program is considered a global variable. Such a variable will be defined in the text where the program appears.

In the program below, the variable n is a global variable.

```
1: function TRUNCATIONMOVEGLOBAL( )
2:     x ~ bin(2, 0.2)
3:     y ← min(x, n)
4:     return y
```

Operators In addition to the operators "~" and "←," we use the addition, subtraction, and multiplication assignment operators "+←," "−←," and "×←," respectively. For example, the statement $x +\!\leftarrow z$ is equivalent to the statement $x \leftarrow x + z$, which replaces the current value of x with the sum of the current values of x and z. The operators "$- \leftarrow$" and "$\times \leftarrow$" are defined similarly. We also use "$+\sim$" as the "addition sampling" operator, which is equivalent to sampling from the right-hand side and adding the result to the left-hand side. For example, $x +\sim \text{bin}(10, 0.2)$ is equivalent to replacing the current value of x with the sum of the current value of x and a realization of a $\text{bin}(10, 0.2)$ random variable. Finally, we use the operator "$\overset{\propto}{\sim}$" to signify a sampling statement where the right-hand side is first normalized to obtain a pmf. For example, suppose that the histogram h is $h(0) = 2$, $h(1) = 3$, and $h(2) = 5$. The statement $x \overset{\propto}{\sim} h$ is equivalent to $x \sim h/10$.

1.6   Graph Theory

This book uses several basic notions from graph theory. Fix some undirected graph $G = (V, E)$, where V denotes its vertex set and E its edge set.

Two vertices are *adjacent* if there is an edge between them, in which case each vertex is said to be a *neighbor* of the other vertex. A vertex sequence $v_0, v_1, \ldots, v_k$ for $k \geq 1$ is called a $v_0 - v_k$ *walk* if $(v_{i-1}, v_i) \in E$ for $i = 1, \ldots, k$. Its *length* is k. A $v_0 - v_k$ walk is called a *path* between v_0 and v_k if there are no vertex repetitions. A $v_0 - v_k$ walk with $v_k = v_0$ is a *cycle* if $k \geq 3$ and $v_0, \ldots, v_{k-1}$ is a path. For disjoint subsets B_1 and B_2 of V, a path between B_1 and B_2 is a path between some $v \in B_1$ and some $w \in B_2$. A graph is *connected* if, for any two vertices, there is a path that contains these two vertices. A graph is *complete* if every pair of distinct vertices is adjacent.

For some subset $S \subseteq V$, the (induced) *subgraph* $G[S]$ is the graph with vertex set S and edge set consisting of all of the edges in E that have both endpoints in S. We also write $G - S = G[V \backslash S]$, where we note that $V \backslash S$ is a set and $G - S$ is a graph. A *component* of a graph is a connected subgraph that is not a subgraph of another connected subgraph.

A set $C \subseteq V$ is called a *clique* if the subgraph $G[C]$ is complete. A *maximal clique* is a clique that is not strictly contained in another clique. The set of all maximal cliques in G is denoted by $\mathcal{C}(G)$. For some subset $S \subseteq V$, we say that we *plant a clique with vertex set S in G* to mean that we add all edges between any two vertices in S to E, if not present already.

1.7   Kullback-Leibler Divergence

Each projection map Π defined in this book is a map on a set of pmfs. We use the concept of *Kullback-Leibler (KL) divergence* to compare a pmf p and its projection $\Pi(p)$. The Kullback-Leibler divergence can be thought of as a distance between two pmfs, albeit without the symmetry property. As our primary interest is to develop a numerical methodology, we define it here only when the first argument is a pmf with finite support to avoid convergence issues for infinite sums.

Definition 1.1 Suppose p_1 and p_2 are pmfs on the same (countable) set and p_1 has finite support. The Kullback-Leibler (KL) divergence between p_1 and p_2 is defined via

$$D(p_1 \parallel p_2) = \sum_y p_1(y) \log\left(\frac{p_1(y)}{p_2(y)}\right).$$

The right-hand side should be interpreted as a sum over the support of p_1 only, and it should be interpreted as ∞ when the denominator is zero for at least one of the summands.

Consider two pmfs p_1 and p_2 such that p_1 has finite support and p_2 is positive on the support of p_1. Clearly, the KL divergence is zero if $p_1 = p_2$. Suppose that $p_1 \neq p_2$. Since the function $-\log(\cdot)$ is convex, it follows from Jensen's inequality that

$$D(p_1 \parallel p_2) = -\sum_y p_1(y) \log\left(\frac{p_2(y)}{p_1(y)}\right) \geq -\log\left(\sum_y p_1(y) \times \frac{p_2(y)}{p_1(y)}\right) = 0.$$

This argument shows that the KL divergence is always nonnegative and that it is zero if and only if $p_1 = p_2$. As a result of the finite-support condition on p_1, we have that $D(p_1 \parallel p_2) < \infty$ if and only if the support of p_2 includes the support of p_1.

1.8 Projection Maps

In this book, we define many maps, some of which we interpret as projection maps. For our purposes, we define a projection map as follows.

Definition 1.2 Given a map $\Pi : X \to X$ and some subset $Y \subseteq X$, we say that Π is a *projection map* from X onto Y if (1) $\Pi(X) \subseteq Y$ and (2) $\Pi(x) = x$ for all $x \in Y$.

The following lemma gives an equivalent formulation of this definition with seemingly stronger versions of its two conditions.

Lemma 1.3 *A map $\Pi : X \to X$ is a projection map from X onto $Y \subseteq X$ if and only if (1) $\Pi(X) = Y$ and (2) $\Pi(x) = x$ if and only if $x \in Y$.*

To establish this lemma, consider a map Π that satisfies the two conditions of Definition 1.2. It suffices to show that Π satisfies conditions (1) and (2) in the lemma. To establish condition (1), it is necessary to argue that the reverse inclusion $Y \subseteq \Pi(X)$ holds. This reverse inclusion follows from condition (2) of the definition since $x \in Y$ implies $x = \Pi(x)$ and therefore $x \in \Pi(X)$. For condition (2), it suffices to show that $\Pi(x) = x$ implies $x \in Y$, which immediately follows from condition (1) in the definition.

For any projection map, we must have that $\Pi \circ \Pi = \Pi$. Indeed, for any $x \in X$, we have that $\Pi(x) \in Y$ by condition (1) in the above definition and therefore that $\Pi(\Pi(x)) = \Pi(x)$ by condition (2). The next lemma provides a converse.

Lemma 1.4 *A map* $\Pi : X \to X$ *is a projection map from* X *onto* $\Pi(X)$ *if and only if* $\Pi \circ \Pi = \Pi$.

To see why the converse holds, suppose that $\Pi \circ \Pi = \Pi$ for some map $\Pi : X \to X$. It suffices to establish condition (2) of Definition 1.2, as condition (1) is trivially satisfied for $Y = \Pi(X)$. Pick some $y \in \Pi(X)$, so that $y = \Pi(x)$ for some $x \in X$. Using $\Pi \circ \Pi = \Pi$, we then find that $\Pi(y) = \Pi(\Pi(x)) = (\Pi \circ \Pi)(x) = \Pi(x) = y$, which establishes the lemma.

Chapter 2
First Look at QPLEX

This chapter develops the QPLEX calculus for the classical $M_t/G/n$ multiserver queueing model in discrete time with more flexibility for arrival process modeling. We focus on this model for four reasons. First, this model has many practical applications in its own right, and it (and its variations) forms a building block for many practical stochastic network models that we will encounter in various forms throughout this book. Second, this model is simple enough to develop and justify the QPLEX calculus from first principles. In subsequent chapters, we will formally derive the calculus for a broad class of models defined by model primitives from which the calculus developed here will be seen as a special case. Third, the reader will see the use of distributional programming, described in general in Sect. 1.5, and the use of our streamlined notation to represent pmfs and conditional pmfs, described in Sect. 1.2, in a concrete context. Fourth, we have undertaken an extensive testbed of experiments to ascertain the quality of the QPLEX approximation for this model, which we summarize at the end of this chapter.

2.1 Multiserver Queueing Model

A facility consists of $n \geq 1$ homogeneous servers that provide some service and a staging area (the "buffer") with unlimited capacity. Customers seeking the service may either be served immediately or must wait in the buffer until a server becomes available. Customers occupy a server while they are in service, and this server can serve no other customer during that time. Customers are served according to the (nonpreemptive, nonidling) first-in, first-out (FIFO) service discipline. All customers exit the system upon completion of service, at which time they release their server.

© The Author(s) 2025   15
A. B. Dieker, S. T. Hackman, *QPLEX: A Computational Modeling and Analysis Methodology for Stochastic Systems*, Springer Series in Operations Research and Financial Engineering, https://doi.org/10.1007/978-3-031-74870-7_2

Applications of this model include healthcare, elections, sports or cultural event planning, and parking. Customers may correspond to patients, voters, event attendees, and (electric) cars, respectively, while servers may correspond to hospital beds, voting machines or staff, security personnel or ticket scanners, and parking spaces, respectively. In this chapter, we use the generic terminology of customers and servers.

We consider a discrete-time model with time indexed by the nonnegative integers. The distribution of the number of customers arriving during a time period can depend on the number of customers in the system at the beginning of this time period, but not on the numbers of customers arriving during prior time periods. We use the symbol A to represent the number of arrivals in some period and, for $t \geq 0$, we write for $\alpha^{(t)}(a|z)$ for the pmf of the number of customers arriving between times t and $t + 1$ if the number of customers in the system at time t is $z \geq 0$. All service durations are independent and identically distributed and are independent of everything else. We use the symbol L for a generic service duration, which is necessarily integer-valued, and write $g(\ell)$ for its pmf. Service durations are assumed to be at least one time period. We use the convention that customers departing or arriving between times t and $t + 1$ do so *immediately before* time $t + 1$, with arrivals immediately being able to use servers that are freed up (if any).

The above setup encompasses several different arrival models, even in the special case where the number of arrivals during each period is independent of the number in the system at the beginning of the period, i.e., when $\alpha^{(t)}(a|z) = \alpha^{(t)}(a)$ for all $t \geq 0$. For instance, it accommodates any independent increment process such as a (discrete-time) Poisson process or compound Poisson process. It also allows for modeling appointments with no-shows, with $\alpha^{(t)}(a)$ being a binomial pmf (if the no-show probability is the same for all arrivals and no-shows are independent across arrivals). Moreover, a superposition of such independent processes can model a combination of appointments with no-shows and walk-ins.

The possibility of the number of arrivals during each period depending on the number in the system at the beginning of the period allows for modeling balking and admission control policies. For instance, a potential arrival may leave before entering the system with a probability that depends on the number of customers in the system and the time period. Similarly, the likelihood of a customer being accepted into a system via a pricing policy can be based on the number of customers in the system and the time period.

The number of servers n, the service duration pmf $g(\ell)$, and the pmfs $\alpha^{(t)}(a|z)$ of the numbers of arrivals can be thought of as a system configuration selected by a system designer, who must balance the cost of technology and labor on the one hand against the benefits of higher service quality on the other. A nonstationary arrival process can make this decision problem quite challenging. One service quality metric commonly used in applications is the waiting time of an arriving customer. One possible objective is to choose the configuration to ensure that the probability of an excessive waiting time for an arriving customer is sufficiently small for all possible time epochs. A proxy for waiting time is the number of customers in the system, which is the number of customers in service plus the number of customers in the buffer waiting for service. So, how might one calculate the distributions of the number of customers in the system over time?

2.2 Markov Chain Model

It is natural to seek a discrete-time Markov chain model to find the distribution of the number of customers in the system over time. For computational purposes, its state space must be as small as possible, and for purposes of our stated performance evaluation goal, the number of customers in the system needs to be a function of the state.

Suppose we include the number of customers in the system in the state. It is not possible to calculate the distribution of the number of customers in the system at the *next* time epoch from the number of customers in the system at the current time epoch alone because the service duration pmf is general; the state must also include some information on the service progress of the customers in service. A natural choice is to track either the elapsed time since each customer began its service or the remaining service duration of each customer. This choice does *not* significantly affect our development, so we will use the remaining service duration. It corresponds to imagining that a customer's service duration is revealed as soon as it starts service. No information about arrivals needs to be included in the state due to the arrival model we have chosen.

As it is immaterial which server serves which customer for this system, it is sufficient to keep track of the number of customers with each possible remaining service duration. In view of our convention that arrivals and departures occur at the end of a period, the lowest possible remaining service duration is 1, and customers with this remaining service duration at the current time epoch will have departed by the next time epoch.

The above discussion leads to the following stochastic process. For $\ell \geq 1$, let $Y_\ell^{(t)}$ represent the number of customers in service with remaining service duration equal to ℓ at time t, and let $Y_0^{(t)}$ represent the total number of customers in the system at time t. Writing $\boldsymbol{Y}^{(t)} = (Y_0^{(t)}, Y_1^{(t)}, \ldots)$, we will see that $\{\boldsymbol{Y}^{(t)} : t \geq 0\}$ is a Markov chain. Since $\sum_{\ell \geq 1} Y_\ell^{(t)}$ represents the number of customers in service at time t, the state space is

$$\left\{ \boldsymbol{y} : y_\ell \in \mathbb{Z}_+ \text{ for all } \ell \geq 0, \sum_{\ell \geq 1} y_\ell = \min(y_0, n) \right\}.$$

We work with a slightly different Markov chain for reasons that will become clear below. We set $Z^{(t)} = Y_0^{(t)}$ and $H^{(t)}(\ell) = Y_\ell^{(t)}$ for $\ell \geq 1$. Working with $(Z^{(t)}, H^{(t)})$ instead of $\boldsymbol{Y}^{(t)}$ is merely a notational choice. We represent a state of the Markov chain $\{(Z^{(t)}, H^{(t)}) : t \geq 0\}$ by a pair (z, h). The first component z is the number of customers in the system. We call z a *counter*. For $\ell \geq 1$, we let $h(\ell)$ represent the number of customers in service with remaining service duration equal to ℓ. The collection $(h(\ell))_{\ell \geq 1}$ can either be viewed as a vector or as a function on $\{1, 2, \ldots\}$. We denote it by h and call it a *histogram of remaining service durations*. This is the second component of a state. (Section 1.3 defines histograms more broadly.) For example, suppose there are ten customers in service, four, five, and one of whom have remaining service durations of 1, 2, and 3, respectively. The associated histogram of remaining service durations h is a histogram on $\{1, 2, 3\}$ and can be represented via the vector $h = (4, 5, 1)$. The state space of the Markov chain $\{(Z^{(t)}, H^{(t)}) : t \geq 0\}$ is

$$\{(z, h) : z \geq 0, |h| = \min(z, n)\}. \tag{2.1}$$

2.3   Transition Probabilities and Kernel Maps

The system description leads to the stochastic process $\{(Z^{(t)}, H^{(t)}) : t \geq 0\}$. We now argue that this process is a Markov chain, describe its transition probabilities, and discuss the computational challenges of working with this Markov chain. Fix time t and consider the period between times t and $t + 1$, which we refer to as *this period*.

Transition Probabilities Suppose we are given realizations of the number of customers in the system z and the histogram of remaining service durations h at time t and of all such quantities at time epochs before t. Given this information, we seek to describe the joint distribution of the number of customers in the system $Z^{(t+1)}$ and the histogram of remaining service durations $H^{(t+1)}$ for time $t + 1$. For notational convenience, we write Z' and H' for $Z^{(t+1)}$ and $H^{(t+1)}$, respectively. We refer to Z' as the *next counter* and H' as the *next histogram*.

A distributional program for the joint pmf of Z' and H' given this information is presented below. We explain and justify the program shortly. Our explanation shows that the process has the Markov property, and we therefore suppress the state history for prior time epochs from the signature of the distributional program in Line 1 as this information is not needed. The number of servers n, the service duration pmf g, and the pmf of the number of arrivals $\alpha^{(t)}(a|z)$ are not made explicit either, as we consider them to be global variables. We denote the output of this distributional program by $p^{(t)}(z', h'|z, h)$ and use the term "transition probabilities" for the output of this distributional program.

1: **function** TRANSITIONPROBABILITY$^{(t)}(z, h)$
2:     $a \sim \alpha^{(t)}(A = \cdot | z)$
3:     $d \leftarrow h(1)$
4:     $z' \leftarrow z - d + a$
5:     $h^{\text{rel}} \leftarrow \overleftarrow{h}$

6: $h^{\text{join}} \sim \text{mult}\left(\min(z', n) - (\min(z, n) - d), g(L' = \cdot)\right)$

7: $h' \leftarrow h^{\text{rel}} + h^{\text{join}}$

8: **return** (z', h')

Sampling the Next Counter z' The number of arrivals sampled in Line 2 uses the fact that its distribution depends on the current number of customers in the system z and is independent of everything else. The independence of these samples across time is implicit, as z is the only variable that appears in the pmf instruction on the right-hand side of this sampling statement. Let d denote the number of system departures (service completions) during this period. Line 3 uses the fact that a customer departs during this period if and only if their remaining service duration equals 1. The value of z' is obtained using standard flow balance (Line 4).

Sampling the Next Histogram h' The next histogram of remaining service durations h' is the sum of two histograms, h^{rel} and h^{join} (Line 7). Here, h^{rel} represents the histogram of remaining service durations at time $t + 1$ of all customers who do not depart during this period, and h^{join} represents the histogram of remaining service durations at time $t + 1$ of all customers who start their service during this period. Line 5 uses the fact that the remaining service duration of each customer who does not depart will decrease by 1 at time $t + 1$. Here, the histogram $\bar{h}$ is defined, for all $\ell' \geq 1$, via

$$\bar{h}(\ell') = h(\ell' + 1).$$

Line 6 uses the fact that there are $\min(z, n) - d$ customers in service at time t who do not depart during this period and therefore $\min(z', n) - (\min(z, n) - d)$ customers who begin their service at time $t + 1$. Since each of these customers independently receives a remaining service duration using pmf g, histogram h^{join} is a multinomial random variable. (Section 1.4 formally defines a histogram-valued multinomial random variable and its pmf.)

Example 2.1 Suppose we have the following setup. There are $n = 10$ servers. At time t, there are $z = 11$ customers in the system, ten customers in service, and one customer waiting in the buffer. The maximum service duration is three periods, so the remaining service durations of the ten customers in service are either 1, 2, or 3. The histogram of remaining service durations at time t is $h = (4, 5, 1)$, where we use vector notation in lieu of writing $h(1) = 4$, $h(2) = 5$, and $h(3) = 1$. Note that $|h| = 10$, which equals $\min(n, z) = \min(10, 11)$, as required for (z, h) to be a state. The number of arrivals a between time t and $t + 1$ equals one. The number of service completions is $d = h(1) = 4$, and we have that $h^{\text{rel}} = (5, 1, 0)$. The latter corresponds to shifting the vector h one position to the left and padding with a zero. The number of customers in the system at time $t + 1$ is

$$z' = z - d + a = 11 - 4 + 1 = 8,$$

of which

$$\min(z', n) - (\min(z, n) - d) = 8 - (10 - 4) = 2$$

customers will begin their service, namely, the one customer in the buffer and the one arriving customer. Note that the buffer is empty at time $t + 1$. If the histogram of remaining service durations for the customers starting their service during this period is $h^{\text{join}} = (0, 1, 1)$, then the histogram of remaining service durations at time $t + 1$ is $h' = (5, 2, 1)$. $\triangle$

Kernel Map Let v denote an arbitrary pmf on the state space defined in (2.1). We define the *kernel map* $\mathcal{K}^{(t)}$ in terms of the transition probabilities via

$$[\mathcal{K}^{(t)}(v)](z', h') = \sum_{z,h} v(z, h) \times p^{(t)}(z', h'|z, h), \qquad (2.2)$$

where the sum is taken over the set in (2.1). The distributional program below represents this map.

1: **function** KERNELMAP$^{(t)}(v)$
2:     $(z, h) \sim v(Z = \cdot, H = \cdot)$
3:     $(z', h') \sim$ TRANSITIONPROBABILITY$^{(t)}(z, h)$
4:     **return** (z', h')

Curse of Dimensionality for Histograms Given some initial pmf $v_{\text{ker}}^{(0)}$ on the state space defined in (2.1), the iterative scheme $v_{\text{ker}}^{(t+1)} = \mathcal{K}^{(t)}(v_{\text{ker}}^{(t)})$ for $t \geq 0$ generates *kernel iterates*

$$v_{\text{ker}}^{(0)}, v_{\text{ker}}^{(1)}, v_{\text{ker}}^{(2)}, \ldots.$$

The kernel iterate $v_{\text{ker}}^{(t)}$ is the pmf of $(Z^{(t)}, H^{(t)})$ for $t \geq 0$ when $(Z^{(0)}, H^{(0)})$ has pmf $v_{\text{ker}}^{(0)}$. Suppose the goal is to calculate these iterates. In that case, one has to contend with the fact that the size of the state space in (2.1) is *enormous*, even when the number of servers and the maximum service duration are small, and that the one-step transition matrix has many nonzero elements per column. For instance, given a state where multiple customers may start their service during the current period, each such customer needs to be assigned a remaining service duration and all combinations of remaining service durations generally have a positive probability. From a computational point of view, this Markov chain is, in general, utterly unworkable due to the familiar *curse of dimensionality*. This is true even for this simple model.

2.4   Re-imagining Information Requirements

Consider the following thought experiment. Again, fix time t. Suppose we are permitted to query only a single customer, chosen (independently of everything else) uniformly at random from the customers in service at time t, to reveal its remaining service duration. With this information alone, one can only assume that each of the other customers has this same remaining service duration, so this does not capture any heterogeneity of service durations across customers. We already know that asking other customers to reveal their remaining service duration leads to enormous state spaces, so this is not an option either. Instead, we proceed exactly as before, except that we repeat this experiment many times, each time with a new collection of remaining service durations ("sample path") and a new independent uniform random variable to select a customer in service, yet *conditional on the value of $Z^{(t)}$ being fixed at some $z > 0$.* If the only information given is that $Z^{(t)} = z$, it would be reasonable to pretend that the remaining service durations of the customers in service are i.i.d. with the "pre-computed" empirical distribution of the remaining service durations from this thought experiment.

This thought experiment leads to the idea of an *iterative scheme* to approximate the pmfs of the $(Z^{(t)}, L^{(t)})$, where $L^{(t)}$ denotes the remaining service duration of a *uniformly chosen at random* customer from among the customers in service. Note that $(Z^{(t)}, L^{(t)})$ is a (random) function of the state $(Z^{(t)}, H^{(t)})$ of the Markov chain at time t. We call $L^{(t)}$ a *label*. We use the symbols Z, H, and L for generic random variables representing a counter, histogram, and label, respectively. *Instead of iteratively calculating joint pmfs of a counter Z and a histogram H, the QPLEX methodology iteratively calculates joint pmfs of a counter Z and a label L.* The latter pmfs have *much* smaller supports.

Suppose a pmf μ of a counter Z and a label L is the iterate at time t and consider the diagram below. The marginal pmf $\mu(z)$ of Z can be taken as the pmf of $Z^{(t)}$, and the *individual* remaining service durations at time t can be taken to be i.i.d. with pmf set to the conditional pmf $\mu(\ell|z)$ of L given $Z = z$, which plays the role of the empirical distribution from the thought experiment. This results in a histogram $H^{(t)}$. Let ν denote the pmf of $(Z^{(t)}, H^{(t)})$, viewed as a pmf ν of (Z, H). By taking a single transition of the Markov chain, recognizing that some customers in service at time t may depart by time $t + 1$ and that some new customers may begin service by time $t + 1$, we obtain $(Z^{(t+1)}, H^{(t+1)})$ and we denote its pmf viewed as a pmf of (Z, H) by ν'. The next iterate μ' is the pmf of $(Z^{(t+1)}, L^{(t+1)})$, where $L^{(t+1)}$ is the remaining service duration of a uniformly chosen at random customer represented in $H^{(t+1)}$, viewed as a pmf of (Z, L).

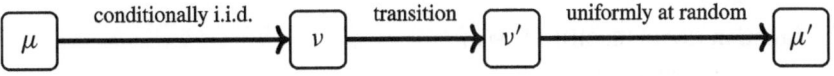

Applying this three-step procedure for each time t generates *QPLEX iterates:*

$$\mu_Q^{(0)}, \mu_Q^{(1)}, \mu_Q^{(2)}, \ldots.$$

We view each QPLEX iterate $\mu_Q^{(t)}$ as a pmf of (Z, L) and each kernel iterate $\nu_{ker}^{(t)}$ as a pmf of (Z, H). Given suitable initial pmfs $\mu_Q^{(0)}$ and $\nu_{ker}^{(0)}$ (e.g., starting with an empty system), the pmf $\mu_Q^{(t)}$ is the QPLEX approximation to the pmf of $(Z^{(t)}, L^{(t)})$ from the Markov chain setting. The marginal counter pmfs $\mu_Q^{(t)}(z)$ approximate the marginal counter pmfs $\nu_{ker}^{(t)}(z)$ of the kernel iterates.

2.5 QPLEX Maps

In the previous section, we described a map that outputs a pmf μ' of (Z, L) given an input pmf μ of (Z, L) in three steps. We call this map the *QPLEX map* and denote it by $\mathcal{Q}^{(t)}$. It is the analog of the kernel map $\mathcal{K}^{(t)}$, and the iterative scheme $\mu_Q^{(t+1)} = \mathcal{Q}^{(t)}(\mu_Q^{(t)})$ is the analog of the iterative scheme $\nu_{ker}^{(t+1)} = \mathcal{K}^{(t)}(\nu_{ker}^{(t)})$. This section works out a distributional program for $\mathcal{Q}^{(t)}$.

Each of the three steps in the diagram corresponds to a map, with the transition step corresponding to $\mathcal{K}^{(t)}$. Choosing uniformly at random leads to what we call the compression map, while the idea of conditionally i.i.d. remaining service durations leads to what we call the multinomial lift map. We take a few liberties in our presentation so that we may place more emphasis on intuition rather than technical precision.

Compression Map In the thought experiment, the label was defined as the remaining service duration of a uniformly chosen at random customer in service. Given some counter $z > 0$ and some histogram h on remaining service durations, since there are $\min(z, n)$ number of customers in service, the normalized histogram $h/\min(z, n)$ is the pmf of the remaining service duration of a uniformly chosen at random customer represented in the histogram. Therefore, it is the pmf of a label L given the counter z. For the corner case $z = 0$, the label L is sampled from the pmf g, which is an arbitrary choice of no consequence for the remainder. For example, if $\min(z, n) = 10$ and the histogram of remaining service durations is $h = (4, 5, 1)$, then $h/\min(z, n) = (0.4, 0.5, 0.1)$ represents the pmf of the remaining service duration of a uniformly chosen at random customer in service.

In view of this observation, the distributional program below outputs the compressed pmf μ of (Z, L) given a pmf ν of (Z, H), the latter being a pmf on the state space defined in (2.1).

1: **function** COMPRESSIONMAP(ν)
2:     $(z, h) \sim \nu(Z = \cdot, H = \cdot)$
3:     **if** $z > 0$ **then**
4:         $\ell \sim h/\min(z, n)$
5:     **else**

6:    $\ell \sim g(L = \cdot)$

7:    **return** (z, ℓ)

Represented in algebraic form, the output pmf μ obtained by compressing the input pmf ν is given by

$$\mu(z, \ell) = \sum_h \frac{h(\ell)}{\min(z, n)} \times \nu(z, h),$$

where the sum is understood to be taken over all histograms h with $|h| = \min(z, n)$. The ratio in this expression should be interpreted as $g(\ell)$ if $z = 0$. As a consequence of the preceding display, we obtain that the conditional pmf

$$\mu(\ell|z) = \sum_h \frac{h(\ell)}{\min(z, n)} \times \nu(h|z)$$

is the weighted average of the pmfs $h/\min(z, n)$ with the weights given by $\nu(h|z)$. Thus, $\mu(\ell|z)$ may also be interpreted as the "mean" histogram under ν given z and normalized by $\min(z, n)$.

Example 2.2 Suppose that the counter Z takes some value z with probability 1 under ν and that $\min(z, n) = 2$. Also suppose each customer's remaining service duration can be 1, 2, or 3. There are two customers in service, so there are six possible histograms on $\{1, 2, 3\}$. The values of $\nu(z, h)$ for this example are displayed in the table below.

h	$(2, 0, 0)$	$(0, 2, 0)$	$(0, 0, 2)$	$(1, 1, 0)$	$(1, 0, 1)$	$(0, 1, 1)$
$\nu(z, h)$	0.1	0.2	0.1	0.3	0.2	0.1

The weighted average $\sum_h h \times \nu(z, h)$ of the six histograms is

$$(2, 0, 0) \times 0.1 + (0, 2, 0) \times 0.2 + (0, 0, 2) \times 0.1 + (1, 1, 0) \times 0.3$$

$$+ (1, 0, 1) \times 0.2 + (0, 1, 1) \times 0.1 = (0.7, 0.8, 0.5).$$

Dividing this weighted average by the sum of its elements, $|h| = \min(z, n) = 2$, yields the normalized vector $(0.35, 0.4, 0.25)$. If we let the pmf μ of (Z, L) be the output of the compression map with input ν, then this vector is $\mu(z, L = \cdot)$. $\triangle$

Multinomial Lift Map The thought experiment suggests a kind of converse to the compression map by letting remaining service durations be conditionally i.i.d. given the number of customers in the system. The distributional program below outputs the pmf ν of (Z, H) given a pmf μ of (Z, L). Under pmf ν, given $Z = z$, the remaining service durations of the $\min(z, n)$ customers in service are assumed to be i.i.d. with pmf $\mu(\ell|z)$, and H is thus conditionally multinomial given $Z = z$. (If

$z = 0$, then we have that $H = 0$ with probability 1 in view of our conventions for multinomial random variables, as there are no customers in service.)

```
1: function MULTINOMIALLIFTMAP(μ)
2:     z ~ μ(Z = ·)
3:     h ~ mult (min(z, n), μ(L = ·|z))
4:     return (z, h)
```

Example 2.3 Suppose that the counter Z takes some value z with probability 1 under μ and that $\min(z, n) = 2$. Suppose, as in Example 2.2, that each customer's remaining service duration can be 1, 2, or 3 and that $\mu(z, L = \cdot) = (0.35, 0.4, 0.25)$. Let the pmf ν of (Z, H) be the output of the multinomial lift map with input μ. The values of $\nu(z, h)$ for this example are displayed in the table below.

h	$(2, 0, 0)$	$(0, 2, 0)$	$(0, 0, 2)$	$(1, 1, 0)$	$(1, 0, 1)$	$(0, 1, 1)$
$\nu(z, h)$	0.1225	0.16	0.0625	0.28	0.175	0.2

For example, the entry for $(1, 1, 0)$ is obtained via

$$\nu\,(z, H = (1, 1, 0)) = \binom{2}{1, 1, 0} \times (0.35)^1 \times (0.4)^1 \times (0.25)^0 = 0.28.$$

Together with Example 2.2, this example shows that first applying the compression map and then the multinomial lift map *changes* a pmf of (Z, H), at least in general. Thus, the composition of the multinomial lift and compression maps is not the identity map; in fact, it will be shown to be an important projection map used repeatedly in the developments in this book. We will see that the reverse composition is the identity map here. Applying the compression map to the pmf ν found in this example yields the weighted average $\sum_h h \times \nu(z, h)$ of the six histograms divided by two, which indeed is the pmf μ we started with.   △

Distributional Program for QPLEX Map The distributional program below outputs the pmf $\mu' = \mathcal{Q}^{(t)}(\mu)$ of (Z', L') given an input pmf μ of (Z, L). A pair (z, h) is sampled from the pmf μ as in the program of the multinomial lift map (Lines 2 and 3), so that its pmf is the image ν of μ under the multinomial lift map. A pair (z', h') is then sampled using the transition probabilities so that the unconditional pmf of (z', h') is $\nu' = \mathcal{K}^{(t)}(\nu)$ (Line 4). The next label ℓ' is then drawn uniformly at random from the $\min(z', n)$ customers represented in h' (Line 6) if $z' > 0$ or drawn from the service duration pmf (Line 8) if $z' = 0$, as in the distributional program of the compression map (applied to time $t + 1$ instead of time t). The output pmf μ' of (Z', L') can also be viewed as a pmf of (Z, L).

```
1: function QPLEXMAP(t)(μ)
2:     z ~ μ(Z = ·)
3:     h ~ mult (min(z, n), μ(L = ·|z))
4:     (z', h') ~ TRANSITIONPROBABILITY(t)(z, h)
```

5:    **if** $z' > 0$ **then**
6:         $\ell' \sim h' / \min(z', n)$
7:    **else**
8:         $\ell' \sim g(L' = \cdot)$
9:    **return** (z', ℓ')

2.6 Efficient Representation of QPLEX Maps

Calculating the QPLEX iterates $\mu_Q^{(t)}$ still requires enumerating an enormous number of histograms as the QPLEX map $Q^{(t)}$ is still expressed in terms of transition probabilities on a massive state space. It may therefore appear, at first blush, that we have not yet circumvented the curse of dimensionality for histograms. In this section, we show that there is an efficient representation of the QPLEX map $Q^{(t)}$ that *no longer* requires enumerating any histograms.

Reproduced below is an expanded version of the equivalent distributional program for the QPLEX map $Q^{(t)}$. It includes the pseudocode from the distributional program for the transition probabilities. The program samples the histograms h, h^{rel}, h^{join}, and h', but these variables are not returned. Thus, calculating the output of this program requires enumerating their possible sample values. Here and throughout this section, we fix time t.

1: **function** QPLEXMAP$^{(t)}(\mu)$
2:      $z \sim \mu(Z = \cdot)$
3:      $h \sim \text{mult}\left(\min(z, n), \mu(L = \cdot | z)\right)$
4:      $a \sim \alpha^{(t)}(A = \cdot | z)$
5:      $d \leftarrow h(1)$
6:      $z' \leftarrow z - d + a$
7:      $h^{\text{rel}} \leftarrow \tilde{h}$
8:      $h^{\text{join}} \sim \text{mult}\left(\min(z', n) - (\min(z, n) - d), g(L' = \cdot)\right)$
9:      $h' \leftarrow h^{\text{rel}} + h^{\text{join}}$
10:     **if** $z' > 0$ **then**
11:          $\ell' \sim h' / \min(z', n)$
12:     **else**
13:          $\ell' \sim g(L' = \cdot)$
14:     **return** (z', ℓ')

Below, we refactor this program, namely, we rewrite it without changing its output. We do this in such a way that the histograms h, h^{rel}, h^{join}, and h' no longer appear. As a consequence, their possible sample values no longer need to be enumerated to calculate the output of this distributional program, which results in computational tractability. The refactoring procedure consists of two steps. In the first step, we show how to eliminate the sample of the histogram h. In the second step, we show how to eliminate the samples of the histograms h^{rel}, h^{join}, and h'.

Step 1: Eliminate h The variable h appears in Lines 3, 5, and 7, and the only other variable used in these instructions is z. Fix some z. We show how to replace Lines 5 and 7 with sampling statements that do not use h yet result in d and h^{rel} having the same joint distribution given z as in the above program. We show how to sample d and h^{rel} without sampling h in such a way that d and h^{rel} have the same joint distribution given z as in the above program. As a consequence, we can replace Lines 3, 5, and 7 with sampling statements for d and h^{rel} only, thereby eliminating h without changing the output of the distributional program.

Since the histogram of remaining service durations h is multinomial, we can think of all customers in service as having i.i.d. remaining service durations at time t with pmf $\mu(\ell|z)$. As a result, each of the $\min(z, n)$ customers in service departs with the same probability $\mu(L = 1|z)$. The number of service completions d given z is therefore binomial with parameters $\min(z, n)$ and $\mu(L = 1|z)$.

Given z and d, the remaining service durations at time t of the $\min(z, n) - d$ customers that do *not* depart during this period are i.i.d. and the probability that their remaining service duration is ℓ is $\mu(\ell|z)/(1 - \mu(L = 1|z))$ if $\ell > 1$ and 0 if $\ell = 1$. Since the remaining service duration of each such customer decreases by 1 at time $t + 1$, their remaining service duration at that time has pmf $\mu_{L-1|Z, L>1}(\ell'|z)$ given by

$$\mu_{L-1|Z, L>1}(\ell'|z) = \frac{\mu(L = \ell' + 1|z)}{1 - \mu(L = 1|z)}.$$

For example, if $\mu(L = \cdot|z) = (0.35, 0.4, 0.25)$ as in Example 2.3, then we have that

$$\mu_{L-1|Z, L>1}(L' = \cdot|z) = \left(\frac{0.4}{0.4 + 0.25}, \frac{0.25}{0.4 + 0.25}, 0\right).$$

The corner case $\mu(L = 1|z) = 1$ leads to $d = \min(z, n)$, in which case the pmf $\mu_{L-1|Z, L>1}(\ell'|z)$ may be arbitrary. This case plays no role in the remainder and relies on our interpretation of a multinomial random variable with the first parameter equal to 0 as the zero histogram.

We conclude that d given z is binomial with parameters $\min(z, n)$ and $\mu(L = 1|z)$ and that h^{rel} given z and d is multinomial with parameters $\min(z, n) - d$ and $\mu_{L-1|Z, L>1}(L' = \cdot|z)$. Of course, it is possible that h^{rel} is the zero histogram. Using these facts, Lines 5 and 7 in the above program can be replaced with the following pseudocode fragment:

```
5:    d ~ bin (min(z, n), μ(L = 1|z))
7:    h^rel ~ mult (min(z, n) − d, μ_{L−1|Z,L>1}(L' = ·|z))
```

Here is the refactored distributional program after the first step.

```
1: function QPLEXMAP^(t)(μ)
2:     z ~ μ(Z = ·)
3:     a ~ α^(t)(A = ·|z)
```

4:    $d \sim \text{bin}\left(\min(z, n), \mu(L = 1 | z)\right)$

5:    $z' \leftarrow z - d + a$

6:    $h^{\text{rel}} \sim \text{mult}\left(\min(z, n) - d, \mu_{L-1|Z, L>1}(L' = \cdot | z)\right)$

7:    $h^{\text{join}} \sim \text{mult}\left(\min(z', n) - (\min(z, n) - d), g(L' = \cdot)\right)$

8:    $h' \leftarrow h^{\text{rel}} + h^{\text{join}}$

9:    **if** $z' > 0$ **then**

10:        $\ell' \sim h' / \min(z', n)$

11:    **else**

12:        $\ell' \sim g(L' = \cdot)$

13:    **return** (z', ℓ')

Step 2: Eliminate h^{rel}, h^{join}, and h' These three remaining histograms appear in Lines 6–8 of the above program. The other variables used in these three instructions are z, d, and z'; in what follows, we fix their values. Below, we show how to sample the next label ℓ' conditionally on z, d, and z' without reference to these histograms. As a consequence, the sampling instructions associated with these variables can be eliminated, and therefore their sample values no longer need to be enumerated to calculate the output of the distributional program.

The conditional pmf of the next label ℓ' given z, d, and z' is

$$\sum_{h'} \frac{h'(\ell')}{\min(z', n)} \times \tilde{v}(h'),$$

where $\tilde{v}$ is the conditional pmf of h' given z, d, and z'. The sum

$$\sum_{h'} h'(\ell') \times \tilde{v}(h')$$

is the mean histogram under $\tilde{v}$. As shown in Lines 6–8, $\tilde{v}$ is the pmf of the sum of two (independent) multinomial histograms. As explained in Sect. 1.4, this mean histogram is therefore given by, for $\ell' \geq 1$,

$$(\min(z, n) - d) \times \mu_{L-1|Z, L>1}(\ell' | z) + (\min(z', n) - (\min(z, n) - d)) \times g(\ell'),$$

and so the next label ℓ' can be sampled from the pmf:

$$\frac{\min(z, n) - d}{\min(z', n)} \times \mu_{L-1|Z, L>1}(\ell' | z) + \left(1 - \frac{\min(z, n) - d}{\min(z', n)}\right) \times g(\ell'). \qquad (2.3)$$

Sampling from this pmf can be viewed as occurring in two steps: first sample the *type* of the uniformly chosen at random customer from service at time $t + 1$, and then sample its remaining service duration at time $t + 1$ given its customer type. There are two possible customer types. Either the customer is continuing its service as it was already in service at time t, in which case we say that their customer type is "old," or the customer is beginning its service, in which case we say that their customer type is "new." If the customer type is "new," then the next label has pmf $g(\ell')$; if it is "old," then the next label has pmf $\mu_{L-1|Z,L>1}(\ell'|z)$. Keep in mind that a customer of type "new" could be a customer who arrived to the system during this period, but it could also be a customer who has been waiting in the buffer. Of the $\min(z', n)$ customers in service at time $t + 1$, there are $\min(z, n) - d$ customers of type "old" and $\min(z', n) - (\min(z, n) - d)$ customers of type "new." The weights in (2.3) can be interpreted as the customer-type probabilities of the uniformly chosen at random customer.

We have argued that the distributional program QPLEXMAP$^{(t)}$ can be refactored as shown below. Note that *there are no longer any histograms*. For aesthetic reasons only, we handle the corner case $z' = 0$ slightly differently in this program to avoid nested if statements. Numerical results can be obtained with a computational engine designed to (efficiently) perform calculations encoded in distributional programs.

```
 1: function QPLEXMAP^(t)(μ)
 2:     z ~ μ(Z = ·)
 3:     d ~ bin (min(z, n), μ(L = 1|z))
 4:     a ~ α^(t)(A = ·|z)
 5:     z' ← z − d + a
 6:     if z' = 0 then
 7:         ℓ' ~ g(L' = ·)
 8:         return (z', ℓ')
 9:     customerIsOld ~ ber((min(z, n) − d)/ min(z', n))
10:     if customerIsOld then
11:         ℓ' ~ μ_{L−1|Z,L>1}(L' = ·|z)
12:     else
13:         ℓ' ~ g(L' = ·)
14:     return (z', ℓ')
```

2.7   Model Variations

Modeling flexibility is built into the QPLEX methodology, and we give many example models in Chaps. 5 and 7. To provide a flavor of what is to come, we modify the above distributional program to accommodate several simple but useful variations. Here are three examples.

Maximum Size of Buffer Suppose that at most $b \geq 0$ customers can be waiting in the buffer to be served. Customers who arrive when the buffer is full are lost. To accommodate this variation, the only change needed is to replace Line 5 with the following sampling instruction:

```
 5:     z' ← min(z − d + a, n + b)
```

Batch Processing Assume that services must be performed in batches of size B, i.e., customers in a group of size B begin and complete their services simultaneously. The number of servers n is an integer multiple of B. For example, there can be four groups of $B = 3$ servers, so that $n = 12$. If the system is empty and there are four arrivals, then three of the four arrivals will begin their service, while the remaining arriving customer must wait in the buffer until there are two more arrivals.

To accommodate this variation, the only changes needed are to replace the number of customers in service in Line 3 with the number of groups in service and to adjust the number of customers completing service this period in Line 5 to reflect that each service completion represents a batch of size B, as shown in the following pseudocode fragment:

3: $\quad d \sim \text{bin}\left(\min(\lfloor z/B \rfloor, n/B), \mu(L = 1|z)\right)$

5: $\quad z' \leftarrow z - B \times d + a$

Here $\lfloor w \rfloor$ denotes the largest integer less than or equal to the real number w.

Time-Varying Number of Servers Creating a time-varying capacity plan $\{n^{(t)} : t \geq 0\}$ can be beneficial when the demand for service varies significantly. Often, servers correspond to labor staff. In many cases, there may be constraints on the degree to which the server capacity can change over time, e.g., it may need to be fixed each shift, day, etc. In a manufacturing or cloud computing setting, there may be a surplus of machines that can be brought back online if necessary. In a healthcare setting, there may be extra rooms that can be readied for use if necessary.

If the number of available servers $n^{(t+1)}$ at time $t + 1$ is less than the number $\min(z, n^{(t)}) - d$ of customers that did not complete their service during this period, then some customers must be removed from service even though their service requirement has not yet been met. One approach is to assume that each possible set of such customers is equally likely to be selected for removal and that the selected customers return to the buffer and have their service restart with a new (independent) service duration when they are taken into service at a later time. To accommodate this variation, the only change needed is to replace Lines 3 and 9 with the following sampling instructions:

3: $\quad d \sim \text{bin}\left(\min(z, n^{(t)}), \mu(L = 1|z)\right)$

9: $\quad \text{customerIsOld} \sim \text{ber}((\min(z, n^{(t)}) - d)/\min(z', n^{(t+1)}))$

A second approach assumes that customers simply leave the system when selected for removal from service. The number of customers leaving the system must then be $\max(d, \min(z, n^{(t)}) - n^{(t+1)})$. In addition to the changes to Lines 3 and 9 above, the flow balance equation in Line 5 needs to be changed to the following instruction:

5: $\quad z' \leftarrow z - \max(d, \min(z, n^{(t)}) - n^{(t+1)}) + a$

2.8 Approximation Quality

To investigate the quality of the QPLEX approximations, we designed an experimental testbed of 180 challenging model instances to put QPLEX through its paces. The details are in Appendix A. For all model instances, the system is initially empty. Here, we describe our results for a single challenging model instance. The quality reported is representative of the quality of the testbed as a whole.

Model Inputs The horizon length is 10 hours, and each period corresponds to a time interval of 0.01 hours, so there are 1000 time periods. Our model instance has four servers and an expected service duration of 0.25 hours. The service duration distribution is (discretized) lognormal, and its density is depicted below. It has a moderately high coefficient of variation (namely, 2) and a long tail, which can pose challenges for standard parametric formulas and methodologies such as matrix-analytic methods; more discussion is in Sect. 2.9.

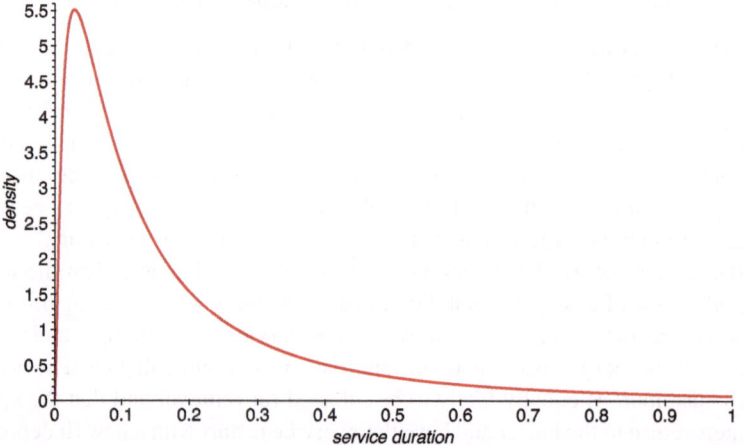

We assume that the number of arrivals during each period is Poisson and depict the hourly rate as a function of time below. The average utilization is 0.95. There are two massive half an hour long spikes, during which 55% of the customers arrive. During these spikes, the arrival rate exceeds the processing capacity, resulting in temporary overload. In particular, the peak instantaneous offered load is 20.9, and the peak over average arrival rate is 5.5.

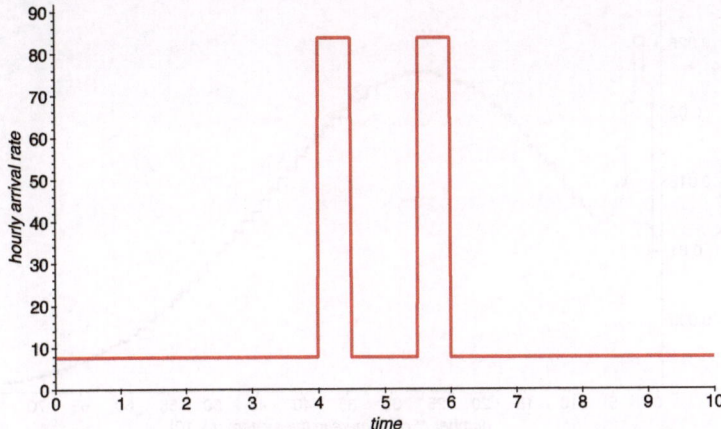

In the pictures that follow, pmfs are shown as outlines of histograms. The pink curve is based on the QPLEX methodology, while the purple curves have been generated using (discrete-event) simulation. Since it is impossible to calculate the distributions of the underlying Markov chain, we use *one million* simulation replications and think of the simulation-based curve as nearly exact.

Number of Customers in the System The picture below shows the pmfs of the number of customers in the system at the end of the time horizon after 10 hours. We include a vertical line at $z = 4$ (the number of servers) as a reference. We make several observations. First, even after going through periods of extreme overload multiple times, and even when the distribution of the number of customers in the system has not settled back into its steady state, the two curves are virtually *indistinguishable*. In fact, the largest absolute difference between the curves is 3.7×10^{-4}. Second, even though part of the curve looks like a bell curve, this distribution looks very different from a common parametric distribution, in particular because it has multiple modes. Third, the QPLEX-based pmf is bimodal, whereas the simulation-based pmf is not (the probability for $z = 30$ exceeds that for $z = 29$). Of course, we do not know if the exact distribution is bimodal; even one million simulation replications are insufficient.

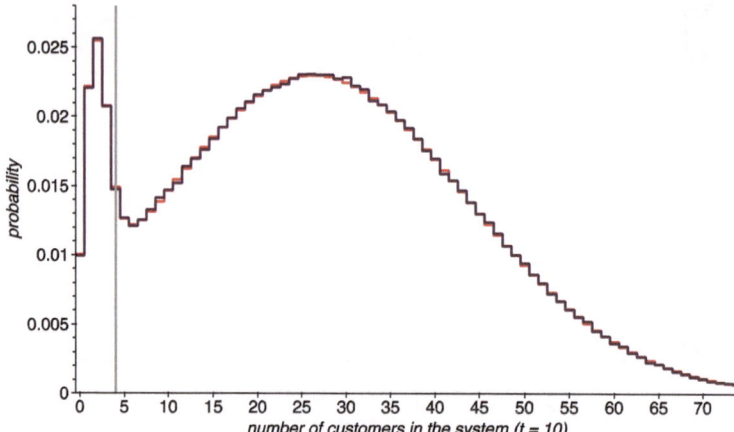

We next consider the distributional dynamics of the number of customers in the system. The pmfs of the number of customers in the system generated by simulation and QPLEX at times 4, 4.5, 5, 5.5, 6, and 6.5 are depicted below. These time epochs correspond to the beginning and end of the arrival rate spikes as well as 0.5 hours after they end. It is clearly visible how the pmf of the number of customers in the system at first responds to the increases in service demand during the arrival rate spikes and subsequently to their decrease. The system is close to steady state at time 4. Again, it is striking that the two curves in each panel are indistinguishable to the naked eye.

For this example, not surprisingly, all servers are busy with probability close to 1, and the pmfs are shaped like bell curves for a time period following each spike. We emphasize, however, that the QPLEX methodology does not impose an a priori parametric form on its output.

Virtual Waiting Time   We define the *virtual waiting time* at time t to be the time an *additional* customer would need to wait before it can enter service if it were to arrive at time t. Appendix A discusses how to calculate such a pmf efficiently given a pmf μ of the counter Z and the label L if the number of customers in the system has pmf $\mu(z)$ and the remaining service durations of the customers in service are conditionally i.i.d. given the number of customers in the system with pmf $\mu(\ell|z)$.

We replace the lognormal service duration with a *constant* service duration (with the same mean). The cumulative distribution functions of the virtual waiting time generated by simulation and QPLEX at times 4, 4.5, 5, 5.5, 6, and 6.5 are depicted below. The "waterfall" shapes should be contrasted with the exponential formulas from textbooks and methodologies built on such formulas. Once again, the curves in each panel are indistinguishable.

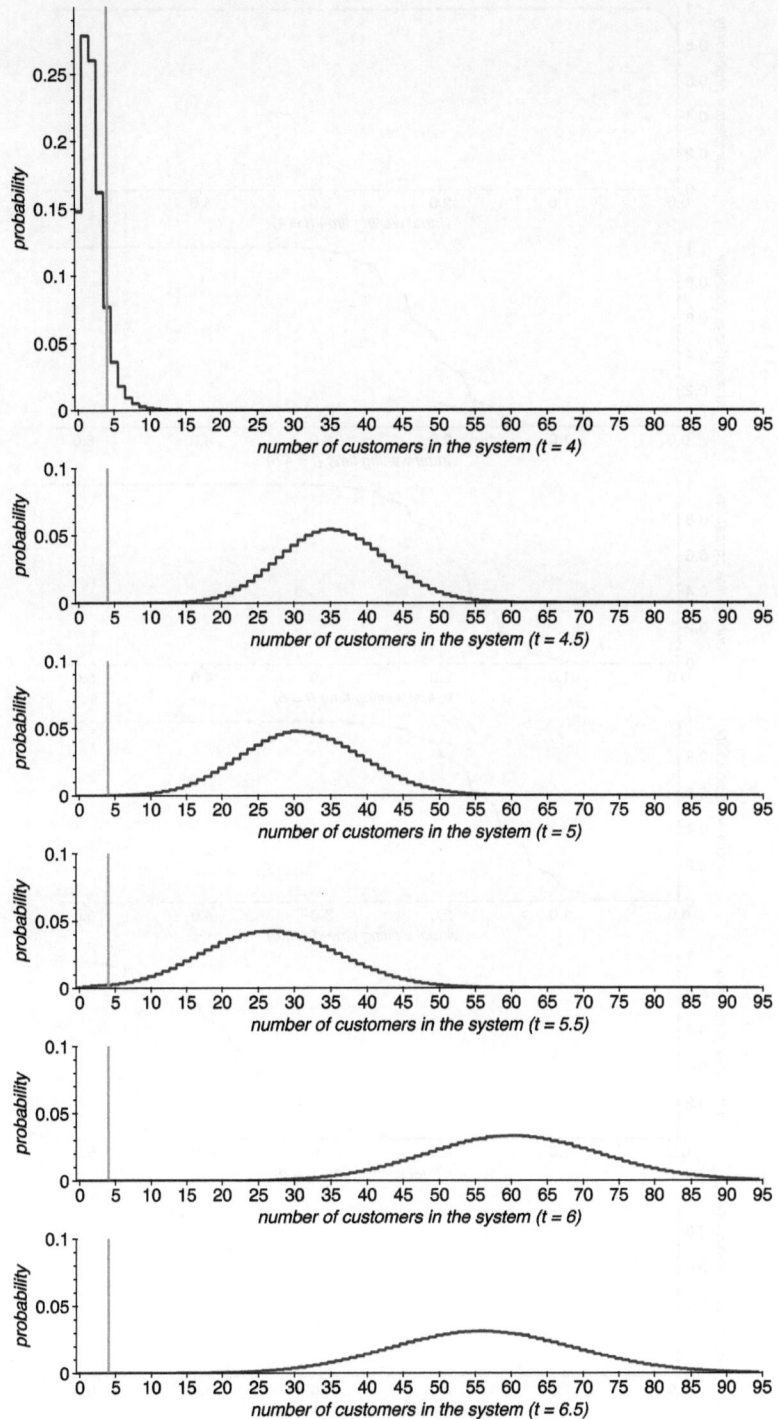

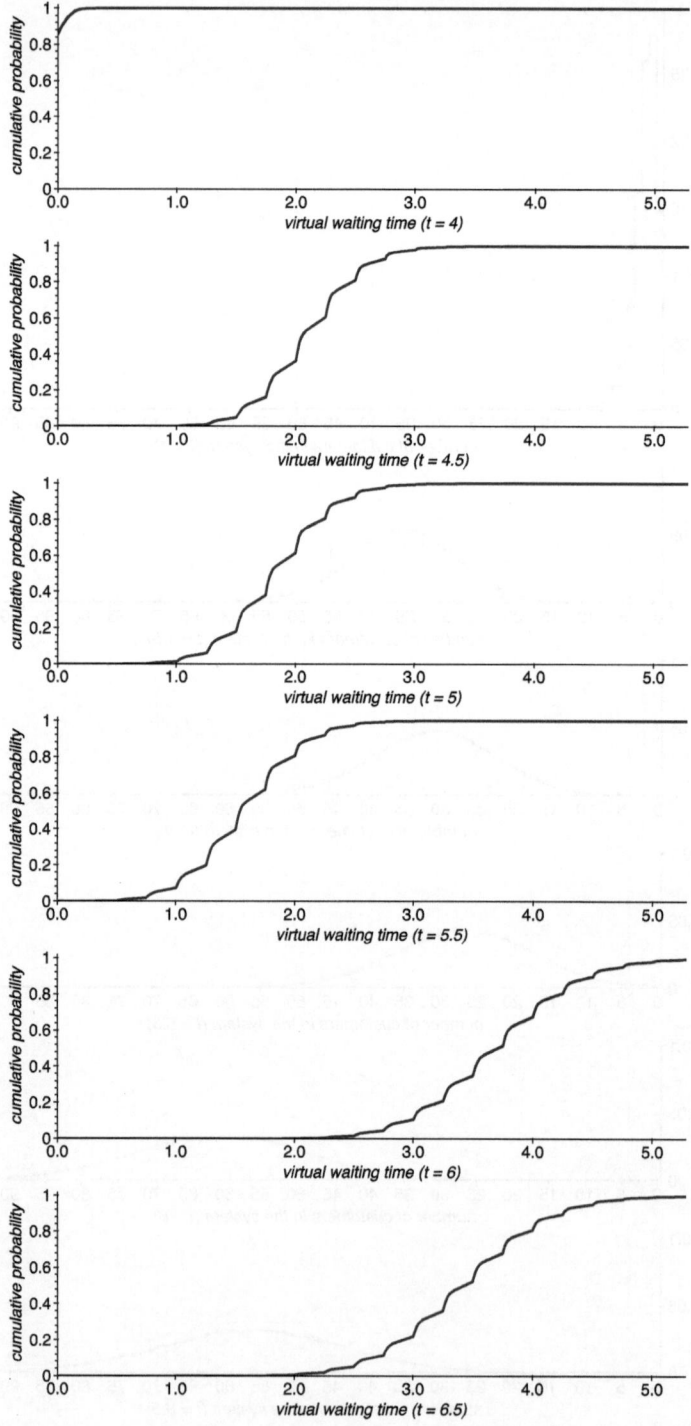

Percentiles In the panels below, we depict the median, 95th percentile, 99th percentile, and 99.5th percentile of the number of customers in the system over time, again for the model with lognormal service durations. The pink and purple curves are again on top of one another, so the quality of the QPLEX approximation in the tails is excellent. In fact, the largest absolute difference is 1.

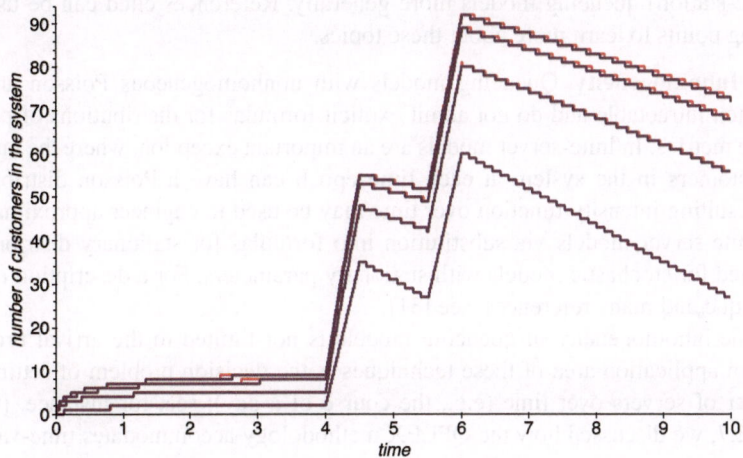

We consider a similar picture for the virtual waiting time. It again conveys that the QPLEX approximation is excellent in the tails. In fact, the largest absolute difference is 0.04 hours.

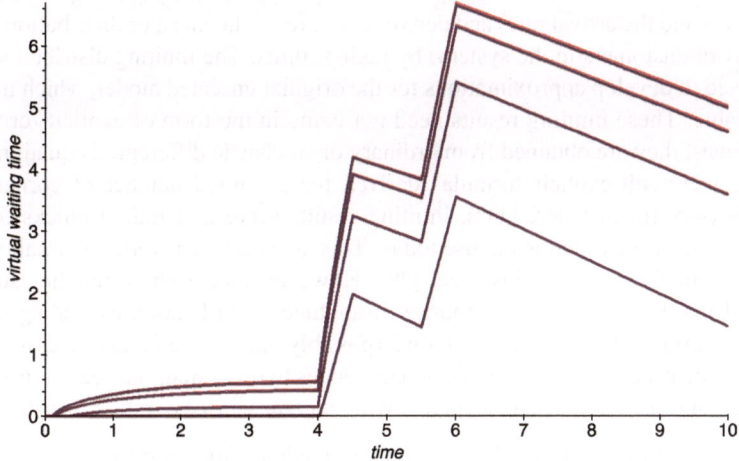

2.9   Bibliographical Notes

The queueing model we used in this chapter to illustrate the QPLEX methodology has been the subject of much research for over a half-century. Many of the techniques developed are more broadly applicable, so we discuss existing work on (single-station) queueing models more generally. References cited can be used as starting points to learn more about these topics.

Time-Inhomogeneity Queueing models with nonhomogeneous Poisson arrivals are often intractable and do not admit explicit formulas for distributions of performance metrics. Infinite-server models are an important exception, where the number of customers in the system at each time epoch can have a Poisson distribution. The resulting intensity function over time may be used to engineer approximations for finite-server models via substitution into formulas for stationary distributions obtained for stochastic models with stationary parameters. For a description of this technique and many references, see [31].

Time-inhomogeneity in queueing models is not limited to the arrival process. A main application area of these techniques is the decision problem of setting the number of servers over time (e.g., the course of a day); see, for instance, [9]. In Sect. 2.7, we discussed how the QPLEX methodology accommodates time-varying numbers of servers.

Scaling Results and Resulting Approximations Much work on queueing systems has been focused on scaling limits for stationary models; see [2, 30] for up-to-date general treatments and [33] for a special case of the model from this chapter. The idea is to consider a relevant scaling regime (e.g., appropriately growing the number of servers and the arrival rate) and derive results (e.g., the mean or distribution of the number of customers in the system) by taking limits. The limiting distributions are then used to develop approximations for the original unscaled model, which may be intractable. These limiting results need not come in the form of explicit formulas; sometimes, they are obtained from ordinary or stochastic differential equations.

Together with explicit formulas derived for a limited number of cornerstone models (see, for instance, [11]), limiting results serve as building blocks for the engineering approximations in use today. This approach can work remarkably well when applicable; see, for instance, [29]. However, model characteristics such as a small number of servers, temporary instantaneous utilization exceeding 100%, general service duration distributions (possibly data-driven), and the need for transient distributions pose significant challenges to this paradigm; see, for instance, [10] for an illustration of the challenge with general service duration distributions.

Matrix-Analytic Methods Matrix-analytic methods are a modeling and analysis framework with applications to a broad class of stochastic models, including queueing models. Like the QPLEX methodology, this framework is primarily geared towards computation. It typically assumes stationary parameters and is generally difficult to adapt to settings with time-inhomogeneity. In the context of the present chapter, these methods use a class of service duration distributions known as *phase-type distributions*. These distributions generalize the exponential distribution and result in matrix analogs of the classical stationary distribution for the number of customers in the system when service durations are exponential. We refer to [3, 12] for recent textbook treatments. We will encounter phase-type distributions again in Sect. 4.1.

Simulation Computer simulation techniques such as discrete-event simulation are powerful tools with conceptually unlimited modeling flexibility; see [20] for a textbook treatment of the standard tools. Today's abundant computational power and storage of data provide new opportunities and pose new challenges [21].

Even in the relatively simple context of the model from this chapter, it is not always straightforward to use discrete-event simulation models for decision problems with a temporal or a spatial component. For instance, when tracking performance metrics over time, [24] argues that the standard summary measures in commercial simulation software are misleading and can lead a user to draw incorrect conclusions. Similarly, [22] points out that simulation has primarily been used to assess long-run average performance but that control and policy-based optimization pose challenges for discrete-event simulation.

Conditionally Independent Remaining Service Durations The papers [17, 26] are related to the ideas presented in this chapter. They develop approximations for *stationary* distributions of the number of customers in the system z by assuming that remaining service durations are conditionally independent given z (only) for some values of z. The pmfs of the remaining service durations are exogenously specified, time-homogeneous functions of the service duration pmf. In contrast, in this chapter, the conditional pmfs $\mu_Q^{(t)}(\ell|z)$ of the remaining service durations are calculated iteratively for *all z*; they are, therefore, *endogenous and vary with time*.

Part I
QPLEX Modeling and Calculus

Chapter 3
Introduction to QPLEX Modeling and Calculus

Chapter 2 used a simple yet highly practical model to introduce the main ideas behind the QPLEX calculus. In Part I, we develop an abstraction of this calculus and show that it applies to a wide variety of models. This chapter presents our terminology and definitions that form the foundation for the developments in this part.

3.1 QPLEX Chains and Kernel Maps

A *QPLEX chain* is a discrete-time, discrete-space Markov chain with a special structure described below. This structure is a necessary starting point to develop the QPLEX calculus that applies to a broad class of models, of which the multiserver queueing model of Chap. 2 is but one simple example. To help motivate the definition of this structure, we revisit the Markov chain representation of this multiserver queueing model to set the stage for the general formulation to follow.

From the model description, we first defined a Markov chain $\{Y^{(t)} : t \geq 0\}$, where $Y^{(t)} = (Y_0^{(t)}, Y_1^{(t)}, \ldots)$, and, for $\ell \geq 1$, $Y_\ell^{(t)}$ represented the number of customers in service with remaining service duration equal to ℓ at time t, while $Y_0^{(t)}$ represented the total number of customers in the system at time t. We then expressed this Markov chain with a slightly different notation. We *grouped* the components $Y_\ell^{(t)}, \ell \geq 1$ into a *histogram* $H^{(t)}$ of remaining service durations, which is a function on $\{1, 2, \ldots\}$ given by $H^{(t)}(\ell) = Y_\ell^{(t)}$. We changed the symbol of the component $Y_0^{(t)}$ left after this procedure to $Z^{(t)}$.

This construction resulted in a Markov chain $\{(Z^{(t)}, H^{(t)}) : t \geq 0\}$ with state space:

© The Author(s) 2025
A. B. Dieker, S. T. Hackman, *QPLEX: A Computational Modeling and Analysis Methodology for Stochastic Systems*, Springer Series in Operations Research and Financial Engineering, https://doi.org/10.1007/978-3-031-74870-7_3

$$\{(z, h) : z \geq 0, |h| = \min(z, n)\}. \tag{3.1}$$

The process $\{(Z^{(t)}, H^{(t)}) : t \geq 0\}$ is our first example of a QPLEX chain. Its state space and the number of transitions are typically enormous, which we referred to as the curse of dimensionality for histograms. We circumvented it in this example by iteratively approximating the pmfs of the $(Z^{(t)}, L^{(t)})$, where $L^{(t)}$ can take the values $1, 2, \ldots$. We constructed $L^{(t)}$ from $H^{(t)}$ by choosing a uniformly at random customer in service at time t and considering its remaining service duration. The distribution of $(Z^{(t)}, L^{(t)})$ has (typically) much smaller support than that of $(Z^{(t)}, H^{(t)})$. (When choosing uniformly at random customers in service *independently* across time, we obtain a stochastic process $\{(Z^{(t)}, L^{(t)}) : t \geq 0\}$. This process is *not* a Markov chain and plays no role in the remainder. If the histograms $H^{(t)}$ are unobservable but $Z^{(t)}$ and $L^{(t)}$ are observable, the "partially" hidden Markov model $\{((Z^{(t)}, H^{(t)}), (Z^{(t)}, L^{(t)})) : t \geq 0\}$ has this process as its fully observable component.)

Counters and Counter Sets We thus consider Markov chains of the form $\{(\mathbf{Z}^{(t)}, \mathbf{H}^{(t)}) : t \geq 0\}$, where

$$(\mathbf{Z}^{(t)}, \mathbf{H}^{(t)}) = (Z_1^{(t)}, \ldots, Z_N^{(t)}, H_1^{(t)}, \ldots, H_M^{(t)})$$

for nonnegative integers N and M. We call the components of $\mathbf{Z}^{(t)}$ *counters* and $\mathbf{Z}^{(t)}$ itself a *counter vector*, and we use the same terminology for (possible) realized values. The components of $\mathbf{H}^{(t)}$ are histograms, and we will shortly explain what they represent. (See Sect. 1.3 for a formal definition of a histogram.) We write $p^{(t)}(z', h'|z, h)$ for the (time-t one-step) transition probabilities, where it should be understood that we use $\mathbf{Z}, \mathbf{H}, \mathbf{Z}'$, and $\mathbf{H}'$ instead of $\mathbf{Z}^{(t)}, \mathbf{H}^{(t)}, \mathbf{Z}^{(t+1)}$, and $\mathbf{H}^{(t+1)}$.

We require that each counter vector $\mathbf{Z}^{(t)}$ takes values in some *counter set* $\Omega_{\mathbf{Z}}$, some Cartesian product of N countable sets. If $N = 1$, we denote the counter set simply by Ω_Z. Counters in this book are taken to be integer-valued, but a counter can also be used to represent categorical information such as the state of the business environment (e.g., robust, average, or recession). Of course, abstractly, it is always possible to work with a single counter. Since we are presenting a modeling rather than a mathematical methodology, using several counters can help make models more expressive. There may be several (potentially equivalent) ways to model, and modeling choices often need to be made.

Entities and Activities Although it is possible to start with an abstract Markov chain $\{\mathbf{Y}^{(t)} : t \geq 0\}$ and group its components as before to construct a QPLEX chain $\{(\mathbf{Z}^{(t)}, \mathbf{H}^{(t)}) : t \geq 0\}$, in this book, we have a specific context in mind, which informs the terminology we have chosen throughout and makes it easier to present the many models we encounter.

Specifically, we consider systems where *entities* (customers, patients, jobs, parts, etc.) undertake *activities* for some random amount of time. They may undertake the same activity any number of times, and several entities may be undertaking the same activity for overlapping time periods. Entities may also arrive to the system or leave the system. Entities need not undertake activities at all times; for example, they may be held (possibly temporarily) until they are able to start an activity. Throughout, we use the symbol m for a generic activity. In Chap. 2, there is a single activity corresponding to undergoing service. We use the word "activity" freely and abstractly. For instance, in Sect. 7.3, we will describe a variation of the model in Chap. 2 with *two* activities corresponding to the undergoing service. This is done by grouping the components of $Y^{(t)}$ into whether they represent "short" or "long" remaining service duration.

Labels and Label Sets Associated with each entity undertaking an activity is a *label*. Typically, an entity's label represents its status with respect to activity progress. Interpretations of labels can be different for different activities; this is a modeling choice. One interpretation is the remaining activity duration; this is the label interpretation we used in Chap. 2. (Shortly, we will discuss other interpretations.) We denote the countable set of possible labels for entities undertaking activity m by Ω_m and call it the *label set* for activity m. Although labels are nonnegative integers for most models in this book, label sets need not consist of adjacent integers or even integers. (We will present some advanced models in which they are pairs of nonnegative integers.)

Here are three canonical examples of label interpretations. The *remaining activity duration* is the number of periods before an entity completes its activity and is applicable when an entity's (random) activity duration is revealed when it starts the activity. (This can often be assumed without loss of generality.) The *elapsed activity duration (age)* is the number of periods since starting the activity (or time spent undertaking the activity if interruptions occur). When we adopt the convention that entities can only start and complete activities at the end of each period, the label set corresponding to the remaining activity duration interpretation can be taken as $\{1, 2, \ldots\}$, while for the age interpretation, it can be taken as $\{0, 1, \ldots\}$ (both can appropriately be truncated from the right if an upper bound on activity durations is known). Indeed, an entity starting its activity at the end of the period between times t and $t + 1$ has an age of 0 at time $t + 1$. An equally valid convention is that entities can only start and complete activities at the beginning of each period, in which case the remaining activity duration can be 0 but an age must be at least 1. Finally, suppose that the activity duration can be represented by a random variable describing the time until absorption of a (discrete-time) finite Markov chain with a single absorbing state. In this case, the activity duration is said to have a (discrete) *phase-type distribution*. Absorption corresponds to completing the activity, and each label represents a nonabsorbing state in this *phase Markov chain*.

Indistinguishability and Histograms The fundamental starting point of the QPLEX methodology is that *entities undertaking the same activity with the same label are indistinguishable*. (This is hardly a conceptual restriction since additional information can be incorporated into the label.) As such, instead of keeping track of each entity's label, keeping track of label frequencies suffices.

We use the symbol h_m for a generic histogram on Ω_m, for which $h_m(\ell_m)$ represents the number of entities undertaking activity m with label ℓ_m for $\ell_m \in \Omega_m$.

As there is one such histogram per activity, the total number of activities equals the number of histograms M. Note that the size $|h_m|$ of h_m represents the number of entities undertaking activity m. We let $\boldsymbol{h} = (h_1, \ldots, h_M)$ denote the vector of these histograms. We let Ω_{H_m} denote the set of histograms on Ω_m and define the set Ω_H of histogram vectors via

$$\Omega_H = \Omega_{H_1} \times \cdots \times \Omega_{H_M}.$$

In the special case $M = 1$, we denote $\boldsymbol{h}$ simply by h and Ω_H by Ω_H. For $\{(\boldsymbol{Z}^{(t)}, \boldsymbol{H}^{(t)}) : t \geq 0\}$ to be a QPLEX chain, we require that each histogram vector $\boldsymbol{H}^{(t)}$ takes values in the set Ω_H, so each $H_m^{(t)}$ is a histogram on Ω_m.

Size Functions For a Markov chain $\{(\boldsymbol{Z}^{(t)}, \boldsymbol{H}^{(t)}) : t \geq 0\}$ on counters and histograms to be a QPLEX chain, we require that *the information contained in the counters is sufficient to determine the number of entities undertaking each activity for each time epoch*. Without this information, it is impossible to work with an Ω_m-valued label instead of a histogram on Ω_m, which is the key to achieving computational efficiency, as we will see shortly.

Returning to our example from Chap. 2 and the state space in (3.1), the number of customers in service can be expressed in terms of the number of customers in the system z via $\min(z, n)$, regardless of the time epoch. More generally, a *size function* $x_m^{(t)}$ for activity m at time t is a function for which $x_m^{(t)}(z)$ represents the number of entities undertaking activity m at time t when the counter vector is z. Thus, by definition, for any possible value $(z, \boldsymbol{h})$ of $(\boldsymbol{Z}^{(t)}, \boldsymbol{H}^{(t)})$, we must always have, for any activity m, that

$$x_m^{(t)}(z) = |h_m|.$$

We write $\boldsymbol{x}^{(t)}(z) = (x_1^{(t)}(z), \ldots, x_M^{(t)}(z))$ and refer to this vector-valued function as a *size function vector*. In the special case $M = 1$, we denote $\boldsymbol{x}^{(t)}$ simply by $x^{(t)}$.

It is possible to include a counter in the model for each histogram to track the histogram size and then order the counters such that $x_m^{(t)}(z) = z_m$ for all m. Instead of requiring that each activity have such a designated counter, we found that the additional flexibility provided by using size functions can make it significantly easier to specify a model and can allow for more expressive models. Of course, if desirable, the modeler is free to introduce these additional counters.

QPLEX Chains We are now in a position to formally define a QPLEX chain. As soon as we specify the counter set Ω_Z, the number of activities M, and the label sets Ω_m (thus also specifying Ω_{H_m}), we can define

$$\Omega_{Z,H} = \Omega_Z \times \Omega_H$$

and, for all $t \geq 0$,

$$\Omega_{Z,H}^{(t)} = \{(z, h) \in \Omega_{Z,H} : |h_m| = x_m^{(t)}(z) \text{ for all } m\}.$$

Definition 3.1 A discrete-time stochastic process $\{(Z^{(t)}, H^{(t)}) : t \geq 0\}$ is a *QPLEX chain* with counter set Ω_Z, M activities, label sets $\Omega_1, \ldots, \Omega_M$, and size function vectors $\{x^{(t)} : t \geq 0\}$ if it satisfies the following requirements:

(R1) $\{(Z^{(t)}, H^{(t)}) : t \geq 0\}$ is a Markov chain with state space included in $\Omega_{Z,H}$.

(R2) The support of $(Z^{(0)}, H^{(0)})$ is included in $\Omega_{Z,H}^{(0)}$.

(R3) For all $(z, h) \in \Omega_{Z,H}^{(t)}$ and $t \geq 0$, the support of $p^{(t)}(z', h'|z, h)$ is included in $\Omega_{Z,H}^{(t+1)}$, where $p^{(t)}(z', h'|z, h)$ are the (time-t one-step) transition probabilities of the Markov chain from (R1).

If the counter set, number of activities, label sets, and size function vectors are clear from the context, we simply say that $\{(Z^{(t)}, H^{(t)}) : t \geq 0\}$ is a QPLEX chain.

Here are three examples to further explain the requirements of this definition. The first two examples examine different state space representations associated with the example system of Chap. 2, and the third one examines a simple variation.

Example 3.1 Consider the stochastic process $\{(Z_B^{(t)}, Z_S^{(t)}, H^{(t)}) : t \geq 0\}$, where $Z_B^{(t)}$ and $Z_S^{(t)}$ represent the number of customers in the buffer and in service, respectively, at time t. Since $(Z_B^{(t)}, Z_S^{(t)}, H^{(t)})$ is in one-to-one correspondence with $(Z^{(t)}, H^{(t)})$, this process is again a Markov chain. It is a QPLEX chain since the histogram size $|H^{(t)}| = \min(Z_B^{(t)} + Z_S^{(t)}, n)$ is a function of the two counters $Z_B^{(t)}$ and $Z_S^{(t)}$, so we can use $x^{(t)}(z_B, z_S) = \min(z_B + z_S, n)$ for the size function for all t. The label set is $\{1, 2, \ldots\}$. △

Example 3.2 Consider the stochastic process $\{(Z_B^{(t)}, H^{(t)}) : t \geq 0\}$, where $Z_B^{(t)}$ is as in the previous example. There is a one-to-one correspondence between $(Z^{(t)}, H^{(t)})$ and $(Z_B^{(t)}, H^{(t)})$ because $Z_B^{(t)} = Z^{(t)} - |H^{(t)}|$, and so this process is also a Markov chain. However, the histogram size $|H^{(t)}|$ cannot be expressed as a function of the counter $Z_B^{(t)}$. As a result, $\{(Z_B^{(t)}, H^{(t)}) : t \geq 0\}$ *cannot* be a QPLEX chain for any choice of size functions. △

Example 3.3 Suppose that all customers arriving to the system are lost if they cannot be immediately served, i.e., there is no buffer. (This is a special case of the first variation from Sect. 2.7.) The process $\{H^{(t)} : t \geq 0\}$ is a Markov chain. However, it is *not* a QPLEX chain. As there are no counters, there is no way to determine how many customers are undergoing service from the counters. Including the variable $Z^{(t)}$, while redundant from the Markov chain perspective, is *not redundant from the QPLEX perspective*, as $\{(Z^{(t)}, H^{(t)}) : t \geq 0\}$ *is* a QPLEX chain, as previously seen. △

Kernel Maps Consider a QPLEX chain $\{(Z^{(t)}, H^{(t)}) : t \geq 0\}$. We use the symbols Z and H for generic random vectors representing a counter vector and a histogram vector, respectively. It is understood that (Z, H) is $\Omega_{Z,H}$-valued. This means that Z is Ω_Z-valued, and each component H_m of H is a histogram on the label set Ω_m.

Fix time t. Let $p^{(t)}(z', h'|z, h)$ denote the *transition probabilities* and recall from Sect. 1.4 that we assume that these probabilities are defined for all (z, h) and (z', h') in the state space. The *kernel map* $\mathcal{K}^{(t)}$ maps a pmf ν of the counter vector Z and the histogram vector H to the pmf $\mathcal{K}^{(t)}(\nu)$ of (Z, H) via

$$[\mathcal{K}^{(t)}(\nu)](z', h') = \sum_{(z,h)} \nu(z, h) \times p^{(t)}(z', h'|z, h), \qquad (3.2)$$

where the sum is taken over the state space. To be able to calculate this sum numerically, the support of ν must be finite. Since $\{(Z^{(t)}, H^{(t)}) : t \geq 0\}$ is a QPLEX chain, the pmf $\mathcal{K}^{(t)}(\nu)$ is a pmf of (Z, H) that is supported on $\Omega_{Z,H}^{(t+1)}$ if ν is a pmf of (Z, H) that is supported on $\Omega_{Z,H}^{(t)}$.

Kernel Iterates Given some initial pmf $\nu_{\text{ker}}^{(0)}$ of (Z, H), the iterative scheme $\nu_{\text{ker}}^{(t+1)} = \mathcal{K}^{(t)}(\nu_{\text{ker}}^{(t)})$ for $t \geq 0$ generates *kernel iterates*:

$$\nu_{\text{ker}}^{(0)}, \nu_{\text{ker}}^{(1)}, \nu_{\text{ker}}^{(2)}, \ldots.$$

Calculating these iterates is generally impossible due to the cardinality of their support. Indeed, if $|B|$ denotes the cardinality of a set B, the number of histograms on B of size x is $\binom{x+|B|-1}{x}$, which is exponential in x. Thus, if the counter vector $Z^{(t)}$ at time t is z, the number of compatible histograms for activity m is exponential in $x_m^{(t)}(z)$. Even if there is only one activity with a handful of possible labels, it is generally utterly unworkable to carry out calculations for each state unless the size functions $x_m^{(t)}$ are uniformly very small. We call this the *curse of dimensionality for histograms*.

3.2   QPLEX Maps

The QPLEX methodology considers a *single* label per activity: the label of an entity *uniformly chosen at random* among the entities undertaking each activity and chosen independently across activities. (Independence across time does not need to be imposed as we do not consider joint distributions of these labels across time.) The QPLEX methodology "replaces" histograms with labels for purposes of calculations, where the replacement should be understood in a probabilistic sense instead of in a state space reduction sense.

Consider a QPLEX chain $\{(Z^{(t)}, H^{(t)}) : t \geq 0\}$. Fix time t and let $L_m^{(t)}$ denote the uniformly chosen label for activity m and write $L^{(t)} = (L_1^{(t)}, \ldots, L_M^{(t)})$, so $(Z^{(t)}, L^{(t)})$ is a random function of the Markov state $(Z^{(t)}, H^{(t)})$. We use the symbol

L for a generic random label vector, and it is understood that each component L_m is an element of the label set Ω_m. We use the symbol ℓ_m to denote a realized or potential label value and write $\boldsymbol{\ell} = (\ell_1, \ldots, \ell_M)$. The QPLEX methodology iteratively calculates approximations to the pmfs of the $(\boldsymbol{Z}^{(t)}, \boldsymbol{L}^{(t)})$. *Instead of iteratively calculating joint pmfs of a counter vector $\boldsymbol{Z}$ and a histogram vector $\boldsymbol{H}$, the QPLEX methodology calculates joint pmfs of a counter vector $\boldsymbol{Z}$ and a label vector $\boldsymbol{L}$.* The latter can be pmfs on *much* smaller sets as their support is unaffected by the size functions, unlike the support of pmfs of $(\boldsymbol{Z}, \boldsymbol{H})$.

Construction We replace the iterative scheme for pmfs of the counter vector $\boldsymbol{Z}$ and the histogram vector $\boldsymbol{H}$ based on the kernel maps $\mathcal{K}^{(t)}$ with an iterative scheme for pmfs of the counter vector $\boldsymbol{Z}$ and the label vector $\boldsymbol{L}$ based on the *QPLEX maps* $\mathcal{Q}^{(t)}$. Refer to the diagram below. There are two "levels": a top level with pmfs of $(\boldsymbol{Z}, \boldsymbol{H})$ and a bottom level with pmfs of $(\boldsymbol{Z}, \boldsymbol{L})$. The pmfs in the top level remain unworkable in practice, and the pmfs on the bottom level have (generally) much smaller supports than their counterparts at the top level.

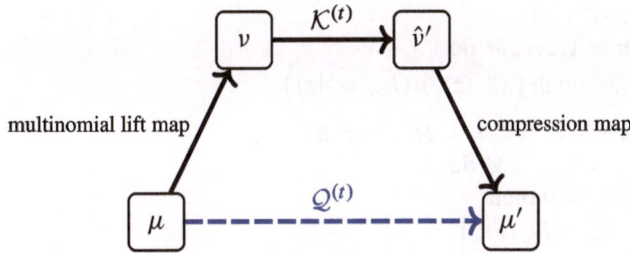

The QPLEX map $\mathcal{Q}^{(t)}$ is the composition of the *compression*, kernel, and *multinomial lift* maps. Starting in the lower left corner of the diagram with some pmf μ on counters and labels, we obtain a pmf ν of $(\boldsymbol{Z}, \boldsymbol{H})$ by applying the multinomial lift map. Next, the kernel map $\mathcal{K}^{(t)}$ is applied to ν to obtain another pmf $\hat{\nu}'$ of $(\boldsymbol{Z}, \boldsymbol{H})$. Finally, the compression map is applied to $\hat{\nu}'$ to obtain the pmf μ' of $(\boldsymbol{Z}, \boldsymbol{L})$.

The multinomial lift and compression maps are the vehicles by which we can "translate" pmfs between the two levels. Both maps arise from the basic idea behind the thought experiment described in Sect. 2.4, but we now work with an arbitrary number of counters and histograms. The mathematical properties of the multinomial lift and compression maps, as well as the compositions of these two maps, play an essential role in the formal analysis presented in Part III. Until then, we will represent these maps as distributional program pseudocode. (See Sect. 16.1 for formal definitions and numerical examples.) For example, the multinomial lift map is *not* linear, which is a key reason why the QPLEX approach is fundamentally different than the Markov chain approach.

Distributional Program Representation The distributional program below represents the QPLEX map. As in Chap. 2, there are three parts to this program, corresponding to the multinomial lift map (Lines 2–4), the kernel map (Line 5), and the compression map (Lines 6–10). The multinomial lift map effectively uses conditional independence to sample the labels of each entity, while the compression map samples the labels of uniformly chosen at random entities.

Given an input pmf μ of (Z, L), the counters are sampled in Line 2 from the marginal pmf of the counters. The histograms for each activity are then sampled from the appropriate multinomial pmf in Lines 3 and 4. These histograms are conditionally *independent* given the counters, which is implicit in these sampling statements. A pair (z', h') is then sampled using the transition probabilities $p^{(t)}(z', h'|z, h)$ (Line 5). The next labels for each activity are sampled from the pmf proportional to the frequencies of the corresponding histograms, which can be interpreted as selecting the label of a *uniformly chosen at random* entity undertaking the activity at time $t + 1$ (Line 8). This sampling is done independently for each activity given h'. Whenever there are no entities undertaking activity m, the next label is sampled from some arbitrary dummy pmf θ_m^0 on Ω_m that plays no significant role (Line 10). (In this abstract setting, there is no analog of the pmf g we used for this purpose in Chap. 2.)

1: **function** QPLEXMAP$^{(t)}(\mu)$
2:      $z \sim \mu(Z = \cdot)$
3:      **for** $m = 1, \ldots, M$ **do**
4:          $h_m \sim \text{mult}\left(x_m^{(t)}(z), \mu(L_m = \cdot|z)\right)$
5:      $(z', h') \sim p^{(t)}(Z' = \cdot, H' = \cdot|z, h)$
6:      **for** $m = 1, \ldots, M$ **do**
7:          **if** $h_m' \neq 0$ **then**
8:              $\ell_m' \sim h_m'/|h_m'|$
9:          **else**
10:             $\ell_m' \sim \theta_m^0(L_m' = \cdot)$
11:     **return** (z', ℓ')

QPLEX Iterates Given some initial pmf $\mu_Q^{(0)}$ of (Z, L), the iterative scheme $\mu_Q^{(t+1)} = \mathcal{Q}^{(t)}(\mu_Q^{(t)})$ for $t \geq 0$ generates *QPLEX iterates*

$$\mu_Q^{(0)}, \mu_Q^{(1)}, \mu_Q^{(2)}, \ldots.$$

Each QPLEX iterate is a pmf of (Z, L). The marginal pmfs of the counters for the QPLEX iterates approximate the marginal pmfs of the counters for the kernel iterates. We stress that these iterates are *deterministic*, unlike distributions obtained from a stochastic simulation.

Although the cardinality of the support of QPLEX iterates is (typically) much smaller than the cardinality of the support of kernel iterates, the QPLEX iterative scheme should *not* be interpreted as having been obtained from a state space reduction of sorts. In fact, we will show in Chap. 8 via a counterexample that the collection of QPLEX approximations of finite-dimensional joint pmfs for counter vectors is, in general, *not* consistent. Consequently, *this collection cannot be associated with a stochastic process* such as a Markov chain or a hidden Markov model. However, we do explore connections with what are known as nonlinear Markov chains.

3.3 Circumventing the Curse of Dimensionality for Histograms

Iteratively calculating the QPLEX iterates $\mu_Q^{(t)}$ remains, in general, impossible due to its reliance on the kernel map $\mathcal{K}^{(t)}$. Indeed, the definition of the QPLEX map $\mathcal{Q}^{(t)}$ involves going "up, over, and down" in the above diagram, which necessitates working with histograms, as the above distributional program makes clear. Even for "small" systems, the number of possible values for the histogram vector h and the number of possible transitions are too large for any practical calculation. To circumvent the curse of dimensionality for histograms, we must be able to evaluate the QPLEX map *without* having to enumerate histograms or transitions.

QPLEX Calculus By imposing structure on the $\mathcal{K}^{(t)}$ maps, specifically by defining model primitives and prescribing underlying transition dynamics in terms of these primitives, we show that it is possible to "go directly across" at the bottom level in the above diagram. By this, we mean that $\mu_Q^{(t+1)}$ can be calculated from $\mu_Q^{(t)}$ by only working with counters and labels, avoiding histograms. We call the resulting methodology the *QPLEX calculus*. Without this structure, there are no computational benefits of working with labels instead of histograms. Choosing appropriate pmf primitives is a balancing act between restricting generality, i.e., constraining which models can be represented, and deriving formulas that allow one to go directly across at the bottom level of the above diagram.

Model Primitives The model primitives we use to define what we call *simple transition dynamics* are presented in Chap. 4. Here is a brief description.

- Each entity undertaking an activity independently determines whether it completes the activity during each period. An entity that does not complete its activity is assigned category "stay" and continues its activity. The probability that an entity completes its activity depends on its current label and the counters. In a queueing system, for example, the completion probability can depend on how many customers are currently being served ("processor sharing"). As another example, a policy can dictate that the completion probability increases if the number of customers in the system exceeds a threshold.
- The next counters are then determined using the *routing* pmf primitive, also incorporating entities arriving from outside sources. This primitive typically involves the most system-dependent logic.
- Each entity that continues its activity or starts an activity is *(re)labeled* using pmf primitives that depend on counters and its current label (if any).

- The labels of all entities undertaking an activity at the next time epoch are aggregated to generate the next histogram for each activity.

Examples of single and multiple activity models that can be represented using these model primitives are presented in Chap. 5.

The model primitives underlying what we call *advanced transition dynamics* are presented in Chap. 6. These primitives extend the ones for simple transition dynamics by permitting more flexible categories, routing, and relabeling. Examples of models that require the flexibility provided by these primitives are presented in Chap. 7.

Multinomial Propagation In Chaps. 4 and 6, we derive the QPLEX calculus by heavily exploiting properties of multinomial random variables in conjunction with the transition dynamics. The dynamics are defined via several stages, each of which preserves multinomial distributions in an appropriate sense. We call this *multinomial propagation*. We chose the model primitives and dynamics in order for the parameters of these multinomial distributions to be expressed and easily calculated in terms of the model primitives, which then results in the sought-after QPLEX calculus.

Chapter 4
Simple Transition Dynamics

This chapter defines and describes four model primitives that impose structure on the transition probabilities of a QPLEX chain. We exploit this structure to derive an "efficient" representation of the QPLEX map that overcomes the curse of dimensionality for histograms, thereby establishing the QPLEX calculus for a broad class of QPLEX chains.

Throughout this chapter, we fix an arbitrary QPLEX chain $\{(\boldsymbol{Z}^{(t)}, \boldsymbol{H}^{(t)}) : t \geq 0\}$ with counter set Ω_Z, M activities, a collection of size function vectors $\{\boldsymbol{x}^{(t)} : t \geq 0\}$, and label sets $\Omega_1, \ldots, \Omega_M$.

4.1 Model Primitives

We define transition probabilities with simple transition dynamics using the following four model primitives for each time t:

- *Completion probability functions* $\gamma_m^{(t)}(z, \ell_m)$,
- *Routing pmf* $\pi^{(t)}(z'|z, \boldsymbol{d})$,
- *Relabeling pmfs* $\pi^{(t),\mathrm{rel}}(\ell_m'|z, \ell_m, z')$, and
- *Join pmfs* $\pi_m^{(t),\mathrm{join}}(\ell_m'|z, z')$.

The superscripts stress that these model primitives may vary over time. Indeed, of primary interest are models where at least one of these model primitives varies over time. It should be understood that the $\gamma_m^{(t)}(z, \ell_m)$ only need to be specified for values of z and ℓ_m that are meaningful in the context of the model, and a similar understanding holds for the other primitives.

Throughout the next three sections, we fix t and consider the period between times t and $t + 1$, which we refer to as *this period*. We speak of labels pertaining to time t as *current labels* and those pertaining to time $t+1$ as *next labels*. Similarly, we

© The Author(s) 2025
A. B. Dieker, S. T. Hackman, *QPLEX: A Computational Modeling and Analysis Methodology for Stochastic Systems*, Springer Series in Operations Research and Financial Engineering, https://doi.org/10.1007/978-3-031-74870-7_4

speak of the *current counter*, *next counter*, *current histogram*, and *next histogram* vectors.

Completion Probability Functions This model primitive is used to determine the number of entities completing their activities during this period. The current label of an entity determines whether or not the entity completes its activity during this period: specifically, each of the entities undertaking activity m at time t independently completes its activity with probability $\gamma_m^{(t)}(z, \ell_m)$ if the current counter vector is z and the entity's current label is ℓ_m.

Here are examples of completion probability functions based on the three label interpretations from Sect. 3.1. Here and throughout all examples in this section, we adopt the convention that entities start and complete activities at the end of a period, as we discussed in Sect. 3.1.

- If the labels for activity m have the remaining activity duration interpretation, then a specification of the completion probability function primitive consistent with this interpretation is

$$\gamma_m^{(t)}(z, \ell_m) = 1(\ell_m = 1).$$

Note that the scope of this completion probability function does not include the current counter vector z. A generalization of this example that *does* use this counter vector is as follows. Suppose at time t it is stipulated, as a matter of policy, that any entity undertaking activity m with a remaining activity duration less than a threshold of $\zeta_m^{(t)}(z)$ is deemed to have completed their activity. (Such a policy may be used to "clear" entities from a congested location.) A specification of the completion probability function primitive consistent with this policy is

$$\gamma_m^{(t)}(z, \ell_m) = 1(\ell_m \leq \zeta_m^{(t)}(z)).$$

- If the labels for activity m have the elapsed activity duration ("age") interpretation, then a specification of the completion probability function consistent with this interpretation is

$$\gamma_m^{(t)}(z, \ell_m) = \frac{g_m^{(t-1-\ell_m)}(\ell_m + 1)}{\sum_{\tilde{\ell}_m > \ell_m} g_m^{(t-1-\ell_m)}(\tilde{\ell}_m)}, \tag{4.1}$$

where $g_m^{(\tau)}$ is the pmf of the activity duration for entities starting their activity between times τ and $\tau + 1$. Each activity duration must take at least one time period. Note that the denominator in (4.1) cannot be zero because this expression is only meaningful for achievable values of the elapsed activity duration ("age") ℓ_m. Large values of ℓ_m may not be relevant for activity m when all entities must have completed their activity well before they could have reached age ℓ_m. Note also that the $g_m^{(\tau)}$ pmfs are *not* among our model primitives; the primitives we have chosen are more flexible.

- If the labels for activity m represent an entity's activity phase, then $\gamma_m^{(t)}(z, \ell_m)$ is the probability that an entity completes its activity given that the current phase is ℓ_m, i.e., that the underlying phase Markov chain reaches the absorbing state.

Routing Pmf Given z and d, the next counter vector z' is sampled using this model primitive. The routing pmf encodes the distributional mechanism by which new entities arrive and move around the system. Each routing pmf should be thought of as a collection of pmfs parameterized by pairs (z, d) that make sense in the context of the model. When specifying this model primitive (e.g., via a distributional program), we do not require that the set of such pairs is stated explicitly. A routing pmf cannot use the label information of entities that complete their activities during this period.

The distributional program below represents the routing pmf $\pi^{(t)}(z'|z, d)$ for the multiserver queueing model of Chap. 2.

1: **function** ROUTING$^{(t)}(z, d)$
2:    $a \sim \alpha^{(t)}(A = \cdot|z)$
3:    $z' \leftarrow z - d + a$
4:    **return** z'

Relabeling Pmfs This primitive is used to relabel entities that do not complete their activities during times t and $t + 1$. Each of the entities with current label ℓ_m undertaking activity m that do not complete the activity during this period independently receives a next label ℓ'_m using the relabeling pmf primitive $\pi_m^{(t),\text{rel}}(\ell'_m|z, \ell_m, z')$.

Here are examples of relabeling pmfs.

- If the labels for activity m have the remaining activity duration interpretation, then a specification of the relabeling pmf primitive consistent with this interpretation is

$$\pi_m^{(t),\text{rel}}(\ell'_m|z, \ell_m, z') = 1(\ell'_m = \ell_m - 1).$$

It may appear that this choice leads to the label $\ell'_m = 0$ if $\ell_m = 1$, which is incompatible with this label interpretation. However, only entities that do not complete their activity are relabeled, and it is therefore implicit that only $\ell_m > 1$ is relevant.

- If the labels for activity m have the elapsed activity duration ("age") interpretation, then a specification of the relabeling pmf primitive consistent with this interpretation is

$$\pi_m^{(t),\text{rel}}(\ell'_m|z, \ell_m, z') = 1(\ell'_m = \ell_m + 1).$$

- If the labels for activity m represent an entity's activity phase, then the relabeling probabilities $\pi_m^{(t),\text{rel}}(\ell'_m|z, \ell_m, z')$ are the phase Markov chain's one-step transition probabilities conditioned on not being absorbed.

Join Pmfs This primitive is used to label entities starting activity m during this period. Each of these entities independently samples a label from the join pmf primitive $\pi_m^{(t),\text{join}}(\ell_m' | z, z')$.

Here are some examples of join pmfs.

- If the labels for activity m have the remaining activity duration interpretation, then a specification of the relabeling pmf primitive consistent with this interpretation is

$$\pi_m^{(t),\text{join}}(\ell_m' | z, z') = g_m^{(t)}(\ell_m'),$$

where we recall that $g_m^{(t)}$ is the pmf of the activity duration for entities starting their activity between times t and $t + 1$.

- If the labels for activity m have the elapsed activity duration ("age") interpretation, then a specification of the relabeling pmf primitive consistent with this interpretation is

$$\pi_m^{(t),\text{join}}(\ell_m' | z, z') = 1(\ell_m' = 0).$$

- If the labels for activity m represent an entity's activity phase, then $\pi_m^{(t),\text{join}}(\ell_m' | z, z')$ is the pmf of the initial phase.

4.2 Transition Probabilities

This section describes how the model primitives parameterize the transition probabilities of a QPLEX chain with simple transition dynamics. We illustrate the construction numerically at the end of this section.

Distributional Program Representation Fix some current counter vector z and current histogram vector h with $|h_m| = x_m^{(t)}(z)$ for each activity m. The distributional program TRANSITIONPROBABILITY$^{(t)}$ below represents the transition probabilities $p^{(t)}(z', h'|z, h)$ from time t to time $t + 1$ in terms of the routing pmf primitive and several distributional programs that use the other model primitives. These programs will be introduced shortly. It generates the next counter and histogram vectors z' and h' with $|h_m'| = x_m^{(t+1)}(z')$ in several stages described below. Implicit in this pseudocode are the definitions $d = (d_m)_m$ and $h' = (h_{m'}')_{m'}$; in this section, we similarly use $h^{\text{stay}} = (h_m^{\text{stay}})_m$, $h^{\text{rel}} = (h_m^{\text{rel}})_m$, and $h^{\text{join}} = (h_{m'}^{\text{join}})_{m'}$.

1: **function** TRANSITIONPROBABILITY$^{(t)}(z, h)$
2:    $h^{\text{stay}} \sim$ STAYHISTOGRAMS$^{(t)}(z, h)$
3:    **for** $m = 1, \ldots, M$ **do**
4:        $d_m \leftarrow |h_m| - |h_m^{\text{stay}}|$
5:    $z' \sim \pi^{(t)}(Z' = \cdot | z, d)$

6:   $h^{\text{rel}} \sim \text{RELABELHISTOGRAMS}^{(t)}(z, h^{\text{stay}}, z')$
7:   $h^{\text{join}} \sim \text{JOINHISTOGRAMS}^{(t)}(z, d, z')$
8:   $h' \leftarrow h^{\text{rel}} + h^{\text{join}}$
9:   **return** (z', h')

Stay Histograms The *stay histogram* h_m^{stay} for activity m comprises the frequencies of all current labels for entities that do not complete the activity during this period. The distributional program below outputs the pmf of the vector of stay histograms given the current counter and histogram vectors z and h using the completion probability function model primitive. Each of the $h_m(\ell_m)$ entities with label ℓ_m undertaking activity m independently completes their activity with the same probability $\gamma_m^{(t)}(z, \ell_m)$, also independently across labels and activities. Thus, the number of entities $h_m^{\text{stay}}(\ell_m)$ with label ℓ_m that do not complete activity m during this period has a binomial pmf with parameters $h_m(\ell_m)$ and $1 - \gamma_m^{(t)}(z, \ell_m)$.

1: **function** $\text{STAYHISTOGRAMS}^{(t)}(z, h)$
2:   **for** $m = 1, \ldots, M$ **do**
3:     **for** $\ell_m \in \Omega_m$ **do**
4:       $h_m^{\text{stay}}(\ell_m) \sim \text{bin}\left(h_m(\ell_m), 1 - \gamma_m^{(t)}(z, \ell_m)\right)$
5:   **return** h^{stay}

Number of Completions and Next Counter Vector The number of entities that complete activity m during this period d_m is simply the difference between the number of entities undertaking activity m and the number of entities that stay (Line 4). The next counter vector z' is sampled from the routing pmf primitive (Line 5).

Relabel Histograms The *relabel histogram* h_m^{rel} for activity m comprises the frequencies of all next labels for entities that do not complete the activity. The distributional program below outputs the pmf of the vector of relabel histograms given the current counter vector z, the vector of stay histograms h^{stay}, and the next counter vector z' using the relabeling pmf primitives. There are $h_m^{\text{stay}}(\ell_m)$ entities undertaking activity m with current label ℓ_m that do not complete the activity during this period. Each of these entities independently receives a next label ℓ_m' using the relabeling pmf primitive $\pi_m^{(t),\text{rel}}(\ell_m' | z, \ell_m, z')$, also independently across different current labels and activities. The histogram h_m^{rel} accumulates the histograms of next labels associated with these entities over all possible choices of ℓ_m, each of which is independent and has a multinomial pmf.

1: **function** $\text{RELABELHISTOGRAMS}^{(t)}(z, h^{\text{stay}}, z')$
2:   **for** $m = 1, \ldots, M$ **do**
3:     $h_m^{\text{rel}} \leftarrow 0$
4:     **for** $\ell_m \in \Omega_m$ **do**
5:       $h_m^{\text{rel}} +\sim \text{mult}\left(h_m^{\text{stay}}(\ell_m), \pi_m^{(t),\text{rel}}(L_m' = \cdot | z, \ell_m, z')\right)$
6:   **return** h^{rel}

Join Histograms　The *join histogram* h_m^{join} for activity m comprises the frequencies of all next labels for entities starting the activity during this period. The distributional program below outputs the pmf of the vector of join histograms given the current counter vector z, the vector d, and the next counter vector z' using the join pmf primitives. There are $x_m^{(t+1)}(z')$ entities undertaking activity m at time $t + 1$. Of these,

$$x_m^{(t+1)}(z') - |h_m^{\text{stay}}| = x_m^{(t+1)}(z') - (x_m^{(t)}(z) - d_m)$$

entities start activity m during this next period. Each of these entities independently obtains a label using the join pmf primitive $\pi_m^{(t),\text{join}}(\ell_m'|z, z')$. The histogram h_m^{join} of the labels associated with these entities therefore has a multinomial pmf.

1: **function** JOINHISTOGRAMS$^{(t)}(z, d, z')$
2: 　　**for** $m = 1, \ldots, M$ **do**
3: 　　　　$h_m^{\text{join}} \sim \text{mult}\left(x_m^{(t+1)}(z') - (x_m^{(t)}(z) - d_m), \pi_m^{(t),\text{join}}(L_m' = \cdot|z, z')\right)$
4: 　　**return** h^{join}

Aggregation of Histograms　The final step, which does not use a model primitive, calculates h' by simply aggregating the components of h^{rel} and h^{join}, as shown in Line 8. The sum is to be interpreted componentwise.

Numerical Example　We suppose that z and h_1 are fixed and given, and we show how a realization of z' and h_1' is generated using the model primitives. We do not work out the probability of the realization we generate. The components of h' are actually jointly sampled using the full current histogram vector h, but for conciseness, we only focus on generating z' and h_1' from z and h_1 here. There are $x_1^{(t)}(z) = |h_1| = 14$ entities undertaking activity 1. The histogram of labels, h_1, the completion probabilities for each of the four labels shown, and a possible realization of h_1^{stay} are given in the table below. The value of d_1 is $|h_1| - |h_1^{\text{stay}}| = 14 - 10 = 4$.

ℓ_1	$h_1(\ell_1)$	$\gamma_1^{(t)}(z, \ell_1)$	$h_1^{\text{stay}}(\ell_1)$
1	4	0.1	4
2	5	0.2	3
3	3	0.4	2
4	2	0.3	1
	$|h_1| = 14$		$|h_1^{\text{stay}}| = 10$

The diagram below shows that h_1^{stay} arises from h_1. Entities with label 1 are depicted by a blue ball, with label 2 by a red ball, etc. We have added the size of each histogram in parentheses.

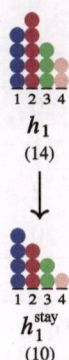

Since we have shown how d_1 has been generated, we imagine this procedure has been similarly done for all activities so that the full vector $\boldsymbol{d}$ has been generated. The next counter vector $\boldsymbol{z}'$ is then sampled using the routing pmf primitive. Suppose that $\boldsymbol{z}'$ satisfies $x^{(t+1)}(\boldsymbol{z}') = 13$. The relabeling probabilities $\pi_1^{(t),\mathrm{rel}}(\ell_1'|\boldsymbol{z}, \ell_1, \boldsymbol{z}')$ are given in the table below for all ℓ_1 and ℓ_1'. For example, the next labels of the $h_1^{\mathrm{stay}}(1) = 4$ entities that had label 1 and did not complete their respective activities are i.i.d., and become 1, 2, 3, 4, with pmf $\pi_1^{(t),\mathrm{rel}}(L_1' = \cdot|\boldsymbol{z}, L_1 = 1, \boldsymbol{z}') = (0.7, 0.1, 0.0, 0.2)$.

ℓ_1 \ ℓ_1'	1	2	3	4
1	0.7	0.1	0.0	0.2
2	0.0	0.3	0.2	0.5
3	0.1	0.2	0.6	0.1
4	0.3	0.4	0.1	0.2

The next table shows a possible realization of the h_1^{rel} generated from h_1^{stay}. For example, of the $h_1^{\mathrm{stay}}(1) = 4$ entities with label 1 that did not complete their activity, the table shows that one of the four entities receives a next label 1, two receive a next label 2, etc. The totals for each of the next labels are represented in the realized histogram $h_1^{\mathrm{rel}} = (2, 3, 3, 2)$. Note that $|h_1^{\mathrm{rel}}| = 10$.

ℓ_1 \ ℓ_1'	1	2	3	4	h_1^{stay}
1	2	1	0	1	4
2	0	1	1	1	3
3	0	0	2	0	2
4	0	1	0	0	1
h_1^{rel}	2	3	3	2	

The diagram below depicts the realization of h_1^{rel} obtained from h_1^{stay} with the intermediate histograms obtained from $h_1^{\text{stay}}(1), \ldots, h_1^{\text{stay}}(4)$. These histograms are shown in the rows of the above table.

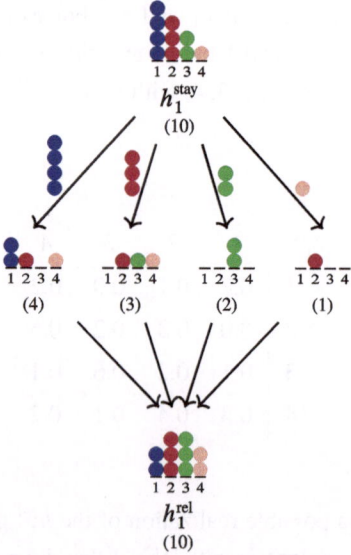

Given that $x_1^{(t+1)}(z') = 13$, there are $x_1^{(t+1)}(z') - |h_1^{\text{rel}}| = 13 - 10 = 3$ entities starting the activity during this period. Suppose that $\pi_1^{(t).\text{join}}(L = \cdot | z, z')$ equals $(0.1, 0.4, 0.2, 0.3)$. If the realization of h_1^{join}, a multinomial random variable with parameters 3 and $(0.1, 0.4, 0.2, 0.3)$, equals $(0, 2, 0, 1)$, then the realization of h_1' is given by $(2, 3, 3, 2) + (0, 2, 0, 1) = (2, 5, 3, 3)$, as depicted below.

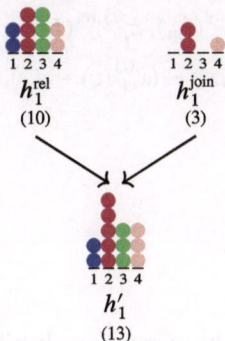

$$h_1^{\text{rel}} \qquad\qquad h_1^{\text{join}}$$
$$(10) \qquad\qquad (3)$$

$$h_1'$$
$$(13)$$

4.3 QPLEX Calculus

The distributional program TRANSITIONPROBABILITY$^{(t)}$ contains sampling statements for each of the components of h^{stay}, h^{rel}, h^{join}. As a result, evaluating the QPLEX map requires enumerating the possible sample values of these histogram vectors, in addition to the possible values of the histogram vectors h and h'. The number of such values is generally prohibitive, even if there were only a single histogram. This section focuses on deriving an alternative representation for the QPLEX map with simple transition dynamics *that does not require enumerating over all possible histogram values*. In terms of the distributional program for the QPLEX map, this amounts to refactoring the distributional program so that histograms no longer appear in the refactored program. At the end of this section, we obtain an "efficient" algebraic representation of the QPLEX map from this refactored distributional program.

The starting point for this section is the distributional program representation of the QPLEX map below. We have substituted all lines associated with the distributional program TRANSITIONPROBABILITY$^{(t)}$ for Line 5 in the distributional program representation of the QPLEX map in Sect. 3.2 and merged for-loops where possible.

1: **function** QPLEXMAP$^{(t)}(\mu)$
2:     $z \sim \mu(\mathbf{Z} = \cdot)$
3:     **for** $m = 1, \ldots, M$ **do**
4:         $h_m \sim \text{mult}\left(x_m^{(t)}(z), \mu(L_m = \cdot|z)\right)$
5:         **for** $\ell_m \in \Omega_m$ **do**
6:             $h_m^{\text{stay}}(\ell_m) \sim \text{bin}\left(h_m(\ell_m), 1 - \gamma_m^{(t)}(z, \ell_m)\right)$
7:         $d_m \leftarrow |h_m| - |h_m^{\text{stay}}|$
8:     $z' \sim \pi^{(t)}(\mathbf{Z}' = \cdot|z, \boldsymbol{d})$
9:     **for** $m = 1, \ldots, M$ **do**
10:         $h_m^{\text{rel}} \leftarrow 0$
11:         **for** $\ell_m \in \Omega_m$ **do**

12:          $h_m^{\mathrm{rel}} +\sim \mathrm{mult}\left(h_m^{\mathrm{stay}}(\ell_m),\, \pi_m^{(t),\mathrm{rel}}(L_m' = \cdot | z, \ell_m, z')\right)$

13:          $h_m^{\mathrm{join}} \sim \mathrm{mult}\left(x_m^{(t+1)}(z') - (x_m^{(t)}(z) - d_m),\, \pi_m^{(t),\mathrm{join}}(L_m' = \cdot | z, z')\right)$

14:          $h_m' \leftarrow h_m^{\mathrm{rel}} + h_m^{\mathrm{join}}$

15:          **if** $h_m' \neq 0$ **then**

16:              $\ell_m' \sim h_m'/|h_m'|$

17:          **else**

18:              $\ell_m' \sim \theta_m^0(L_m' = \cdot)$

19:          **return** (z', ℓ')

We derive the sought-after histogram-free distributional program representing the QPLEX map by eliminating histograms from the program without changing its output. We proceed in three steps, each of which preserves multinomial distributions in an appropriate sense, a concept we call *multinomial propagation*.

Step 1: Eliminate the h_m The variable h_m in the above program is used to generate h_m^{stay} and d_m but is subsequently unused. We replace Lines 4–7

4:          $h_m \sim \mathrm{mult}\left(x_m^{(t)}(z),\, \mu(L_m = \cdot | z)\right)$

5:          **for** $\ell_m \in \Omega_m$ **do**

6:              $h_m^{\mathrm{stay}}(\ell_m) \sim \mathrm{bin}\left(h_m(\ell_m),\, 1 - \gamma_m^{(t)}(z, \ell_m)\right)$

7:          $d_m \leftarrow |h_m| - |h_m^{\mathrm{stay}}|$

with pseudocode that results in the *same* joint distribution of h_m^{stay} and d_m but *without sampling* h_m. We do so by *first* sampling d_m and then h_m^{stay}.

Fix some z with $\mu(z) > 0$ and some activity m. We can think of the labels of the $x_m^{(t)}(z)$ entities undertaking activity m as being i.i.d. with pmf $\mu(L_m = \cdot | z)$ since the pmf of the histogram of labels h_m is multinomial with parameters $x_m^{(t)}(z)$ and $\mu(L_m = \cdot | z)$. These entities independently sample whether they complete their activity during this period given their label. Unconditionally, each such entity completes its activity with the *same* probability, so the number of activity completions has a binomial pmf. The number of trials parameter is $x_m^{(t)}(z)$. The success probability parameter is obtained by conditioning on the label ℓ_m and using the completion probability function $\gamma_m^{(t)}$, so we can sample d_m in Line 7 directly using the distributional program NUMBEROFCOMPLETIONS$_{\mu,m}^{(t)}$ below.

1: **function** NUMBEROFCOMPLETIONS$_{\mu,m}^{(t)}(z)$

2:     $d_m \sim \mathrm{bin}\left(x_m^{(t)}(z),\, \sum_{\ell_m} \mu(\ell_m | z) \times \gamma_m^{(t)}(z, \ell_m)\right)$

3:     **return** d_m

It remains to sample h_m^{stay} given d_m. If $x_m^{(t)}(z) - d_m = 0$, then we have that $h_m^{\mathrm{stay}} = 0$; otherwise, the labels of each of the $x_m^{(t)}(z) - d_m$ entities that do not complete their activity during this period are i.i.d. with pmf $\xi_{\mu,m}^{(t)}(\ell_m | z)$ given by, for $\ell_m \in \Omega_m$,

$$\xi_{\mu,m}^{(t)}(\ell_m|z) = \frac{\mu(\ell_m|z) \times (1 - \gamma_m^{(t)}(z, \ell_m))}{\sum_{\tilde{\ell}_m} \mu(\tilde{\ell}_m|z) \times (1 - \gamma_m^{(t)}(z, \tilde{\ell}_m))}, \tag{4.2}$$

as can be seen using Bayes' rule. Indeed, the distribution of the label of such an entity changes as it is known that it does not complete its activity. Thus, h_m^{stay} has a multinomial pmf with number of trials parameter $x_m^{(t)}(z) - d_m$ and pmf parameter $\xi_{\mu,m}^{(t)}(L_m = \cdot|z)$.

We conclude that we can replace Lines 4–7 from the distributional program QPLEXMAP$^{(t)}$ with the following pseudocode fragment:

$$d_m \sim \text{NUMBEROFCOMPLETIONS}_{\mu,m}^{(t)}(z)$$
$$h_m^{\text{stay}} \sim \text{mult}\left(x_m^{(t)}(z) - d_m, \xi_{\mu,m}^{(t)}(L_m = \cdot|z)\right)$$

The histogram h_m has been eliminated. Note that the case $x_m^{(t)}(z) - d_m = 0$ need not be separated out because of our convention that the multinomial random variable equals zero in that case.

Step 2: Eliminate the h_m^{stay} The variable h_m^{stay} is only subsequently used in Line 12 to sample histograms that accumulate into h_m^{rel}. In this step, we rewrite the sampling statement for h_m^{stay} from the previous step together with Lines 10–12

$$h_m^{\text{stay}} \sim \text{mult}\left(x_m^{(t)}(z) - d_m, \xi_{\mu,m}^{(t)}(L_m = \cdot|z)\right)$$

10:      $h_m^{\text{rel}} \leftarrow 0$

11:      **for** $\ell_m \in \Omega_m$ **do**

12:          $h_m^{\text{rel}} +\sim \text{mult}\left(h_m^{\text{stay}}(\ell_m), \pi_m^{(t),\text{rel}}(L_m' = \cdot|z, \ell_m, z')\right)$

and replace it with pseudocode that results in the *same* pmf of h_m^{rel} but *without sampling* h_m^{stay}. Since h_m^{stay} is multinomial, we can think of the labels of the $x_m^{(t)}(z) - d_m$ entities that do not complete activity m as being i.i.d. with pmf $\xi_{\mu,m}^{(t)}(\ell_m|z)$. If the label of such an entity is ℓ_m, its next label has pmf $\pi_m^{(t),\text{rel}}(\ell_m'|z, \ell_m, z')$. Consequently, the (unconditional) next labels of such entities are i.i.d. with pmf:

$$\sum_{\ell_m} \xi_{\mu,m}^{(t)}(\ell_m|z) \times \pi_m^{(t),\text{rel}}(\ell_m'|z, \ell_m, z'). \tag{4.3}$$

It follows that h_m^{rel} is multinomial with number of trials parameter $x_m^{(t)}(z) - d_m$ and pmf parameter given by (4.3). We thus find that the above pseudocode fragment can be replaced with the following single sampling instruction:

$$h_m^{\text{rel}} \sim \text{mult}\left(x_m^{(t)}(z) - d_m, \sum_{\ell_m} \xi_{\mu,m}^{(t)}(\ell_m|z) \times \pi_m^{(t),\text{rel}}(L_m' = \cdot|z, \ell_m, z')\right)$$

Step 3: Eliminate the h_m^{rel}, h_m^{join}, and h_m' In this step, we rewrite the sampling statement for h_m^{rel} from the previous step together with Lines 13–18

$$h_m^{\text{rel}} \sim \text{mult}\left(x_m^{(t)}(z) - d_m, \textstyle\sum_{\ell_m} \xi_{\mu,m}^{(t)}(\ell_m|z) \times \pi_m^{(t),\text{rel}}(L_m' = \cdot|z, \ell_m, z')\right)$$

13: $\quad h_m^{\text{join}} \sim \text{mult}\left(x_m^{(t+1)}(z') - (x_m^{(t)}(z) - d_m), \pi_m^{(t),\text{join}}(L_m' = \cdot|z, z')\right)$

14: $\quad h_m' \leftarrow h_m^{\text{rel}} + h_m^{\text{join}}$

15: $\quad$ **if** $h_m' \neq 0$ **then**

16: $\quad\quad \ell_m' \sim h_m'/|h_m'|$

17: $\quad$ **else**

18: $\quad\quad \ell_m' \sim \theta_m^0(L_m' = \cdot)$

and replace it with pseudocode that results in the *same* pmf of ℓ_m' but *without sampling* h_m^{rel}, h_m^{join}, and h_m'. Recall that the sample in Line 16 of a next label, which is drawn from a pmf that is proportional to the frequencies in h_m', can be interpreted as selecting the label of a "uniformly chosen at random" entity among the $x_m^{(t+1)}(z')$ entities represented in h_m', each of which undertakes activity m at time $t + 1$. Since h_m' in turn is the sum of h_m^{rel} and h_m^{join} (and these histograms are independent), we can alternatively sample the next label in two stages: first sample the *entity type* of the uniformly chosen at random entity, i.e., whether this entity is represented in h_m^{rel} or h_m^{join}, and then sample its next label given its entity type.

To sample the entity type, we need the probabilities that a uniformly chosen at random entity from among the $x_m^{(t+1)}(z')$ entities either did not complete its activity during this period, in which case we say that its entity type is "old," or just started its activity, in which case we say that its entity type is "new." We let k_m' denote this (random) entity type. We say that the entity type of the uniformly chosen at random entity is "n/a" when there are no entities undertaking activity m. The distributional program below outputs the entity type given the current counter vector z, the number of activity completions d_m, and the next counter vector z'.

1: **function** ENTITYTYPE$_m^{(t)}(z, d_m, z')$

2: $\quad$ **if** $x_m^{(t+1)}(z') = 0$ **then**

3: $\quad\quad$ **return** n/a

4: $\quad$ probabilityOld $\leftarrow (x_m^{(t)}(z) - d_m)/x_m^{(t+1)}(z')$

5: $\quad$ entityIsOld $\sim$ ber(probabilityOld)

6: $\quad$ **if** entityIsOld **then**

7: $\quad\quad$ **return** old

8: $\quad$ **else**

9: $\quad\quad$ **return** new

For the second stage, we need to sample a next label ℓ_m' given the entity type k_m'. There are three cases to consider:

- *Case* $k_m' = $ new. Since h_m^{join} is (conditionally) multinomial with parameters $x_m^{(t+1)}(z') - (x_m^{(t)}(z) - d_m)$ and $\pi_m^{(t),\text{join}}(L_m' = \cdot|z, z')$, we can think of all entities with type "new" as having i.i.d. next labels with pmf $\pi_m^{(t),\text{join}}(\ell'_m | z, z')$. The next label ℓ'_m is sampled from this pmf.

- *Case k'_m = old.* Since h_m^{rel} is (conditionally) multinomial with parameters $x_m^{(t)}(z) - d_m$ and pmf parameter given by (4.3), we can think of all entities with type "old" as being i.i.d. with this pmf. The next label ℓ'_m is sampled from this pmf, which is equivalent to first sampling ℓ_m from the pmf $\xi_{\mu,m}^{(t)}(\ell_m | z)$ in (4.2) and then sampling ℓ'_m from the pmf $\pi_m^{(t),\text{rel}}(\ell'_m | z, \ell_m, z')$.
- *Case k'_m = n/a.* The next label ℓ'_m is sampled from the pmf θ_m^0.

This second stage is summarized in the distributional program below. The right-hand side of the sampling statement in Line 6 of this program is proportional to the pmf $\xi_{\mu,m}^{(t)}(\ell_m | z)$ in (4.2), and in its current form, it makes the model primitives explicit.

1: **function** NEXTLABEL$_{\mu,m}^{(t)}(z, z', k'_m)$
2:    **switch** k'_m **do**
3:       **case** new
4:          $\ell'_m \sim \pi_m^{(t),\text{join}}(L'_m = \cdot | z, z')$
5:       **case** old
6:          $\ell_m \propto \mu(L_m = \cdot | z) \times (1 - \gamma_m^{(t)}(z, \cdot))$
7:          $\ell'_m \sim \pi_m^{(t),\text{rel}}(L'_m = \cdot | z, \ell_m, z')$
8:       **case** n/a
9:          $\ell'_m \sim \theta_m^0(L'_m = \cdot)$
10:   **return** ℓ'_m

We thus find that the pseudocode fragment we started with in this step can be replaced with the following two sampling instructions:

$$k'_m \sim \text{ENTITYTYPE}_m^{(t)}(z, d_m, z')$$
$$\ell'_m \sim \text{NEXTLABEL}_{\mu,m}^{(t)}(z, z', k'_m)$$

Histogram-Free Refactored Distributional Program Putting it all together, we have refactored the distributional program QPLEXMAP$^{(t)}$ from the beginning of this section as the distributional program below. Note that there are no longer any sampling statements for histograms, neither in the program itself nor in the distributional programs called within the program. As such, this distributional program can be thought of as an "efficient" representation of the QPLEX map.

1: **function** QPLEXMAP$^{(t)}(\mu)$
2:   $z \sim \mu(Z = \cdot)$
3:   **for** $m = 1, \ldots, M$ **do**
4:      $d_m \sim \text{NUMBEROFCOMPLETIONS}_{\mu,m}^{(t)}(z)$
5:   $z' \sim \pi^{(t)}(Z' = \cdot | z, d)$
6:   **for** $m = 1, \ldots, M$ **do**
7:      $k'_m \sim \text{ENTITYTYPE}_m^{(t)}(z, d_m, z')$

8:        $\ell'_m \sim \text{NEXTLABEL}_{\mu,m}^{(t)}(z, z', k'_m)$

9:    **return** (z', ℓ')

Algebraic Representation of $\mathcal{Q}^{(t)}$ We can use this program to define several pmfs. Let $q_{\mu,m}^{(t)}(d_m | z)$ be the output of the distributional program $\text{NUMBEROFCOMPLE-TIONS}_{\mu,m}^{(t)}$, i.e.,

$$q_{\mu,m}^{(t)}(d_m | z) = \Pr\left[\text{bin}\left(x_m^{(t)}(z), \sum_{\ell_m} \mu(\ell_m | z) \times \gamma_m^{(t)}(z, \ell_m) \right) = d_m \right].$$

Let $q_m^{(t)}(k'_m | z, d_m, z')$ be the output of the distributional program $\text{ENTITYTYPE}_m^{(t)}$, i.e.,

$$q_m^{(t)}(k'_m | z, d_m, z') = \begin{cases} 1 - \frac{x_m^{(t)}(z) - d_m}{x_m^{(t+1)}(z')} & \text{if } k'_m = \text{new and } x_m^{(t+1)}(z') > 0 \\ \frac{x_m^{(t)}(z) - d_m}{x_m^{(t+1)}(z')} & \text{if } k'_m = \text{old and } x_m^{(t+1)}(z') > 0 \\ 1 & \text{if } k'_m = \text{n/a and } x_m^{(t+1)}(z') = 0 \\ 0 & \text{otherwise.} \end{cases}$$

Let $q_{\mu,m}^{(t)}(\ell'_m | z, z', k'_m)$ be the output of the distributional program $\text{NEXTLABEL}_{\mu,m}^{(t)}$, i.e.,

$$q_{\mu,m}^{(t)}(\ell'_m | z, z', k'_m) = \begin{cases} \pi_m^{(t),\text{join}}(\ell'_m | z, z') & \text{if } k'_m = \text{new} \\ \sum_{\ell_m} \xi_{\mu,m}^{(t)}(\ell_m | z) \times \pi_m^{(t),\text{rel}}(\ell'_m | z, \ell_m, z') & \text{if } k'_m = \text{old} \\ \theta_m^0(\ell'_m) & \text{if } k'_m = \text{n/a,} \end{cases}$$
$$(4.4)$$

where (4.2) defines $\xi_{\mu,m}^{(t)}$ in terms of μ and $\gamma_m^{(t)}(z, \ell_m)$.

We now define

$$q_\mu^{(t)}(z, d, z', k', \ell') = \mu(z) \times \prod_m \left[q_{\mu,m}^{(t)}(d_m | z) \right] \times \pi^{(t)}(z' | z, d)$$

$$\times \prod_m \left[q_m^{(t)}(k'_m | z, d_m, z') \times q_{\mu,m}^{(t)}(\ell'_m | z, d_m, z') \right]$$

and note that we can express the QPLEX map $\mathcal{Q}^{(t)}$ in terms of this pmf.

Theorem 4.1 *Consider a QPLEX chain with simple transition dynamics. For every $t \geq 0$, the QPLEX map $\mathcal{Q}^{(t)}$ may be represented via*

$$[\mathcal{Q}^{(t)}(\mu)](z', \boldsymbol{\ell}') = \sum_{z,d,k'} q_{\mu}^{(t)}(z, d, z', k', \boldsymbol{\ell}').$$

4.4 Numerical Illustration

We consider the following example of a multiserver queueing model from Chap. 2 for two periods:

- The number of servers is $n = 2$. The size function $x^{(t)}(z)$ therefore equals $\min(z, 2)$ for $z \geq 0$ and $t = 0, 1$.
- Service is completed in three periods or less. The service duration pmf is taken to be the pmf on $\{1, 2, 3\}$ given by $g(L = \cdot) = (0.6, 0.3, 0.1)$.
- There are either two or three arrivals at the end of the first period, and there are no arrivals at the end of the second period. The probability of two arrivals at the end of the first period is 0.3. The numbers of arrivals are independent of the number of customers in the system at time 0.
- The system is initially empty.
- Labels represent remaining service durations, so the label set is given by $\{1, 2, 3\}$.
- We may set $\theta^0 = g$, but this pmf does not appear in our calculations.

Throughout this section, we drop the subscript "Q" from QPLEX iterates. Our goal is to calculate $\mu^{(2)}(Z = 2, \ell)$, which is part of the QPLEX iterate $\mu^{(2)}$.

The Pmf $\mu^{(1)}$ The pmf $\mu^{(1)}$ is readily found from the input data and label interpretation to be

$$\mu^{(1)}(Z = 2, L = \cdot) = 0.3 \times (0.6, 0.3, 0.1)$$

$$\mu^{(1)}(Z = 3, L = \cdot) = 0.7 \times (0.6, 0.3, 0.1),$$

so in particular, of course, $\mu^{(1)}(Z = 2) = 0.3$ and $\mu^{(1)}(Z = 3) = 0.7$.

QPLEX Calculus We next consider the second period, between times 1 and 2. Due to the imposed label interpretation, NUMBEROFCOMPLETIONS$_{\mu}^{(1)}(z)$ outputs a binomial pmf with parameters $x^{(1)}(z)$ and $\mu(L = 1|z)$. We therefore have that

$$\mu^{(2)}(Z = 2, L = \cdot)$$

$$= \sum_{(z,d)\in\{(2,0),(3,1)\}} \mu^{(1)}(z) \times \Pr[\text{bin}(x^{(1)}(z), \mu^{(1)}(L = 1|z)) = d]$$

$$\times \Big\{ [\text{ENTITYTYPE}^{(1)}(z, d, 2)](\text{old}) \times \text{NEXTLABEL}_{\mu^{(1)}}^{(1)}(z, 2, \text{old})$$

$$+ [\text{ENTITYTYPE}^{(1)}(z, d, 2)](\text{new}) \times \text{NEXTLABEL}_{\mu^{(1)}}^{(1)}(z, 2, \text{new}) \Big\}.$$

The sum over (z, d) reflects the fact that there are only two possible ways ("paths") for the next counter z' to equal 2 given that the current counter z can only be 2 or 3:

- $z = 2$ and neither of the two customers in service completes their service this period ($d = 0$), and
- $z = 3$ and exactly one of the two customers in service completes their service this period ($d = 1$).

As we consider these paths only, the routing pmf term

$$\pi^{(1)}(z'|z, d) = \alpha^{(1)}(A = z' - z + d|z) = 1(z' = z - d)$$

is omitted in the above equation, as there are no arrivals in the second period. Note that z and z' are the numbers of customers in the system at time t and $t + 1$, respectively.

Number of Completions For $z = 2, 3$, the relevant probabilities for the numbers of completions are given by

$$\Pr[\text{bin}(x^{(1)}(z), \mu^{(1)}(L = 1|z)) = d] = \Pr[\text{bin}(2, 0.6) = d].$$

Entity Type Pmfs Thus, the relevant output of ENTITYTYPE$^{(1)}$ satisfies, for $k' = $ old, new,

$$[\text{ENTITYTYPE}^{(1)}(2, 0, 2)](k') = 1(k' = \text{old})$$

$$[\text{ENTITYTYPE}^{(1)}(3, 1, 2)](k') = 0.5.$$

Thus, no probability mass is assigned to $k' = \text{n/a}$.

Next Label Pmfs If the entity type is "new," the next label pmf is the service duration pmf. If the entity type is "old," the next label pmf is the pmf of $L - 1$ given $Z = z$ and $L > 1$ under $\mu^{(1)}$ as we use the remaining service duration label interpretation. Thus, the relevant outputs of NEXTLABEL$_{\mu^{(1)}}^{(1)}$ are given by, for $z = 2, 3$,

$$\text{NEXTLABEL}_{\mu^{(1)}}^{(1)}(3, 2, \text{new}) = (0.6, 0.3, 0.1).$$

$$\text{NEXTLABEL}_{\mu^{(1)}}^{(1)}(z, 2, \text{old}) = \left(\frac{0.3}{1 - 0.6}, \frac{0.1}{1 - 0.6}, 0\right) = (0.75, 0.25, 0).$$

Putting it all together, we have that

$$\mu^{(2)}(Z = 2, L = \cdot)$$
$$= \mu^{(1)}(Z = 2) \times \Pr[\text{bin}(2, 0.6) = 0] \times (0.75, 0.25, 0)$$
$$+ \mu^{(1)}(Z = 3) \times \Pr[\text{bin}(2, 0.6) = 1]$$
$$\times \left\{0.5 \times (0.6, 0.3, 0.1) + 0.5 \times (0.75, 0.25, 0)\right\}.$$

Upon substitution, we find that

$$\mu^{(2)}(Z = 2, L = \cdot)$$
$$= 0.3 \times 0.4^2 \times (0.75, 0.25, 0)$$
$$+0.7 \times (2 \times 0.6 \times 0.4) \times \Big\{0.5 \times (0.6, 0.3, 0.1) + 0.5 \times (0.75, 0.25, 0)\Big\}$$
$$= 0.048 \times (0.75, 0.25, 0) + 0.336 \times (0.675, 0.275, 0.05)$$
$$= (0.2628, 0.1044, 0.0168).$$

Note that $\mu^{(2)}(Z = 2) = 0.384$.

4.5 Endogenous Model Primitives

The current QPLEX iterate $\mu_Q^{(t)}$ or even the history $\{\mu_Q^{(\tau)} : \tau \leq t\}$ of QPLEX iterates that have already been calculated can be freely used in the specification of the model primitives. With this understanding, the QPLEX calculus admits policies with *endogenous* parameters in the sense that they use information that is not revealed until the QPLEX iterates are calculated over time. *Such QPLEX calculus need no longer be tied to a QPLEX chain; for instance, the Markov property may fail.* Here are two examples.

Example 4.1 (Accelerated Service Policies) Consider the multiserver queueing model from Chap. 2 and recall that we use the remaining service duration label interpretation. In this model, each customer with remaining service duration $\ell = 1$ is certain to complete its service during this period, and each customer with remaining service duration $\ell \geq 2$ is certain not to complete its service during this period.

One example of an "accelerated" service policy is represented by integers $\tilde{z}$ and $\tilde{r} > 1$ and a positive probability β: if the current number of customers in the system z exceeds the pre-determined threshold $\tilde{z}$, then each customer with remaining service duration ℓ greater than 1 but less than the threshold $\tilde{r}$ completes its service during this period with probability β. The choice of $\tilde{z}$ could be based on, say, the 95th percentile of the marginal QPLEX iterate $\mu_Q^{(t)}(z)$ or of prior marginal QPLEX iterates.

Another way to endogenously accelerate service is to base the join pmf primitive $\pi_m^{(t),\mathrm{join}}(\ell_m | z, z')$ on, say, the 95th percentile of the marginal QPLEX iterate $\mu_Q^{(t)}(z)$ or of prior marginal QPLEX iterates. $\triangle$

Example 4.2 (Queue Selection or Load Balancing) Each arriving customer chooses among heterogeneous multiserver stations and joins the one with the lowest, say, median cycle time. The cycle time pmf of each station uses the latest available QPLEX iterate; see Appendix A.4 for the details on this calculation. In this setting, the routing pmf is no longer exogenous as it depends on the current QPLEX iterate. This example illustrates that the stochastic process of counters and histograms becomes non-Markovian in nature, and we will work out this example in Sect. 5.6.   △

Chapter 5
Models with Simple Transition Dynamics

This chapter provides examples of practical models that can be represented as QPLEX chains with simple transition dynamics via suitable choices of model primitives. These examples include generalizations and variations of the multiserver queueing model presented in Chap. 2, a stochastic infectious disease model, a production-inventory model with backlogs and finished goods capacity, and several variations of a queueing network of multiserver stations.

Our selection of models strives to illustrate different aspects of QPLEX modeling, but we do not present models in their fullest generality. We make two simplifications. First, we only use the three standard label interpretations from Sect. 4.1 along with our standard convention that both arrivals and activity completions occur at the end of each period. Any of these label interpretations can be chosen for each activity in each of the models, and this choice thus determines the label set, the completion probability function, relabeling pmfs, and join pmfs corresponding to that activity. Second, all activity duration pmfs in this chapter are taken to be time-homogeneous, and we denote the pmf for activity m by g_m.

Each of the models in this chapter uses only time-homogeneous size functions. Therefore, we write x and x_m for $x^{(t)}$ and $x_m^{(t)}$, respectively. We simply write x and x, respectively, when there is a single activity.

Given a model description, we specify counters, histograms, counter sets, size functions, and the routing pmf primitive. These elements define a QPLEX chain after choosing one of the three standard label interpretations for each activity, assuming that an appropriate initial distribution is given. The routing pmf requires the most model-specific logic, and we represent it via a distributional program ROUTING$^{(t)}$.

Suppose, for instance, that we choose the remaining activity duration label interpretation in each of the models in this chapter. The program QPLEXMAP$^{(t)}$ below represents the QPLEX map $\mathcal{Q}^{(t)}$ corresponding to the QPLEX chain we have thus defined. This program can be obtained by appropriately refactoring the

© The Author(s) 2025
A. B. Dieker, S. T. Hackman, *QPLEX: A Computational Modeling and Analysis Methodology for Stochastic Systems*, Springer Series in Operations Research and Financial Engineering, https://doi.org/10.1007/978-3-031-74870-7_5

histogram-free distributional program for $\mathcal{Q}^{(t)}$ from Sect. 4.3 after substituting the specification of the primitives from Sect. 4.1 for the remaining activity duration label interpretation. (A similar program can be obtained when the labels have the elapsed activity duration interpretation, which only requires changes in Lines 4, 12, and 14.) We arbitrarily use $\theta_m^0 = g_m$; see Line 8. The next label is sampled from the pmf proportional to the function defined via $\mu(L_m = \ell_m' + 1|z)$ for $\ell_m' \geq 1$; see Line 12.

1: **function** QPLEXMAP$^{(t)}(\mu)$
2:     $z \sim \mu(\mathbf{Z} = \cdot)$
3:     **for** $m = 1, \ldots, M$ **do**
4:         $d_m \sim \text{bin}(x_m(z), \mu(L_m = 1|z))$
5:     $z' \sim \text{ROUTING}^{(t)}(z, \boldsymbol{d})$
6:     **for** $m = 1, \ldots, M$ **do**
7:         **if** $x_m(z') = 0$ **then**
8:             $\ell_m' \sim g_m(L_m' = \cdot)$
9:             **continue**
10:        entityIsOld $\sim \text{ber}((x_m(z) - d_m)/x_m(z'))$
11:        **if** entityIsOld **then**
12:            $\ell_m' \overset{\propto}{\sim} \mu(L_m - 1 = \cdot|z)$
13:        **else**
14:            $\ell_m' \sim g_m(L_m' = \cdot)$
15:    **return** $(z', \boldsymbol{\ell}')$

Each example starts with a model description, after which we consider a fixed period between time t and $t + 1$, which we call *this period*.

5.1   Multiserver Queueing Models

This section considers several variations of the multiserver queuing model from Chap. 2.

Reentrant Flow   Consider the multiserver queueing model from Chap. 2 where each customer, upon completing service, will have to complete another service with a probability $\beta \geq 0$, independent of any other information. Such customers are treated as new customers. (A model with several classes of customers will be presented in Sect. 7.2.) Of course, in the special case $\beta = 0$, we recover the model from Chap. 2. Recall that n denotes the number of servers, $g(\ell)$ denotes the service duration pmf, and $\alpha^{(t)}(a|z)$ denotes the pmf of the number of arrivals between times t and $t + 1$ if the number of customers in the system at time t is z.

Counters and Histograms   We define the following random variables:

- $Z^{(t)}$ is the number of customers in the system at time t.
- $H^{(t)}$ is the histogram of remaining service durations, service ages, or service phases of the customers undergoing service at time t.

We take the counter set as $\Omega_Z = \mathbb{Z}_+$. The size function is given by $x(z) = \min(z, n)$. The activity duration pmf is the service duration pmf $g(\ell)$.

Routing Pmf Primitive This routing pmf is a simple variation of the one given in Sect. 4.1 for the model in Chap. 2 but must reflect the more general flow dynamics. In the distributional program below, the variable ϕ represents the number of reentrant customers.

1: **function** ROUTING$^{(t)}(z, d)$
2: $\quad a \sim \alpha^{(t)}(A = \cdot | z)$
3: $\quad \phi \sim \mathrm{bin}(d, \beta)$
4: $\quad z' \leftarrow z - d + \phi + a$
5: $\quad$ **return** z'

Renewal Arrivals Consider the multiserver queueing model from Chap. 2 but with arrivals governed by a renewal process. We let η denote the pmf of the *interrenewal times*, namely, the pmf of the number of periods to the next arrival given that an arrival has just occurred. Interrenewal times are also independent and assumed to be at least one time period.

Counters and Histograms We define the following random variables:

- $Z^{(t)}$ is the number of customers in the system at time t.
- $Z_A^{(t)}$ is the *interrenewal age* at time t, namely, the number of periods since the last arrival.
- $H^{(t)}$ is the histogram of remaining service durations, service ages, or service phases of the customers undergoing service at time t.

We take the counter set as $\Omega_Z = \mathbb{Z}_+^2$. The size function is given by $x(z, z_A) = \min(z, n)$. The activity duration pmf is the service duration pmf $g(\ell)$.

Possible alternate choices for interpreting the counter $Z_A^{(t)}$ include the remaining interrenewal duration or the phase of the interrenewal distribution. Choosing such an interpretation affects the routing pmf primitive, which we only work out for the choice given above.

Routing Pmf Primitive The routing pmf must reflect the renewal arrival process. The distributional program below outputs the routing pmf for this model. The probability of an arrival is a function of the interrenewal age z_A and is given in Line 2. Line 4 uses the fact that the interrenewal age increases by one if there is no arrival; otherwise, it is zero.

1: **function** ROUTING$^{(t)}(z, z_A, d)$
2: $\quad a \sim \mathrm{ber}\left(\dfrac{\eta(z_A+1)}{\sum_{\tilde{z}_A \geq z_A+1} \eta(\tilde{z}_A)} \right)$
3: $\quad z' \leftarrow z - d + a$
4: $\quad z_A' \leftarrow 1(a = 0) \times (z_A + 1)$
5: $\quad$ **return** (z', z_A')

Batch Renewal Arrivals Consider a generalization of the previous example model with arrivals occurring in batches. If an arrival batch of size b occurred between times t and $t+1$, the probability that the next arrival batch will occur between times $t+\tau$ and $t+\tau+1$ and be of size $b' > 0$ is given by the exogenously specified pmf $\beta^{(t)}(\tau, b'|b)$, with $\tau > 0$. We insert a dummy batch of size 0 right before time 0 and define the pmf $\beta^{(0)}(\tau, b'|b)$ only for $b = 0$. The service durations, once begun, are independent of past or future arrivals.

Counters and Histograms We define the following random variables:

- $Z^{(t)}$ is the number of customers in the system at time t.
- $Z_A^{(t)}$ is the *arrival age* at time t, namely, the number of periods since the last arrival batch.
- $Z_B^{(t)}$ is the size of the last arrival batch, which came between times $t - Z_A^{(t)} - 1$ and $t - Z_A^{(t)}$.
- $H^{(t)}$ is the histogram of remaining service durations, service ages, or service phases of the customers undergoing service at time t.

We take the counter set as $\Omega_Z = \mathbb{Z}_+^3$. The size function is given by $x(z, z_A, z_B) = \min(z, n)$. The activity duration pmf is the service duration pmf $g(\ell)$.

Routing Pmf Primitive The routing pmf must change to reflect the different arrival process. The routing pmf builds on the one for the model with renewal arrivals. The distributional program below outputs the routing pmf for this model. The probability of an arrival batch of size b' during this period given $Z_A^{(t)} = z_A$ and $Z_B^{(t)} = z_B$ is

$$\beta^{(t-z_A-1)}(\mathcal{T} = z_A + 1, b'|\mathcal{T} > z_A, B = z_B).$$

We think of b' as the number of arrivals to the system, so in Line 3, we use the symbol a to be consistent with the notation we have used in the routing pmf of prior models. The logic for the next counter z_A' remains the same as in the renewal arrival model. The next counter z_B' for the size of the last arrival batch is the same as the current counter z_B if there are no arrivals; otherwise, it equals the number of arrivals a (Line 6).

```
1: function ROUTING(t)(z, zA, zB, d)
2:     (τ, b') ~ β(t−zA−1)(𝒯 = ·, B' = ·|𝒯 > zA, B = zB)
3:     a ← 1(τ = zA + 1) × b'
4:     z' ← z − d + a
5:     zA' ← 1(a = 0) × (zA + 1)
6:     zB' ← 1(a = 0) × zB + 1(a > 0) × a
7:     return (z', zA', zB')
```

Superposition of Renewal Arrivals This model generalizes the model with renewal arrivals in a different direction than the model with batch renewal arrivals. Here, the arrival process is a superposition of n_R identical renewal processes, i.e., the interrenewal pmf is the same for each of the renewal processes. Even if $n_R = 1$, the model we present here is technically different from the model with renewal arrivals since here we reclassify the counter z_A as a histogram. The two QPLEX chains that result are in one-to-one correspondence.

Counters and Histograms We define the following random variables:

- $Z^{(t)}$ is the number of customers in the system at time t.
- $H_R^{(t)}$ is the histogram of times since the last renewal, remaining times to a renewal, or phases of the n_R renewal processes at time t.
- $H_S^{(t)}$ is the histogram of remaining service durations, service ages, or service phases of the customers undergoing service at time t.

Note that for this model, there are *two* activities: one corresponding to the renewal arrival process (activity R) and the other corresponding to undergoing service (activity S). Completing activity R is synonymous with supplying a new arrival to the system.

We take the counter set as $\Omega_Z = \mathbb{Z}_+$. The size functions for activities R and S are given by $x_R(z) = n_R$ and $x_S(z) = \min(z, n)$, respectively. The activity duration pmfs associated with activities R and S are the interrenewal pmf and the service duration pmf, respectively.

Routing Pmf Primitive The distributional program below represents the routing pmf for this model. Note that z' is a *deterministic* function of z, d_R, and d_S, so the output of this distributional program corresponds to a point mass.

```
1: function ROUTING^(t)(z, d_R, d_S)
2:     z' ← z − d_S + d_R
3:     return z'
```

5.2 Stochastic Infectious Disease Model

Each individual in a population of K individuals is in one of three compartments: susceptible "S," infectious "I," or recovered "R." An individual progresses from the S to the I and then to the R compartment. Each individual in the S compartment becomes infectious (transitions to the I compartment) independently with probability $1 - e^{-\beta \times \zeta}$, where ζ is the infectious proportion of the population and $\beta > 0$ is a model parameter that controls the infectiousness of the disease. Once an individual becomes infectious, (s)he remains infectious for a random amount of time. The numbers of periods needed to recover from the disease after becoming infectious are assumed to be independent across individuals and to be at least one time period. The R compartment contains individuals who have recovered, and individuals stay in this compartment as soon as they enter it. We assume there is at least one infectious individual initially (to avoid trivial dynamics).

Counters and Histograms   We define the following random variables:

- $Z_S^{(t)}$ is the number of susceptible individuals at time t.
- $Z_I^{(t)}$ is the number of infectious individuals at time t.
- $H^{(t)}$ is the histogram of remaining infectious durations, infectious ages, or infectious phases of the infectious individuals at time t.

We take the counter set as $\Omega_Z = \mathbb{Z}_+^2$. The size function is given by $x(z_S, z_I) = z_I$. The activity duration pmf is the pmf of the number of periods needed to recover from the disease after becoming infectious.

Routing Pmf Primitive   The distributional program below outputs the routing pmf for this model. Lines 3 and 4 use standard flow balance.

```
1: function ROUTING^(t)(z_S, z_I, d_I)
2:     a ~ bin(z_S, 1 − e^{−β×z_I/K})
3:     z'_S ← z_S − a
4:     z'_I ← z_I − d_I + a
5:     return (z'_S, z'_I)
```

5.3   Production-Inventory Model

Consider the multiserver queueing model from Chap. 2 but with the number of arrivals in each period independent of the number of customers in the system at the beginning of the period. In this example, a customer corresponds to a unit of product, which begins as raw material and then is transformed into a finished good after undergoing the production process. (The arrivals of raw material units are outside the control of the production process.) The number of periods needed to complete the production process is assumed to be independent for each unit and to be at least one time period. There is a finished goods warehouse with finite capacity n_F that houses completed units. Demand for finished goods is exogenous and independent across periods. We let $\delta_F^{(t)}(d_F)$ denote the pmf of the number of units demanded between times t and $t + 1$. Demand is backlogged. If there is no space to store a finished unit, it continues to occupy production capacity until space at the finished goods warehouse becomes available. Thus, it is blocked from moving to the finished goods warehouse. Completed units are equally likely to be moved to the finished goods warehouse if some but not all such units can be moved.

Counters and Histograms   We define the following random variables:

- $Z_R^{(t)}$ is the number of units in production or buffer at time t.
- $Z_I^{(t)}$ is the inventory position of finished goods at time t, namely, the number of completed units in the facility minus the number of units on backorder. If the inventory position is positive, units are on hand to fulfill future demand. On

the other hand, if the inventory position is negative, then finished goods to be
completed are committed to satisfy past demand.

- $H^{(t)}$ is the histogram of remaining production durations, production ages, or
  production phases for units undergoing the production process at time t.

The value of $Z_I^{(t)}$ cannot exceed $n + n_F$, the sum of the number of servers plus
the capacity of the finished goods warehouse. If $Z_I^{(t)} \geq 0$, then the number of
completed units in the finished goods warehouse is $\min\{Z_I^{(t)}, n_F\}$, and the number
of completed units occupying servers is $\max\{Z_I^{(t)} - n_F, 0\}$. If $Z_I^{(t)} \leq 0$, then the
number of units on backorder is $-Z_I^{(t)}$.

We take the counter set as $\Omega_Z = \mathbb{Z}_+ \times \mathbb{Z}$. The size function is not the familiar
one. The *available* production capacity is the total capacity n *less* the number of
completed units blocked from being transferred to the finished goods warehouse.
Since z_R is the number of units in production and in the buffer and $\max(z_I - n_F, 0)$
is the number of completed units that occupy production capacity, the size function
is given by

$$x(z_R, z_I) = \min(z_R, n - \max(z_I - n_F, 0))$$
$$= \min(z_R, n, n + n_F - z_I),$$

where $n + n_F - z_I$ is the number of finished goods that can be added if the inventory
position is z_I. The activity duration pmf is the pmf of the time to complete the
production process.

Routing Pmf Primitive The distributional program below outputs the routing pmf
primitive for this model. After sampling the values for the numbers of raw material
units and exogenous demand (Lines 2 and 3), the next values of the counters are
obtained using standard flow balance (Lines 4 and 5).

```
1: function ROUTING^(t)(z_R, z_I, d)
2:     a ~ α^(t)(A = ·)
3:     d_F ~ δ_F^(t)(D_F = ·)
4:     z'_R ← z_R − d + a
5:     z'_I ← z_I + d − d_F
6:     return (z'_R, z'_I)
```

5.4 Flow Line Model

The remaining models in this chapter are networks of heterogeneous stations. Each
station is modeled as a pool of homogeneous servers and a buffer serving as a staging
area, as in the multiserver queueing model of Chap. 2. The model description from
Chap. 2 applies to each station, except that the arrival dynamics are different from

example to example. We let n_m and g_m denote, respectively, the number of servers and the service duration pmf at station m.

We start with a simple flow line, as it illustrates the basic "flow balance" logic often required to describe a routing pmf primitive for networks. There are M stations. Customers arrive only to station 1. We let $\alpha_1^{(t)}(a)$ denote the pmf of the number of arrivals A between times t and $t+1$, and the numbers of such arrivals are assumed to be independent across time periods. After completing service at station $m = 1, \ldots, M - 1$, a customer visits station $m + 1$. If a server is available, the customer immediately begins service at this station; otherwise, the customer is held in that station's buffer. Customers leave the system after completing their service at station M.

Counters and Histograms We define the following random variables:

- $Z_m^{(t)}$ is the number of customers at station m at time t.
- $H_m^{(t)}$ is the histogram of remaining service durations, service ages, or service phases of customers undergoing service at station m at time t.

We take the counter set as $\Omega_Z = \mathbb{Z}_+^M$ and write $(z_1, \ldots, z_M)$ for an element of this set. The size functions are, for $m = 1, \ldots, M$, given by

$$x_m(z_1, \ldots, z_M) = \min(z_m, n_m).$$

The activity duration pmf for undergoing service at station m is g_m.

Routing Pmf Primitive The distributional program below outputs the routing pmf primitive for this model. The flow balance equations are used in Line 4, where d_0 represents the number of arrivals to station 1.

```
1: function ROUTING(t)(z1, ..., zM, d1, ..., dM)
2:     d0 ~ α1(t)(A = ·)
3:     for m = 1, ..., M do
4:         z′m ← zm − dm + dm−1
5:     return (z′1, ..., z′M)
```

5.5   Queueing Network Model with Probabilistic Routing

We consider a model for queueing networks with probabilistic (or Markovian) routing, which contains the flow line as a special case. Once again, we consider a network of M stations, with two notable changes relative to the flow line model. First, customers may arrive to any of the M stations. We let $\alpha_m^{(t)}(a)$ denote the pmf of the number of external arrivals to station m between times t and $t + 1$, and the numbers of such arrivals are assumed to be independent across time periods and stations. Second, each customer who completes their service at station m is routed to their next destination independently of the other customers. The probability that a customer exits the system right before the end of this period is denoted by $\beta^{(t)}_{m \to 0}$, and the probability that it goes to station $\tilde{m}$ is denoted by $\beta^{(t)}_{m \to \tilde{m}}$. We allow $\tilde{m} = m$, and thus we allow a customer to return to the same station where it is currently being served. Such a customer may need to wait until it can enter service again and, if so, will receive a new independent service duration as in the reentrant model. We let

$$\beta^{(t)}_{m \to \cdot} = (\beta^{(t)}_{m \to 0}, \beta^{(t)}_{m \to 1}, \ldots, \beta^{(t)}_{m \to M})$$

denote the pmf on destinations for customers that complete their service at station m.

Counters and Histograms The definitions given for the flow line model apply to this example verbatim.

Routing Pmf Primitive The distributional program below outputs the routing pmf primitive for this model. It reflects the more general flow dynamics relative to the flow line model. The number of external arrivals to station m, $\phi_{0 \to m}$, is sampled from $\alpha^{(t)}_m(a)$ (Line 3). We let $\phi_{m \to \tilde{m}}$ denote the number of customers that move from station m to destination $\tilde{m}$ during this period, where it is understood that $\tilde{m} = 0$ corresponds to the exit. Given the assumed independence of the destination locations for each of the d_m customers who complete their activity at station m, the flow vector $(\phi_{m \to 0}, \phi_{m \to 1}, \ldots, \phi_{m \to M})$ out of station m is multinomial with parameters d_m and $\beta^{(t)}_{m \to \cdot}$. (Line 4). The total number of arrivals to station m, both external and from other stations but not those that come back to m, is a_m (Line 6). The calculation of z'_m (Line 7) uses standard flow balance associated with the multiserver queueing model with reentrant flow, as described in Sect. 5.1.

```
1:  function ROUTING(t)(z1, ..., zM, d1, ..., dM)
2:      for m = 1, ..., M do
3:          φ0→m ~ α(t)m(A = ·)
4:          (φm→0, ..., φm→M) ~ mult (dm, β(t)m→.)
5:      for m = 1, ..., M do
6:          am ← φ0→m + Σm̃∉{0,m} φm̃→m
7:          z'm ← zm − dm + φm→m + am
8:      return (z'1, ..., z'M)
```

5.6 Load Balancing Model

We consider a network of $M > 1$ stations, where customers arrive to the system as a whole instead of individual stations. A *load balancing (queue selection) policy* determines how many system arrivals are sent to each station, and we will discuss two such policies. Each customer exists the system upon completing service at the station to which it has been routed. In this model, $\alpha^{(t)}(a)$ now denotes the pmf of the number of system arrivals between times t and $t + 1$, and the numbers of such arrivals are independent across different time periods.

Counters and Histograms The definitions given for the flow line model apply to this example verbatim.

Routing Pmf Primitive The specification of this pmf requires a load balancing policy, which we represent via a distributional program with the numbers of customers $z_1, \ldots, z_M$ at each of the M stations and the number of system arrivals a as input. Such a distributional program outputs a pmf of $(a_1, \ldots, a_M)$, where a_m represents the number of system arrivals routed to station m. Line 5 updates the counters using standard flow balance.

1: **function** ROUTING$^{(t)}(z_1, \ldots, z_M, d_1, \ldots, d_M)$
2:      $a \sim \alpha^{(t)}(A = \cdot)$
3:      $(a_1, \ldots, a_M) \sim$ LOADBALANCINGPOLICY$(z_1, \ldots, z_M, a)$
4:      **for** $m = 1, \ldots, M$ **do**
5:          $z'_m \leftarrow z_m - d_m + a_m$
6:      **return** $(z'_1, \ldots, z'_M)$

We consider the well-known "join the shortest queue" policy and a policy that routes arrivals to the station with the lowest cycle time percentile. For each policy, we consider two variations: assign all arrivals to the same station or sequentially assign arrivals to stations.

We use the following notational conventions in the distributional programs below. Given a vector of M nonnegative numbers $(i_1, \ldots, i_M)$, we write

$$\arg\min(i_1, \ldots, i_M) = \{m : i_m = \min(i_1, \ldots, i_M)\}.$$

For an arbitrary nonempty finite set I, we furthermore let unif(I) denote a uniform random variable on I.

Example 5.1 A policy that routes arriving customers to the station with the fewest customers with ties broken arbitrarily is called the *join the shortest queue* policy. We first consider the policy of sending *all* arrivals to the station with the fewest customers each period with ties broken uniformly at random. The distributional program below outputs the load balancing pmf under this policy.

1: **function** ALLARRIVALSJOINSHORTESTQUEUE$(z_1, \ldots, z_M, a)$
2:      $m^* \sim$ unif$(\arg\min(z_1, \ldots, z_M))$
3:      **for** $m = 1, \ldots, M$ **do**
4:          $a_m = a \times 1(m = m^*)$
5:      **return** $(a_1, \ldots, a_M)$

The station m^* that receives all arrivals is chosen uniformly at random from among the stations having the fewest customers (Line 2). The number of system arrivals to each station is then updated accordingly (Lines 3 and 4).

Next, we consider the policy of sending *each successive* arrival to the station with the fewest customers and ties broken uniformly at random. The distributional program below outputs the load balancing pmf under this policy.

1: **function** SEQUENTIALARRIVALSJOINSHORTESTQUEUE($z_1, \dots, z_M, a$)
2:      **for** $m = 1, \dots, M$ **do**
3:          $a_m \leftarrow 0$
4:      **for** $j = 1, \dots, a$ **do**
5:          $m_j^* \sim \text{unif}(\arg\min(z_1 + a_1, \dots, z_M + a_M))$
6:          $a_{m_j^*} \leftarrow a_{m_j^*} + 1$

7:      **return** $(a_1, \dots, a_M)$

The number of customers at a station will change after an arrival is assigned to it. As a result, the number of customers at one of the stations must be updated after each assignment. We first initialize counters that keep track of how many arrivals have been assigned to each station (Lines 2 and 3). For each subsequent arrival, we then calculate the station with the fewest customers, taking the number of assigned arrivals into account (Line 5), and update the appropriate assignment counter (Line 6).   △

Example 5.2 We now consider the policy that routes arriving customers to the station with the shortest cycle time percentile with ties broken uniformly at random. For presentational purposes, we work with the median cycle time. We assume that the cycle time pmf calculated for each station is based on the number of customers at the station at the beginning of the period. A special feature of this model is that the logic required to define the load balancing pmf uses the cycle time pmfs, which in turn can be approximated using the marginal pmfs $\mu_Q^{(t)}(\ell_m | z_1, \dots, z_M)$ of the QPLEX iterate for each station m; see Appendix A.4 for the details of this calculation. Thus, the signature of the distributional program that outputs the load balancing pmf must change to the following:

1: **function** LOADBALANCINGPOLICY$_\mu(z_1, \dots, z_M, a)$

It is understood that here μ is assigned the value $\mu_Q^{(t)}$. Since the routing pmf depends on the current QPLEX iterate, the QPLEX calculus is no longer tied to a QPLEX chain as the Markov property fails; see also Sect. 4.5.

In what follows, we let $t_{\theta_m, m}(z_m)$ denote the median cycle time at station m when there are $z_m \geq 0$ customers at time t and the labels at time t in station m are i.i.d. with common pmf θ_m.

We first consider the policy that routes all arrivals to the station with the lowest median cycle time with ties broken uniformly at random. The distributional program that outputs the load balancing pmf under this policy is identical to the program ALLARRIVALSJOINSHORTESTQUEUE above, except that the sampling instruction

2:      $m^* \sim \text{unif}(\arg\min(t_{\mu(L_1=\cdot|z),1}(z_1), \ldots, t_{\mu(L_M=\cdot|z),M}(z_M)))$

replaces Line 2.

Next, we consider the policy of sequentially routing arrivals to the station with the (updated) lowest median cycle time with ties broken uniformly at random. The distributional program that outputs the load balancing pmf under this policy is identical to the program SEQUENTIALARRIVALSJOINSHORTESTQUEUE above, except that the sampling instruction

5:      $m_j^* \sim \text{unif}(\arg\min(t_{\mu(L_1=\cdot|z),1}(z_1 + a_1), \ldots, t_{\mu(L_M=\cdot|z),M}(z_M + a_M)))$

replaces Line 5.   △

5.7   Queueing Network Model with Blocking

We consider a network of $M + 1$ stations, where station $M + 1$ has no external customer arrivals and a finite buffer of size b_{M+1}. We let $\alpha_m^{(t)}(a)$ denote the pmf of the number of external arrivals A at station $m = 1, \ldots, M$ between times t and $t + 1$, and the numbers of such arrivals are assumed to be independent across time periods and stations. When customers complete service at any of the first M stations, they are eligible to move to station $M + 1$, but they may not be able to do so due to a lack of buffer capacity. Customers leave the system after completing service at station $M + 1$. Given the number of customers eligible to move from each of the first M stations, an allocation policy determines how many of these customers get to move to station $M + 1$. We present two such policies at the end of this section. Eligible customers who are not selected remain at their station, continue to occupy a server, and wait until they are selected to be moved at a later time. Space freed up due to system departures from station $M + 1$ during this period is available for allocation during this period. Customers who finish their service during this period in any of the first M stations can be selected to move during that period, in addition to blocked customers who completed their service before time t.

Counters and Histograms We define the following random variables:

- $Z_{m,S}^{(t)}$ is the total number of customers at station $m = 1, \ldots, M$ that are either in service or in the buffer at time t.
- $Z_{m,C}^{(t)}$ is the number of customers who have completed their service but remain at station $m = 1, \ldots, M$ because they are blocked at time t. These customers may have completed their service at any point before time t.
- $Z_{M+1}^{(t)}$ is the total number of customers at station $M + 1$ at time t.
- $H_m^{(t)}$ is the histogram of remaining service durations, service ages, or service phases of customers undergoing service at station $m = 1, \ldots, M + 1$ at time t.

Note that here there are $M + 1$ and not M activities.

We take the counter set as $\Omega_Z = \mathbb{Z}_+^{2M+1}$. The size function for station $M + 1$ is given by

$$x_{M+1}(z_{1,S}, z_{1,C}, \ldots, z_{M+1}) = \min(z_{M+1}, n_{M+1}).$$

The size functions for the first M stations resemble the size function in Sect. 5.3, which also incorporates blocking. Here, the number of *available* servers is the number of servers n_m *less* the number of servers currently being occupied by customers who have completed their service but who are blocked from being transferred to station $M + 1$. As a result, the size functions x_m for $m = 1, 2, \ldots, M$ are given by

$$x_m(z_{1,S}, z_{1,C}, \ldots, z_{M+1}) = \min(z_{m,S}, n_m - z_{m,C}).$$

The activity duration pmf for undergoing service at station m is g_m.

Routing Pmf Primitive The specification of this pmf requires an *allocation policy*, which we represent as a distributional program with the demands $\hat{d}_1, \ldots, \hat{d}_M$ for slots at station $M + 1$ by each of the M stations and the supply s of available slots as input. Such a distributional program outputs a pmf on M-tuples of pairs

$$((\phi_1, u_1), \ldots, (\phi_M, u_M)),$$

where ϕ_m represents the "met" demand and $u_m = \hat{d}_m - \phi_m$ represents the "unmet" demand. (We use the symbol ϕ_m since met demand represents flow.) The allocation policy need not be randomized, so the output of the distributional program may correspond to a point mass.

The distributional program below outputs the routing pmf primitive for this model. All $z_{M+1} - d_{M+1}$ customers at station $M + 1$ who did not complete their service continue to occupy slots in station $M + 1$, and so the available supply of space is $s = n_{M+1} + b_{M+1} - (z_{M+1} - d_{M+1})$ (Line 2). Customers who complete their service at station m are no longer counted in the variable $z'_{m,S}$, as they are either counted in z'_{M+1} or in $z'_{m,C}$, and so $z'_{m,S}$ is found from familiar flow balance (Line 5). The demand for space from station m is calculated in Line 6. The met and unmet demands for each of the stations are calculated using a distributional program ALLOCATIONPOLICY, which represents the chosen allocation policy. The variable z'_{M+1} is then calculated using standard flow balance (Line 8).

1: **function** ROUTING$^{(t)}(z_{1,S}, z_{1,C}, \ldots, z_{M,S}, z_{M,C}, z_{M+1}, d_1, \ldots, d_{M+1})$
2:      $s \leftarrow n_{M+1} + b_{M+1} - (z_{M+1} - d_{M+1})$
3:      **for** $m = 1, \ldots, M$ **do**
4:          $a_m \sim \alpha_m^{(t)}(A = \cdot)$
5:          $z'_{m,S} \leftarrow z_{m,S} - d_m + a_m$
6:          $\hat{d}_m \leftarrow d_m + z_{m,C}$
7:          $((\phi_1, z'_{1,C}), \ldots, (\phi_M, z'_{M,C})) \sim$ ALLOCATIONPOLICY$(\hat{d}_1, \ldots, \hat{d}_M; s)$

8:     $z'_{M+1} \leftarrow z_{M+1} - d_{M+1} + \sum_{m=1}^{M} \phi_m$
9:     **return** $(z'_{1,S}, z'_{1,C}, \ldots, z'_{M,S}, z'_{M,C}, z'_{M+1})$

The model could be modified to only allow blocked customers to move to station $M + 1$ during this period (instead of also allowing customers that complete their service during this period to move). The routing pmf primitive for this modification is obtained from the above routing pmf primitive by removing Line 6 and replacing $\hat{d}_m$ with $z_{m,C}$ in Line 7.

Here are two examples of an allocation policy.

Example 5.3 In a *station-based priority* allocation policy, the stations are ordered according to their priority, where station 1 has the highest priority. The available space at station $M + 1$ for customers who have completed their service at one of the first M stations is sequentially allocated first to the customers in station 1 that are eligible to be moved, then to eligible customers in station 2, etc.

The distributional program below outputs the allocations for this station-based priority allocation policy. There are no sampling instructions, so its output is a point mass. The variable s_m stands for the available space (supply) at station $M + 1$ before allocating demand for space from station $m + 1$, so it is initialized as s before allocating slots to station 1 (Line 2). The met demand for station m is the minimum of its demand for slots and the available supply (Line 4), and the unmet demand is simply the difference between the demand and unmet demand (Line 5). The available supply is sequentially updated in Line 6.

1: **function** PRIORITYDEMANDSPLIT($\hat{d}_1, \ldots, \hat{d}_M; s$)
2:     $s_0 \leftarrow s$
3:     **for** $m = 1, \ldots, M$ **do**
4:         $\phi_m \leftarrow \min(\hat{d}_m, s_{m-1})$
5:         $u_m \leftarrow \hat{d}_m - \phi_m$
6:         $s_m \leftarrow s_{m-1} - \phi_m$
7:     **return** $((\phi_1, u_1), \ldots, (\phi_M, u_M))$

For example, PRIORITYDEMANDSPLIT(1, 3, 4; 5) outputs a pmf with a point mass at $((1,0), (3,0), (1,3))$.   △

Example 5.4 In a *randomized* allocation policy, each customer from the first M stations is equally likely to be given an available slot at station $M + 1$ if not all demand can be met. The distributional program below outputs the pmf of the allocations associated with this routing policy.

1: **function** RANDOMDEMANDSPLIT($\hat{d}_1, \ldots, \hat{d}_M; s$)
2:     $(\phi_1, \ldots, \phi_M) \sim$ multivariateHypergeometric($\hat{d}_1, \ldots, \hat{d}_M; s$)
3:     **for** $m = 1, \ldots, M$ **do**
4:         $u_m \leftarrow \hat{d}_m - \phi_m$
5:     **return** $((\phi_1, u_1), \ldots, (\phi_M, u_M))$

This program uses the multivariate hypergeometric pmf; see Sect. 1.4 for its definition.   △

Chapter 6
Advanced Transition Dynamics

This chapter extends the QPLEX calculus to a larger universe of abstract models by using model primitives different from those used for simple transition dynamics. We say that such models have *advanced transition dynamics*. The model primitives extend those for simple transition dynamics in three key ways.

- *General completion categories.* Under simple transition dynamics, each entity undertaking an activity either does or does not complete the activity during each period, which can be thought of as assigning either "stay" or "done" to each such entity. We now allow an entity to be assigned any (finite) number of *categories*. This extension can be used to model a system with reentrant flow and preemptive service or one in which servers may fail and need to be repaired while an entity waits.
- *Flexible routing.* Entity routing can incorporate the information of the completion categories the entities have been assigned. It will no longer be required that an entity assigned category "stay" remain at the activity. This can be used to model a queueing system in which entities may abandon if the time spent waiting in the buffer exceeds a threshold.
- *General entity types.* Entity types "old" and "new" are now replaced with general entity types. All entities undertaking an activity at time $t + 1$ are assigned an entity type, of which there are two kinds. If an entity was undertaking an activity at time t, the entity type identifies this activity in conjunction with its assigned category. If an entity was not undertaking an activity at time t, the entity type is one of several possible *origins*. Multiple origins can be used to model a system with multiple classes of entities. The next label an entity receives can depend on its entity type, the activity it undertakes at time $t + 1$, and the counters at times t and $t + 1$.

After defining the requisite notation and terminology and describing the model primitives for models with advanced transition dynamics, we present the transition probabilities of the resulting QPLEX chain via a distributional program.

© The Author(s) 2025

A. B. Dieker, S. T. Hackman, *QPLEX: A Computational Modeling and Analysis Methodology for Stochastic Systems*, Springer Series in Operations Research and Financial Engineering, https://doi.org/10.1007/978-3-031-74870-7_6

We then derive an efficient representation of the QPLEX map with advanced transition dynamics, thus establishing the QPLEX calculus for *any* QPLEX chain with advanced transition dynamics. In this chapter, we also recast the stochastic infectious disease model of Sect. 5.2 as a QPLEX chain with advanced model primitives to illustrate the connection between the two sets of model primitives.

Throughout this chapter, we consider a fixed QPLEX chain with M activities, counter set Ω_z, size functions $x_m^{(t)}$ for $t \geq 0$, $m = 1, \ldots, M$, and label sets Ω_m, $m = 1, \ldots, M$. We fix time t and consider the period between times t and $t + 1$, which we again refer to as *this period*.

6.1 Categories and Entity Types

This section introduces the notions of categories and entity types for QPLEX chains with advanced transition dynamics. Categories are assigned to all entities undertaking an activity at time t, while entity types are assigned to all entities undertaking an activity m' at time $t + 1$. Assigned categories and entity types only pertain to this period, so the assigned categories of entities undertaking an activity in past or future time periods play no role here. We fix counter vectors z and z' at time t and $t + 1$, respectively.

Categories The finite set of categories is arbitrary, but it need not be possible to assign each category to entities undertaking a given activity. We use the symbol k_m to denote a generic feasible category for activity m, where feasibility means that category k_m can be assigned to an entity undertaking activity m.

The symbol d_m now denotes the histogram on feasible categories for the $x_m^{(t)}(z)$ entities undertaking activity m at time t. In particular, $d_m(k_m)$ denotes the total number of entities undertaking activity m assigned category k_m. We call d_m the *category assignment histogram* for activity m and write $\boldsymbol{d} = (d_m)_m$. Since all entities must be assigned a category, we have, for each activity m, the identity

$$|d_m| = x_m^{(t)}(z). \tag{6.1}$$

It is possible to let all categories be feasible for each activity and then define the routing pmf primitive (to be discussed in the next section) to ensure that some (if any) components of the category assignment histograms are always zero. Instead, we work with feasible categories to make models more expressive.

Example 6.1 The set of categories for a QPLEX chain with simple transition dynamics, recast as a QPLEX chain with advanced transition dynamics, contains two categories: "done" and "stay." Both categories are feasible for each activity. Therefore, each d_m is a histogram on these two categories and can be written as $d_m = (d_m(\text{done}), d_m(\text{stay}))$. In view of (6.1), each d_m can be identified with its first component $d_m(\text{done})$ only. We emphasize that in contrast to the symbol

d_m representing a scalar in the setting with simple transition dynamics, it now represents a histogram on categories in the more general context of this chapter. The two perspectives are equivalent if we have simple transition dynamics through this identification.   △

Entity Types Loosely speaking, the entity type of an entity undertaking an activity at the time $t + 1$ represents where it was at time t. Consider an entity undertaking activity m' at time $t + 1$. There are two possibilities. Either this entity was undertaking some activity m (quite possibly $m = m'$) at time t, or it was not undertaking any activity at time t but it starts activity m' during this period.

If this entity was undertaking some activity m at time t, then it must have been assigned some category k_m during this period, in which case we denote its entity type by $m{:}k_m$.

If this entity was not undertaking any activity at time t, then it could be represented in the counter vector z (e.g., it could be waiting in some buffer at time t) or it could have arrived from outside the system. In this case, such an entity has no assigned category. We think of such an entity as having come to activity m' from some *origin* and denote its entity type by $o_{m'}$. The finite set of origins is arbitrary, and we use the symbol $o_{m'}$ to denote a generic feasible origin for activity m', where feasibility means that activity m' may receive entities from origin $o'_{m'}$. Often, the set of origins is a singleton, in which case we denote its one element by o.

We use the symbol $k'_{m'}$ for a generic entity type associated with activity m', so $k'_{m'}$ is of the form $o_{m'}$ or $m{:}k_m$. Not all entity types may be feasible for each activity in a given model. For instance, it may not be possible for an entity undertaking activity m' at time $t + 1$ to have been undertaking activity m at time t and be assigned category k_m.

For each activity m', we let $f_{m'}$ denote the histogram on entity types for the $x_{m'}^{(t+1)}(z')$ entities undertaking activity m' at time $t + 1$. In particular, $f_{m'}(k'_{m'})$ denotes the number of entities undertaking activity m' at the time $t + 1$ with entity type $k'_{m'}$. We call $f_{m'}$ the *entity-type flow histogram* for activity m' and write $f = (f_{m'})_{m'}$. By construction, we have the identity

$$|f_{m'}| = x_{m'}^{(t+1)}(z'). \qquad (6.2)$$

Often, few entity types are feasible for each activity, so it is easier to specify models by making the feasibility structure explicit.

We allow for the possibility that an entity undertaking an activity at time t does not undertake an activity at time $t+1$ regardless of their assigned category, although the routing pmf primitive (to be discussed in the next section) can preclude this. The number of such entities can be found from the category assignment histograms and entity-type flow histograms as follows. The total number of entities with entity type $m{:}k_m$ at any of the activities cannot exceed $d_m(k_m)$, the number of entities currently undertaking activity m assigned category k_m. Consequently, for each activity m and category k_m, we must have the flow balance constraint:

$$\sum_{m'} f_{m'}(m{:}k_m) \le d_m(k_m). \tag{6.3}$$

The difference between the right-hand and left-hand sides in this inequality is the number of entities that are currently undertaking activity m, have been assigned category k_m, and are not undertaking an activity at time $t + 1$.

Example 6.2 The feasible entity types for each activity m' for a QPLEX chain with simple transition dynamics, recast as a QPLEX chain with advanced transition dynamics, are m':stay and o, so each entity-type flow histogram $f_{m'} = (f_{m'}(o), f_{m'}(m'{:}\text{stay}))$ is a histogram on the entity types o and m':stay. In Chap. 4, the entity type o is called "new" and all entity types of the form m':stay are called "old."

Note that entities assigned category "done" during this period no longer undertake an activity at time $t + 1$, which can be deduced from the fact that "m:done" is not a feasible entity type for any activity m. However, we *cannot* deduce from the set of feasible entity types that entities assigned category "stay" always undertake an activity at time $t + 1$ for models with simple transition dynamics; this will be enforced by the routing pmf primitive to be discussed in the next section. $\triangle$

We now illustrate all concepts introduced in this section with the following example.

Example 6.3 There are two activities, a single origin o, and three possible categories labeled a, b, and c. Both activities can receive entities from the origin. The sets of feasible categories and entity types are shown in the table below.

	Activity 1	Activity 2
Feasible categories	$\{a, b\}$	$\{b, c\}$
Feasible entity types	$\{1{:}b, 2{:}b, o\}$	$\{1{:}b, 2{:}c, o\}$

Categories Entities undertaking activity 1 can either be assigned category a or b but not c. Therefore, the histogram $d_1 = (d_1(a), d_1(b))$ on categories for the $x_1^{(t)}(z)$ entities undertaking activity 1 at time t must satisfy the identity

$$d_1(a) + d_1(b) = x_1^{(t)}(z).$$

Entities undertaking activity 2 can either be assigned category b or c but not a. Therefore, the histogram $d_2 = (d_2(b), d_2(c))$ on categories for the $x_2^{(t)}(z)$ entities undertaking activity 2 at time t must satisfy the identity

$$d_2(b) + d_2(c) = x_2^{(t)}(z).$$

Entity Types The feasible entity types for activity 1 are 1:b, 2:b, and o. Therefore, the histogram $f_1 = (f_1(1{:}b), f_1(2{:}b), f_1(o))$ on entity types for the $x_1^{(t+1)}(z')$ entities undertaking activity 1 at time $t+1$ must satisfy the identity

$$f_1(1{:}b) + f_1(2{:}b) + f_1(o) = x_1^{(t+1)}(z').$$

The feasible entity types for activity 2 are 1:b, 2:c, and o. Therefore, the histogram $f_2 = (f_2(1{:}b), f_2(2{:}c), f_2(o))$ on entity types for the $x_2^{(t+1)}(z')$ entities undertaking activity 2 at time $t+1$ must satisfy the identity

$$f_2(1{:}b) + f_2(2{:}c) + f_2(o) = x_2^{(t+1)}(z').$$

Flow Balance Constraints The flow balance constraint (6.3) for each entity type can be deduced from the table. Entities undertaking activity 1 at time t assigned category a do not undertake an activity at time $t+1$, as the entity type 1:a is not feasible for either activity. Thus, the flow balance constraint for such entities is automatically satisfied as the left-hand side of (6.3) is zero.

Entities from activity 1 assigned category b can undertake either activity at time $t+1$, but could also undertake no activity at all. Thus, the flow balance constraint for such entities is $f_1(1{:}b) + f_2(1{:}b) \leq d_1(b)$.

Entities undertaking activity 2 at time t assigned category b either undertake activity 1 at time $t+1$ or will not undertake any activity. Thus, the flow balance constraint for such entities is $f_1(2{:}b) \leq d_2(b)$.

Entities undertaking activity 2 at time t assigned category c will either continue to undertake activity 2 at time $t+1$ or will not undertake any activity. Thus, the flow balance constraint for such entities is $f_2(2{:}c) \leq d_2(c)$. $\triangle$

6.2   Pmf Primitives

We define the transition probabilities for a QPLEX chain with advanced transition dynamics using the following model primitives for each time t:

- *Category pmfs* $\pi_m^{(t),\mathrm{cat}}(k_m | z, \ell_m)$,
- *Routing pmf* $\pi^{(t)}(z', f | z, d)$,
- *Relabeling pmfs* $\pi_{m:k_m \to m'}^{(t),\mathrm{rel}}(\ell'_m | z, \ell_m, z')$, and
- *Join pmfs* $\pi_{o_{m'} \to m'}^{(t),\mathrm{join}}(\ell'_m | z, z')$.

The feasible entity types for each activity dictate which relabeling and join pmfs need to be specified. Specifically, it is only necessary to specify the relabeling pmf $\pi_{m:k_m \to m'}^{(t),\mathrm{rel}}(\ell'_m | z, \ell_m, z')$ if $m{:}k_m$ is a feasible entity type for activity m', and the join pmf $\pi_{o_{m'} \to m'}^{(t),\mathrm{join}}(\ell'_m | z, z')$ only needs to be specified if $o_{m'}$ is a feasible entity type for activity m'.

As in Chap. 4, we speak of labels pertaining to time t as *current labels* and those pertaining to time $t + 1$ as *next labels*. Similarly, we speak of the *current counter, next counter, current histogram,* and *next histogram* vectors.

Category Pmfs Given the current counter vector z, each of the $h_m(\ell_m)$ entities undertaking activity m at time t with label ℓ_m is independently assigned a category using the *category pmf*

$$\pi_m^{(t),\text{cat}}(k_m | z, \ell_m).$$

This category assignment is also done independently across activities m.

This primitive extends the completion probability function $\gamma_m^{(t)}(z, \ell_m)$ used in the simple transition dynamics setting. Indeed, if the only possible categories are "done" and "stay," then the category pmf primitive for each activity may be identified with the probability of an entity assigned category "done." Thus, we have that

$$\pi_m^{(t),\text{cat}}(\text{done} | z, \ell_m) = \gamma_m^{(t)}(z, \ell_m), \tag{6.4}$$

$$\pi_m^{(t),\text{cat}}(\text{stay} | z, \ell_m) = 1 - \gamma_m^{(t)}(z, \ell_m). \tag{6.5}$$

Routing Pmf Given the current counter vector z and the vector of category assignment histograms d, the *routing pmf*

$$\pi^{(t)}(z', f | z, d)$$

determines the joint pmf of the next counter vector z' and the vector of entity-type flow histograms f. As is the case for a QPLEX chain with simple transition dynamics, a routing pmf cannot use entity labels.

The routing pmf primitive can only assign positive probability to entity-type flow histograms satisfying identity (6.2), i.e., $|f_{m'}| = x_{m'}^{(t+1)}(z')$, and flow balance constraint (6.3), i.e., $\sum_{m'} f_{m'}(m:k_m) \leq d_m(k_m)$.

For a QPLEX chain with simple transition dynamics recast as a QPLEX chain with advanced transition dynamics as in Example 6.2, the flow balance constraint becomes $f_m(m:\text{stay}) \leq d_m(\text{stay})$; however, since entities that do not complete their activity will continue to undertake the activity, this inequality becomes the equality:

$$f_m(m:\text{stay}) = d_m(\text{stay}). \tag{6.6}$$

Identity (6.3) becomes $f_{m'}(o) + f_{m'}(m':\text{stay}) = x_{m'}^{(t+1)}(z')$, and so

$$f_{m'}(o) = x_{m'}^{(t+1)}(z') - f_{m'}(m':\text{stay}) = x_{m'}^{(t+1)}(z') - d_{m'}(\text{stay}). \tag{6.7}$$

Rewriting the routing pmf primitive via the chain rule as

$$\pi^{(t)}(z', f | z, d) = \pi^{(t)}(z' | z, d) \times \pi^{(t)}(f | z, d, z'),$$

the second term on the right-hand side is a *canonical* degenerate pmf determined by (6.6) and (6.7) in this case.

QPLEX chains with advanced transition dynamics permit the second term $\pi^{(t)}(f|z, d, z')$ to be general. The particulars of the system under consideration dictate its form, so it is not canonical as for QPLEX chains with simple transition dynamics recast as QPLEX chains with advanced transition dynamics. This explains why we did not include entity-type flow histograms in the routing pmf primitive in the setting with simple transition dynamics. With advanced transition dynamics, the pmf of these histograms needs to be specified via the routing pmf primitive. Of course, it is still possible for the vector of entity-type flow histograms f to be a deterministic function of z, d, and z'.

Relabeling Pmfs Fix some current and next counter vectors z and z'. Consider an entity undertaking activity m at time t with label ℓ_m assigned category k_m that undertakes activity m' at time $t + 1$. Each such entity receives a new label using the *relabeling pmf*:

$$\pi_{m:k_m \to m'}^{(t),\text{rel}}(\ell'_{m'}|z, \ell_m, z').$$

The sampling is done independently across activities m, categories k_m, activities m', and labels ℓ_m.

This primitive only needs to be specified if $m{:}k_m$ is a feasible entity type for activity m'. We use the subscript $m{:}k_m \to m'$ to highlight that such entities were undertaking activity m at time t, were assigned category k_m, and are undertaking activity m' at time $t + 1$. Of course, it is very well possible that $m = m'$. Relabeling pmfs are not required to be distinct.

For a QPLEX chain with simple transition dynamics recast as a QPLEX chain with advanced transition dynamics as in Example 6.2, the one relabeling pmf for each activity m is given by the simple transition dynamics relabeling pmf, i.e.,

$$\pi_{m:\text{stay} \to m}^{(t),\text{rel}}(\ell'_m|z, \ell_m, z') = \pi_m^{(t),\text{rel}}(\ell'_m|z, \ell_m, z'). \tag{6.8}$$

Join Pmfs Given the current and next counter vectors z and z', respectively, each new entity at activity m' with origin $o_{m'}$ receives an independent label using the *join pmf*

$$\pi_{o_{m'} \to m'}^{(t),\text{join}}(\ell'_{m'}|z, z').$$

This labeling is also done independently across activities m' and origins $o_{m'}$.

This primitive only needs to be specified for feasible origins $o_{m'}$ for activity m'. We use the subscript $o_{m'} \to m'$ to highlight that such entities come from origin $o_{m'}$ and are undertaking activity m' at time $t + 1$.

For a QPLEX chain with simple transition dynamics recast as a QPLEX chain with advanced transition dynamics as in Example 6.2, the one join pmf for each activity m' is given by the simple transition dynamics join pmf, i.e.,

$$\pi^{(t),\text{join}}_{o_{m'}\to m'}(\ell'_{m'}|z, z') = \pi^{(t),\text{join}}_{m'}(\ell'_{m'}|z, z'). \tag{6.9}$$

Example 6.4 We recast the stochastic infectious disease model presented in Sect. 5.2 using advanced model primitives. The specification of the category, relabeling, and join pmf primitives here revert to their simple transition dynamics counterparts as described above, so we will not repeat them here.

Recall that each individual in a population of K individuals is in one of three compartments: susceptible "S," infectious "I," or recovered "R." An individual progresses from the S to the I and then to the R compartment. Each individual in the S compartment becomes infectious (transitions to the I compartment) independently with probability $1 - e^{-\beta \times \zeta}$, where ζ is the infectious proportion of the population and $\beta > 0$ is a model parameter that controls the infectiousness of the disease. Once an individual becomes infectious, (s)he remains infectious for a random amount of time. The numbers of periods needed to recover from the disease after becoming infectious are assumed to be independent across individuals and to be at least one time period. The R compartment contains individuals who have recovered, and individuals stay in this compartment as soon as they enter it. We assume there is at least one infectious individual initially (to avoid trivial dynamics).

Counters and Histograms The three random variables $Z_S^{(t)}$, which represents the number of susceptible individuals at time t, $Z_I^{(t)}$, which represents the number of infectious individuals at time t, and $H^{(t)}$, which represents the histogram of remaining infectious times, ages, or phases of the entities recovering from infection at time t are the same as before.

The counter set is $\Omega_Z = \mathbb{Z}_+^2$, the size function is $x^{(t)}(z_S, z_I) = z_I$, and the label set is the same as before.

Categories and Entity Types The two categories are denoted by "done" and "stay." The feasible categories and entity types for activity I are listed and interpreted in the tables below.

Activity	Category	Interpretation
I	done	Individual recovers from infection
I	stay	Individual does not recover from infection

Activity	Entity Type	Description
I	I:stay	Individual continues to recover from their infection
I	o	A susceptible individual becomes infectious

Routing Pmf The distributional program below outputs the routing pmf for this model. (For convenience, we reuse the name ROUTING$^{(t)}$.) This program takes as input the current counters z_S, z_I and the category assignment histogram $d_I =$

$(d_I(\text{done}), d_I(\text{stay}))$, and outputs the pmf of the next counters z'_S, z'_I and the entity-type flow histogram $f_I = (f_I(I:\text{stay}), f_I(o))$.

All individuals who do not recover from their infection during this period will remain infectious (Line 2). The number of new infectious individuals is drawn from the binomial pmf (Line 3). Standard flow balance determines the new values z'_S (Line 4). Line 5 uses the fact that $x_I^{(t+1)}(z') = |f_I|$ and that in this model $x_I^{(t+1)}(z') = z'_I$.

1: **function** ROUTING$^{(t)}(z_S, z_I, d_I)$
2:     $f_I(I:\text{stay}) \leftarrow d_I(\text{stay})$
3:     $f_I(o) \sim \text{bin}\left(z_S, 1 - e^{-\beta z_I/K}\right)$
4:     $z'_S \leftarrow z_S - f_I(o)$
5:     $z'_I \leftarrow |f_I|$
6:     **return** (z'_S, z'_I, f_I)

We see that the number of entities $d_I(\text{done})$ assigned category "done" during this period plays no role in calculating f_I. Therefore, those entities are no longer undertaking activity I at time $t + 1$. On the other hand, the number of entities $d_I(\text{stay})$ assigned category "stay" during this period all continue to undertake activity I; see Line 2. This results from recasting a model with simple transition dynamics as a model with advanced transition dynamics.

The logical flow in the distributional program above is natural for this model, as it first specifies the entity-type flow histogram for activity I and then specifies the next counters in terms of this entity-type flow histogram. Below is an equivalent distributional program.

1: **function** ROUTING$^{(t)}(z_S, z_I, d_I)$
2:     $z'_I \leftarrow d_I(\text{stay}) + \text{bin}\left(z_S, 1 - e^{-\beta z_I/K}\right)$
3:     $z'_S \leftarrow z_S - (z'_I - d_I(\text{stay}))$
4:     $f_I(I:\text{stay}) \leftarrow d_I(\text{stay})$
5:     $f_I(o) \leftarrow z'_I - d_I(\text{stay})$
6:     **return** (z'_S, z'_I, f_I)

In this program, the entity-type flow histogram is calculated after the next counters, resulting in identities (6.6) and (6.7) becoming explicit. $\triangle$

6.3 Transition Probabilities

Fix some counter vector z and histogram vector h with $|h_m| = x_m^{(t)}(z)$ for each activity m. The distributional program TRANSITIONPROBABILITY$^{(t)}$ outputs the transition probabilities $p^{(t)}(z', h'|z, h)$ of the QPLEX chain for the period between times t and $t + 1$ in terms of the routing pmf primitive and two distributional programs that use the other model primitives, which will be presented shortly.

It generates the next counter vector z' and the next histogram vector h' with $|h'_{m'}| = x_{m'}^{(t+1)}(z')$ for all m' in several stages, as follows.

We first construct, for each m, the *current label category histogram* h_m^{clc} on pairs of current labels and categories for activity m (Line 2). We then calculate the category assignment histograms d_m from h_m^{clc} by marginalizing out the current label (Line 4). We generate the next counter vector z' and the vector of entity-type flow histograms f from the current counter vector z and the vector of category assignment histograms d using the routing pmf primitive (Line 5). We subsequently construct, for each activity m', the *next label-type histogram* $h_{m'}^{\text{nlt}}$ on pairs of next labels and entity types (Line 6). Finally, we construct the histograms $h'_{m'}$ from $h_{m'}^{\text{nlt}}$ by marginalizing out the entity types (Line 8). Implicit in this pseudocode are the definitions $h^{\text{clc}} = (h_m^{\text{clc}})_m$, $d = (d_m)_m$, $h^{\text{nlt}} = (h_{m'}^{\text{nlt}})_{m'}$, and $h' = (h'_{m'})_{m'}$.

The remainder of this chapter uses the following terminology for histograms on the Cartesian product of two sets. We refer to a two-dimensional histogram with one of its arguments fixed as a *marginal histogram*. For instance, $h_m^{\text{clc}}(\ell_m, \cdot)$ and $h_{m'}^{\text{nlt}}(\cdot, m{:}k_m)$ appearing in this program are marginal histograms on categories and next labels, respectively.

1: **function** TRANSITIONPROBABILITY$^{(t)}(z, h)$
2:     $h^{\text{clc}} \sim$ CURRENTLABELCATEGORYHISTOGRAMS$^{(t)}(z, h)$
3:     **for** $m = 1, \ldots, M$ **do**
4:         $d_m \leftarrow \sum_{\ell_m} h_m^{\text{clc}}(\ell_m, \cdot)$
5:     $(z', f) \sim \pi^{(t)}(Z' = \cdot, F = \cdot | z, d)$
6:     $h^{\text{nlt}} \sim$ NEXTLABELTYPEHISTOGRAMS$^{(t)}(z, h^{\text{clc}}, z', f)$
7:     **for** $m' = 1, \ldots, M$ **do**
8:         $h'_{m'} \leftarrow \sum_{o_{m'}} h_{m'}^{\text{nlt}}(\cdot, o_{m'}) + \sum_{m{:}k_m} h_{m'}^{\text{nlt}}(\cdot, m{:}k_m)$
9:     **return** (z', h')

We use Example 6.3 as a running example to facilitate the understanding of the details of how the current label category histograms and next label-type histograms are generated. For convenience, we have reproduced the table showing the sets of feasible categories and entity types for each activity.

	Activity 1	Activity 2
Feasible categories	$\{a, b\}$	$\{b, c\}$
Feasible entity types	$\{1{:}b, 2{:}b, o\}$	$\{1{:}b, 2{:}c, o\}$

The diagram below presents one possible realization of the four vectors of histograms h, h^{clc}, h^{nlt}, and h' for this example. The current and new counter vectors z and z' are not depicted. The vectors of category assignment histograms d and entity-type flow histograms f are not shown explicitly either, but shortly we will obtain these from the diagram.

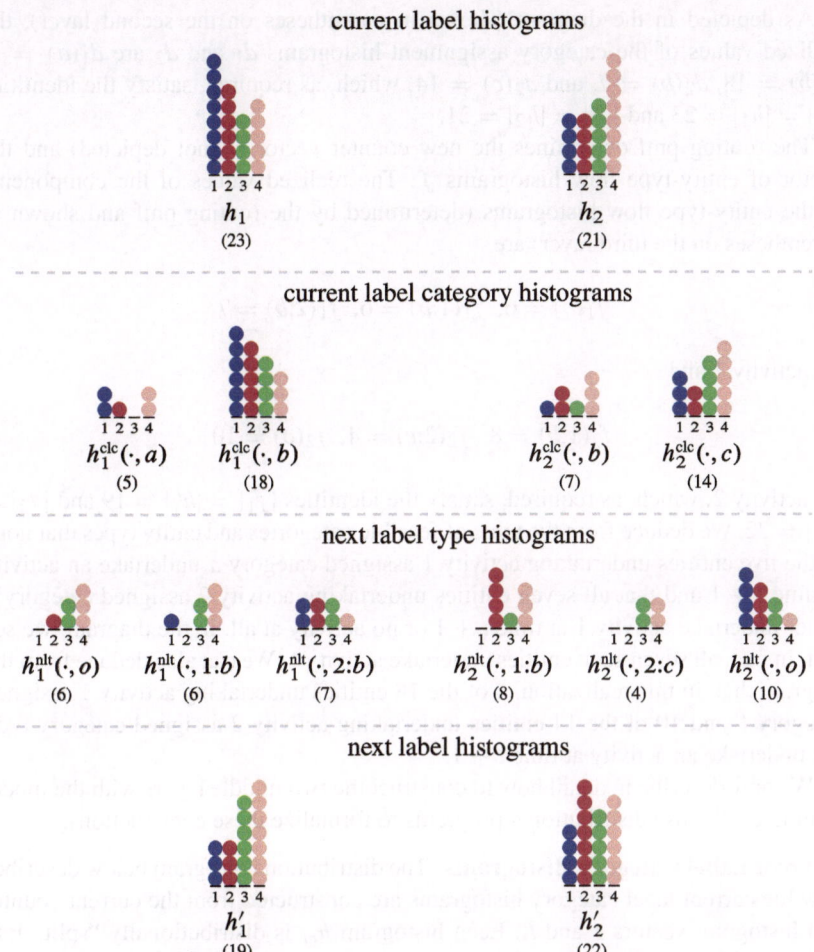

current label histograms

h_1
(23)

h_2
(21)

current label category histograms

$h_1^{\text{clc}}(\cdot, a)$
(5)

$h_1^{\text{clc}}(\cdot, b)$
(18)

$h_2^{\text{clc}}(\cdot, b)$
(7)

$h_2^{\text{clc}}(\cdot, c)$
(14)

next label type histograms

$h_1^{\text{nlt}}(\cdot, o)$
(6)

$h_1^{\text{nlt}}(\cdot, 1{:}b)$
(6)

$h_1^{\text{nlt}}(\cdot, 2{:}b)$
(7)

$h_2^{\text{nlt}}(\cdot, 1{:}b)$
(8)

$h_2^{\text{nlt}}(\cdot, 2{:}c)$
(4)

$h_2^{\text{nlt}}(\cdot, o)$
(10)

next label histograms

h_1'
(19)

h_2'
(22)

The first layer displays the current label histograms h_1 and h_2 associated with activities 1 and 2. The label sets are $\Omega_1 = \Omega_2 = \{1, 2, 3, 4\}$. The numbers of entities undertaking activities 1 and 2 are, respectively, $x_1^{(t)}(z) = |h_1| = 23$ and $x_2^{(t)}(z) = |h_2| = 21$, as shown below the histograms in parentheses. The second layer represents the current label category histograms h_1^{clc} and h_2^{clc} via their marginal histograms for each feasible category. These are histograms on current labels. Similarly, we show the marginal histograms of the next label-type histograms h_1^{nlt} and h_2^{nlt} for each feasible entity type. These are histograms on next labels. Throughout, an entity with label 1, 2, 3, or 4 is identified with a blue, red, green, or beige ball, respectively, but keep in mind that an entity's label (and therefore the color of its corresponding ball) may change when moving from the second layer to the third layer.

As depicted in the diagram (shown in parentheses on the second layer), the realized values of the category assignment histograms d_1 and d_2 are $d_1(a) = 5$, $d_1(b) = 18$, $d_2(b) = 7$, and $d_2(c) = 14$, which, as required, satisfy the identities $|d_1| = |h_1| = 23$ and $|d_2| = |h_2| = 21$.

The routing pmf determines the new counter vector z' (not depicted) and the vector of entity-type flow histograms f. The realized values of the components of the entity-type flow histograms (determined by the routing pmf and shown in parentheses on the third layer) are

$$f_1(o) = 6, \ f_1(1{:}b) = 6, \ f_1(2{:}b) = 7$$

for activity 1 and

$$f_2(1{:}b) = 8, \ f_2(2{:}c) = 4, \ f_2(o) = 10$$

for activity 2, which, as required, satisfy the identities $|f_1| = |h'_1| = 19$ and $|f_2| = |h'_2| = 22$. We deduce from the table of feasible categories and entity types that none of the five entities undertaking activity 1 assigned category a undertake an activity at time $t + 1$ and that all seven entities undertaking activity 2 assigned category b either undertake activity 1 at time $t + 1$ or no activity at all. In the diagram, we see that, in fact, all seven such entities undertake activity 1. We can also deduce from the diagram that, in this realization, 4 of the 18 entities undertaking activity 1 assigned category b and 10 of the 14 entities undertaking activity 2 assigned category c do not undertake an activity at time $t + 1$.

We now describe in detail how to construct the two middle layers with the model primitives. We use distributional programs to formalize these constructions.

Current Label Category Histograms The distributional program below describes how the current label category histograms are constructed from the current counter and histogram vectors z and h. Each histogram h_m is distributionally "split" into h_m^{clc}, which takes a current label ℓ_m and a feasible category k_m associated with activity m for its argument. This splitting is achieved by assigning each entity with label ℓ_m represented in h_m an i.i.d. category with pmf $\pi_m^{(t),\text{cat}}(k_m | z, \ell_m)$.

1: **function** CURRENTLABELCATEGORYHISTOGRAMS$^{(t)}(z, h)$
2:     **for** $m = 1, \ldots, M$ **do**
3:         **for** $\ell_m \in \Omega_m$ **do**
4:             $h_m^{\text{clc}}(\ell_m, \cdot) \sim \text{mult}\left(h_m(\ell_m), \pi_m^{(t),\text{cat}}(K_m = \cdot | z, \ell_m) \right)$
5:     **return** h^{clc}

Consider the entities in activity 1 of our example. Recall that the support of the category pmf primitive $\pi_1^{(t),\text{cat}}(k_1 | z, \ell_1)$ is such that an entity undertaking activity 1 can be assigned category a or b but not c. (The actual probabilities are irrelevant here.) After categories have been assigned, two of the $h_1(1) = 8$ entities with label 1 are assigned category a and the other six are assigned category b, and so $h_1^{\text{clc}}(1, a) = 2$ and $h_1^{\text{clc}}(1, b) = 6$; one of the $h_1(2) = 6$ entities with label 2 is assigned category a

and the other five are assigned category b, and so $h_1^{\text{clc}}(2, a) = 1$ and $h_1^{\text{clc}}(2, b) = 5$; and so on.

Next Label-Type Histograms The distributional program below describes how the next label-type histograms are constructed from the current counter vector z, the vector h^{clc} of current label category histograms, the next counter vector z', and the vector of entity-type flow histograms f.

Using the pmf primitive $\pi_{o \to m'}^{(t),\text{join}}(\ell'_m | z, z')$, Lines 3 and 4 construct the marginal histograms on next labels corresponding to entity types of the form $o_{m'}$. The construction of the marginal histograms on next labels corresponding to entity types of the form $m{:}k_m$ comprises two steps. The first step generates what we call *route histograms*, and the second step generates the requisite marginal histograms from these route histograms.

Fix activity m and category k_m. There are $d_m(k_m)$ entities undertaking activity m at time t assigned category k_m, and their current labels are represented by the marginal histogram $h_m^{\text{clc}}(\cdot, k_m)$. Of these $d_m(k_m)$ entities, $f_{m'}(m{:}k_m)$ are undertaking activity m' at time $t + 1$. By successive sampling without replacement, Lines 6 and 10 produce a *route histogram* $h_{m'}^{\text{rte}}$ on pairs of current labels and non-origin entity types such that

$$\sum_{\ell_m} h_{m'}^{\text{rte}}(\ell_m, m{:}k_m) = f_{m'}(m{:}k_m).$$

All $d_m(k_m) - \sum_{m'} f_{m'}(m{:}k_m)$ entities that remain after this sampling without replacement play no further role during this period. The inequalities in (6.3) ensure that the sampling instruction in Line 6 is feasible. Entity labels are not changed in this process; this is done in Lines 7–9. Each entity represented in the marginal histogram $h_{m'}^{\text{rte}}(\cdot, m{:}k_m)$ with current label ℓ_m receives an i.i.d. next label $\ell'_{m'}$ with pmf primitive $\pi_{m:k \to m'}^{(t),\text{rel}}(\ell'_{m'} | z, \ell_m, z')$. The marginal histogram $h_{m'}^{\text{nlt}}(\cdot, m{:}k_m)$ on next labels accumulates the resulting histograms for all possible current labels ℓ_m, each of which has a multinomial pmf.

1: **function** NEXTLABELTYPEHISTOGRAMS$^{(t)}(z, h^{\text{clc}}, z', f)$
2:     **for** $m' = 1, \ldots, M$ **do**
3:         **for all** $o_{m'}$ **do**
4:             $h_{m'}^{\text{nlt}}(\cdot, o_{m'}) \sim \text{mult}\left(f_{m'}(o_{m'}), \pi_{o_{m'} \to m'}^{(t),\text{join}}(L'_{m'} = \cdot | z, z') \right)$
5:         **for all** $m{:}k_m$ feasible for m' **do**
6:             $h_{m'}^{\text{rte}}(\cdot, m{:}k_m) \sim \text{SAMPLEW/OREPLACEMENT}(f_{m'}(m{:}k_m), h_m^{\text{clc}}(\cdot, k_m))$
7:             $h_{m'}^{\text{nlt}}(\cdot, m{:}k_m) \leftarrow 0$
8:             **for** $\ell_m \in \Omega_m$ **do**
9:                 $h_{m'}^{\text{nlt}}(\cdot, m{:}k_m) +\sim \text{mult}\left(h_{m'}^{\text{rte}}(\ell_m, m{:}k_m), \pi_{m:k_m \to m'}^{(t),\text{rel}}(L'_{m'} = \cdot | z, \ell_m, z') \right)$
10:            $h_m^{\text{clc}}(\cdot, k_m) -\leftarrow h_{m'}^{\text{rte}}(\cdot, m{:}k_m)$
11:     **return** h^{nlt}

Returning to our running example, we illustrate Lines 5–10 in the diagram below. Entities currently undertaking activity 1 assigned category b are represented either in $h_1^{\text{nlt}}(\cdot, 1{:}b)$ or in $h_2^{\text{nlt}}(\cdot, 1{:}b)$, both marginal histograms on next labels, unless they no longer undertake an activity at time $t + 1$. Since $f_1(1{:}b) = 6$ and $f_2(1{:}b) = 8$, the sampling sizes are given by

$$\sum_{\ell_1} h_1^{\text{rte}}(\ell_1, 1{:}b) = \sum_{\ell_1'} h_1^{\text{nlt}}(\ell_1', 1{:}b) = 6$$

$$\sum_{\ell_1} h_2^{\text{rte}}(\ell_1, 1{:}b) = \sum_{\ell_2'} h_2^{\text{nlt}}(\ell_2', 1{:}b) = 8.$$

We construct the marginal histogram $h_1^{\text{rte}}(\cdot, 1{:}b)$ on current labels by sampling $f_1(1{:}b) = 6$ entities without replacement from the entities represented in marginal histogram $h_1^{\text{clc}}(\cdot, b)$ on current labels. (The distributional program SAMPLEW/OREPLACEMENT should be thought of as a built-in program.) We similarly construct marginal histogram $h_2^{\text{rte}}(\cdot, 1{:}b)$ on current labels by sampling $f_2(1{:}b) = 8$ entities without replacement from the 12 remaining entities represented in the residual marginal histogram $h_1^{\text{clc}}(\cdot, b) - h_1^{\text{rte}}(\cdot, 1{:}b)$ on current labels. The remaining $d_1(b) - f_1(1{:}b) - f_2(1{:}b) = 4$ entities assigned category b play no further role during this period. The marginal histograms $h_1^{\text{nlt}}(\cdot, 1{:}b)$ and $h_2^{\text{nlt}}(\cdot, 1{:}b)$ on next labels are constructed from the marginal histograms $h_1^{\text{rte}}(\cdot, 1{:}b)$ and $h_2^{\text{rte}}(\cdot, 1{:}b)$ on current labels, respectively, by giving each entity a new label according to the appropriate relabeling pmf.

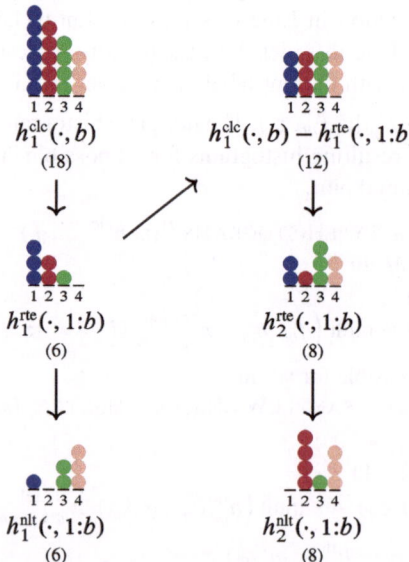

We make two remarks about this program. First, the for-loops over m' and $m{:}k_m$ in Lines 2 and 5 of the program can be traversed in any order since the pmf represented by this program is not affected by the order in which the sampling without replacement is done. Second, the argument h^{clc} of this program is modified in Line 10. We do not specify how arguments are passed in our pseudocode (e.g., by value or by reference), and so h^{clc} technically becomes undefined in the TRANSITIONPROBABILITY$^{(t)}$ program after calling NEXTLABELTYPEHISTOGRAMS$^{(t)}$. However, this does not pose a problem, as this variable is no longer used.

6.4 QPLEX Calculus

The distributional program QPLEXMAP$^{(t)}$ below is the starting point for this section. In this distributional program, we have expanded the pseudocode for the multinomial lift and compression maps but not for the transition probabilities.

1: **function** QPLEXMAP$^{(t)}(\mu)$
2:     $z \sim \mu(Z = \cdot)$
3:     **for** $m = 1, \ldots, M$ **do**
4:         $h_m \sim \text{mult}\left(x_m^{(t)}(z), \mu(L_m = \cdot | z)\right)$
5:     $(z', h') \sim \text{TRANSITIONPROBABILITY}^{(t)}(z, h)$
6:     **for** $m' = 1, \ldots, M$ **do**
7:         **if** $h'_{m'} \neq 0$ **then**
8:             $\ell'_{m'} \sim h'_{m'}/|h'_m|$
9:         **else**
10:             $\ell'_{m'} \sim \theta^0_{m'}(L'_{m'} = \cdot)$
11:     **return** (z', ℓ')

The transition probabilities in the above program encode the advanced transition dynamics. As such, it contains (implicit) sampling statements for each of the components of h, h^{clc}, h^{rte}, h^{nlt}, and h'. As a result, evaluating the QPLEX map involves enumerating all possible values of these histograms. The number of such values is generally prohibitive, even for a single histogram.

We now refactor this distributional program to eliminate these histograms. We proceed in four steps. Each step heavily exploits properties of multinomial random variables and preserves, in an appropriate sense, multinomial pmfs for subsequent use, which we once again refer to as *multinomial propagation*. The derivation below generalizes the multinomial propagation we derived to establish the QPLEX calculus using simple transition dynamics. At the end of this section, we convert the refactored program into an equivalent algebraic formulation for the QPLEX map.

Step 1: Eliminate the h_m Expanding the above program using the distributional programs TRANSITIONPROBABILITY$^{(t)}$, CURRENTLABELCATEGORYHISTOGRAMS$^{(t)}$, and NEXTLABELTYPEHISTOGRAMS$^{(t)}$, the variable h_m in the above program is used to generate the current label category histogram h_m^{clc} but is subsequently unused. We replace the sampling instructions

$$h_m \sim \text{mult}\left(x_m^{(t)}(z), \mu(L_m = \cdot|z)\right)$$

for $\ell_m \in \Omega_m$ **do**

$$h_m^{\text{clc}}(\ell_m, \cdot) \sim \text{mult}\left(h_m(\ell_m), \pi_m^{(t),\text{cat}}(K_m = \cdot|z, \ell_m)\right)$$

$$d_m \leftarrow \sum_{\ell_m} h_m^{\text{clc}}(\ell_m, \cdot)$$

with pseudocode that results in the *same* joint pmf of h_m^{clc} and category assignment histogram d_m, but *without sampling* h_m. We do so by *first* sampling d_m and then h_m^{clc} by sampling marginal histograms $h_m^{\text{clc}}(\cdot, k_m)$ on current labels for each category k_m.

Since z is obtained by sampling from the marginal pmf of $\mathbf{Z}$ under μ (Line 2), it must satisfy $\mu(z) > 0$. Fix some activity m. Since the pmf of the histogram of current labels h_m is multinomial with parameters $x_m^{(t)}(z)$ and $\mu(L_m = \cdot|z)$, we can think of the current labels of the $x_m^{(t)}(z)$ entities undertaking activity m as being i.i.d. with pmf $\mu(\ell_m|z)$. Conditional on their current label, these entities independently sample their assigned categories. Unconditionally, the categories of these entities are independent with the *same* pmf, so the category assignment histogram d_m has a multinomial pmf. The number of trials parameter is $x_m^{(t)}(z)$, and the pmf parameter is obtained by conditioning on the label ℓ_m and using the category pmf $\pi_m^{(t),\text{cat}}(k_m|z, \ell_m)$. Therefore, we can sample d_m directly using the distributional program CATEGORYHISTOGRAM$_{\mu,m}^{(t)}$ below.

1: **function** CATEGORYHISTOGRAM$_{\mu,m}^{(t)}(z)$

2:    $d_m \sim \text{mult}\left(x_m^{(t)}(z), \sum_{\ell_m} \mu(\ell_m|z) \times \pi_m^{(t),\text{cat}}(K_m = \cdot|z, \ell_m)\right)$

3:    **return** d_m

It remains to sample the marginal histogram $h_m^{\text{clc}}(\cdot, k_m)$ on current labels given $\sum_{\ell_m} h_m^{\text{clc}}(\ell_m, k_m) = d_m(k_m)$ for each category k_m. In view of the above construction, the labels of each of the $d_m(k_m)$ entities currently at activity m that have been assigned category k_m are i.i.d. with pmf $\xi_{\mu,m:k_m}^{(t)}(\ell_m|z)$ given by, for $\ell_m \in \Omega_{L_m}$,

$$\xi_{\mu,m:k_m}^{(t)}(\ell_m|z) = \frac{\mu(\ell_m|z) \times \pi_m^{(t),\text{cat}}(k_m|z, \ell_m)}{\sum_{\tilde{\ell}_m} \mu(\tilde{\ell}_m|z) \times \pi_m^{(t),\text{cat}}(k_m|z, \tilde{\ell}_m)}, \tag{6.10}$$

as can be seen using Bayes' rule. Indeed, the pmf of the current label of such an entity changes from $\mu(\ell_m|z)$ to $\xi_{\mu,m:k_m}^{(t)}(\ell_m|z)$ upon revealing that its category is k_m, but independence is preserved. Thus, the marginal histogram $h_m^{\text{clc}}(\cdot, k_m)$ on current labels has a multinomial pmf with number of trials parameter $d_m(k_m)$ and pmf parameter $\xi_{\mu,m:k_m}^{(t)}(L_m = \cdot|z)$.

We conclude that we can replace the above pseudocode fragment with the following pseudocode fragment, where the histogram h_m no longer appears:

$$d_m \sim \text{CATEGORYHISTOGRAM}_{\mu,m}^{(t)}(z)$$

for all k_m **do**

$$h_m^{\text{clc}}(\cdot, k_m) \sim \text{mult}\left(d_m(k_m), \xi_{\mu,m:k_m}^{(t)}(L_m = \cdot|z)\right)$$

Note that the case $d_m(k_m) = 0$ is not separated out due to our convention that a multinomial random variable equals zero in that case, just like the case $x^{(t)}(z) = 0$ is not separated out in the original program.

Step 2: Eliminate the h_m^{clc} The marginal histogram $h_m^{\text{clc}}(\cdot, k_m)$ on current labels is only used to sample the marginal histograms $h_{m'}^{\text{rte}}(\cdot, m:k_m)$ on current labels for all m' for which entity type $m:k_m$ is feasible. The starting point of this step is the sampling statement for the marginal histogram $h_{m'}^{\text{clc}}(\cdot, k_m)$ from the previous step and the construction of these marginal histograms of the route histograms. Specifically, we start with the pseudocode fragment

$$h_m^{\text{clc}}(\cdot, k_m) \sim \text{mult}\left(d_m(k_m), \xi_{\mu,m:k_m}^{(t)}(L_m = \cdot|z)\right)$$

for all m' for which entity type $m:k_m$ is feasible **do**

$$h_{m'}^{\text{rte}}(\cdot, m:k_m) \sim \text{SAMPLEW/OREPLACEMENT}(f_{m'}(m:k_m), h_m^{\text{clc}}(\cdot, k_m))$$
$$h_m^{\text{clc}}(\cdot, k_m) -\!\leftarrow h_{m'}^{\text{rte}}(\cdot, m:k_m)$$

and replace it with a pseudocode fragment that *directly* samples the marginal histogram $h_{m'}^{\text{rte}}(\cdot, k_m)$ on current labels *without sampling the marginal histograms* $h_m^{\text{clc}}(\cdot, k_m)$ *on current labels.*

Since the marginal histogram $h_m^{\text{clc}}(\cdot, k_m)$ on current labels is multinomial, we can think of the labels of the $d_m(k_m)$ entities assigned category k_m as being i.i.d. with pmf $\xi_{\mu,m:k_m}^{(t)}(\ell_m|z)$. Since the sampling of the marginal histogram $h_{m'}^{\text{rte}}(\cdot, m:k_m)$ is done without replacement, the probability of the entity labels remains unchanged, and so we can replace the above pseudocode fragment with the following pseudocode fragment:

for all m' for which entity type $m:k_m$ is feasible **do**

$$h_{m'}^{\text{rte}}(\cdot, m:k_m) \sim \text{mult}\left(f_{m'}(m:k_m), \xi_{\mu,m:k_m}^{(t)}(L_m = \cdot|z)\right)$$

Step 3: Eliminate the $h_{m'}^{\text{rte}}$ Each marginal $h_{m'}^{\text{rte}}(\cdot, m:k_m)$ of $h_{m'}^{\text{rte}}$ is only used to sample the corresponding marginal $h_{m'}^{\text{nlt}}(\cdot, m:k_m)$ of the next label-type histogram. The starting point of this step is the sampling statement for the marginal histogram $h_{m'}^{\text{rte}}(\cdot, m:k_m)$ on current labels from the previous step and the construction of the corresponding marginal histograms $h_{m'}^{\text{nlt}}(\cdot, m:k_m)$ on next labels. Specifically, we start with the pseudocode fragment

$$h_{m'}^{\text{rte}}(\cdot, m:k_m) \sim \text{mult}\left(f_{m'}(m:k_m), \xi_{\mu,m:k_m}^{(t)}(L_m = \cdot|z)\right)$$
$$h_{m'}^{\text{nlt}}(\cdot, m:k_m) \leftarrow 0$$

for $\ell_m \in \Omega_m$ **do**

$$h_{m'}^{\text{nlt}}(\cdot, m:k_m) +\!\sim \text{mult}\left(h_{m'}^{\text{rte}}(\ell_m, m:k_m), \pi_{m:k_m \to m'}^{(t),\text{rel}}(L'_{m'} = \cdot|z, \ell_m, z')\right)$$

and replace it with a pseudocode fragment that *directly* samples the marginal histogram $h_{m'}^{\text{nlt}}(\cdot, m{:}k_m)$ on next labels *without sampling the marginal histogram* $h_{m'}^{\text{rte}}(\cdot, m{:}k_m)$ *on current labels*.

If the label of an entity undertaking activity m, assigned category k_m and routed to activity m' is ℓ_m, its next label pmf is $\pi_{m:k_m \to m'}^{(t),\text{rel}}(\ell_m' | z, \ell_m, z')$. Consequently, the (unconditional) next label of each such entity is i.i.d. with pmf

$$\sum_{\ell_m} \xi_{\mu,m:k_m}^{(t)}(\ell_m | z) \times \pi_{m:k_m \to m'}^{(t),\text{rel}}(L_m' = \cdot | z, \ell_m, z'). \tag{6.11}$$

It follows that the marginal histogram $h_{m'}^{\text{nlt}}(\cdot, m{:}k_m)$ on next labels is multinomial with number of trials parameter $f_{m'}(m{:}k)$ and pmf parameter given by (6.11). We thus find that the above pseudocode fragment can be replaced by the following single sampling instruction:

$$h_{m'}^{\text{nlt}}(\cdot, m{:}k_m) \sim \text{mult}\left(f_{m'}(m{:}k_m), \sum_{\ell_m} \xi_{\mu,m:k_m}^{(t)}(\ell_m | z) \times \pi_{m:k_m \to m'}^{(t),\text{rel}}(L_m' = \cdot | z, \ell_m, z') \right)$$

Step 4: Eliminate the $h_{m'}^{\text{nlt}}$ and $h_{m'}'$ The next label-type histogram $h_{m'}^{\text{nlt}}$ is only used to sample the next label histogram $h_{m'}'$. This step starts from the sampling statements for the marginal histograms $h_{m'}^{\text{nlt}}(\cdot, k_{m'}')$ on next labels, which are obtained for $k_{m'}' = m{:}k_m$ from the previous step, and the construction of $\ell_{m'}'$ by first marginalizing the next label-type histogram to obtain $h_{m'}'$ and subsequently sampling $\ell_{m'}'$ using the lines representing the compression map. Specifically, we start with the pseudocode fragment

for all $o_{m'}$ **do**
$$h_{m'}^{\text{nlt}}(\cdot, o_{m'}) \sim \text{mult}\left(f_{m'}(o_{m'}), \pi_{o_{m'} \to m'}^{(t),\text{join}}(L_{m'}' = \cdot | z, z') \right)$$
for all $m{:}k_m$ **do**
$$h_{m'}^{\text{nlt}}(\cdot, m{:}k_m) \sim \text{mult}\left(f_{m'}(m{:}k_m), \sum_{\ell_m} \xi_{\mu,m:k_m}^{(t)}(\ell_m | z) \times \pi_{m:k_m \to m'}^{(t),\text{rel}}(L_m' = \cdot | z, \ell_m, z') \right)$$
$h_{m'}' \leftarrow \sum_{o_{m'}} h_{m'}^{\text{nlt}}(\cdot, o_{m'}) + \sum_{m:k_m} h_{m'}^{\text{nlt}}(\cdot, m{:}k_m)$
if $h_{m'}' \neq 0$ **then**
$\quad \ell_{m'}' \sim h_{m'}' / |h_{m'}'|$
else
$\quad \ell_{m'}' \sim \theta_{m'}^0(L_{m'}' = \cdot)$

and replace it with pseudocode that *directly* samples ℓ_m' *without sampling* $h_{m'}^{\text{nlt}}$ and *calculating* $h_{m'}'$.

As shown in the above pseudocode fragment, the next label is drawn from a pmf that is proportional to the frequencies in $h_{m'}'$. This can be interpreted as selecting the label of a "uniformly chosen at random" entity among the $x_{m'}^{(t+1)}(z')$ entities represented in $h_{m'}'$, each of which will be undertaking activity m' at time $t + 1$. Since $h_{m'}'$ is a sum of histograms with multinomial distributions, the same argument we used for simple transition dynamics shows that we can alternatively sample the next label in two stages: first sample the entity's type and then sample the next label given its entity type.

For the first stage, we let $k'_{m'}$ denote this (random) entity type. Its value is "n/a" when there are no entities undertaking activity m'; otherwise, its value is $o_{m'}$ with probability $\frac{f_{m'}(o_{m'})}{x^{(t+1)}_{m'}(z')}$ and $m{:}k_m$ with probability $\frac{f_{m'}(m{:}k_m)}{x^{(t+1)}_{m'}(z')}$. Note that these probabilities add up to 1 in view of (6.2). The distributional program below outputs the pmf of this (random) entity type.

1: **function** ENTITYTYPE$^{(t)}_{m'}(f_{m'})$
2:     **if** $|f_{m'}| > 0$ **then**
3:         $k'_{m'} \sim f_{m'}/|f_{m'}|$
4:     **else**
5:         $k'_{m'} \leftarrow$ n/a
6:     **return** $k'_{m'}$

For the second stage, we sample a next label $\ell'_{m'}$ given the entity type $k'_{m'}$. There are three cases to consider:

- *Case $k'_{m'} = o_{m'}$.* Since $h^{\mathrm{nlt}}_{m'}(\cdot, o_{m'})$ is multinomial with parameters $f_{m'}(o_{m'})$ and $\pi^{(t),\mathrm{join}}_{o \to m'}(L'_{m'} = \cdot|z, z')$, we can think of all entities that join activity m' from origin $o_{m'}$ as having i.i.d. next labels with pmf $\pi^{(t),\mathrm{join}}_{o_{m'} \to m'}(\ell'_{m'}|z, z')$. The next label $\ell'_{m'}$ is sampled from this pmf.

- *Case $k'_{m'} = m{:}k_m$.* Since $h^{\mathrm{nlt}}_{m'}(\cdot, m{:}k_m)$ is multinomial with parameters $f_{m'}(m{:}k_m)$ and pmf parameter given by (6.11), we can think of all entities who come to activity m' with entity type $m{:}k_m$ as being i.i.d. with this pmf. The next label $\ell'_{m'}$ is sampled from this pmf, which is equivalent to first sampling ℓ_m from the pmf $\xi^{(t)}_{\mu,m{:}k_m}(\ell_m|z)$ and then sampling $\ell'_{m'}$ from the pmf $\pi^{(t),\mathrm{rel}}_{m{:}k_m \to m'}(\ell'_{m'}|z, \ell_m, z')$.

- *Case $k'_{m'} =$ n/a.* The next label $\ell'_{m'}$ is sampled from the pmf $\theta^0_{m'}$.

This second stage is summarized in the distributional program below. The right-hand side of the sampling statement in Line 6 of this program is proportional to the pmf $\xi^{(t)}_{\mu,m{:}k_m}(\ell_m|z)$ in (6.10), and in its current form, the model primitives are made explicit.

1: **function** NEXTLABEL$^{(t)}_{\mu,m'}(z, z', k'_{m'})$
2:     **switch** $k'_{m'}$ **do**
3:         **case** $o_{m'}$
4:             $\ell'_{m'} \sim \pi^{(t),\mathrm{join}}_{o_{m'} \to m'}(L'_{m'} = \cdot|z, z')$
5:         **case** $m{:}k_m$
6:             $\ell_m \propto \mu(L_m = \cdot|z) \times \pi^{(t),\mathrm{cat}}_m(k_m|z, L_m = \cdot)$
7:             $\ell'_{m'} \sim \pi^{(t),\mathrm{rel}}_{m{:}k_m \to m'}(L'_{m'} = \cdot|z, \ell_m, z')$
8:         **case** n/a

9:           $\ell'_{m'} \sim \theta^0_{m'}(L'_{m'} = \cdot)$

10:    **return** $\ell'_{m'}$

We thus find that the pseudocode fragment we started with in this step can be replaced with the following two sampling instructions:

$$k'_{m'} \sim \text{ENTITYTYPE}^{(t)}_{m'}(f_{m'})$$
$$\ell'_{m'} \sim \text{NEXTLABEL}^{(t)}_{\mu,m'}(z, z', k'_{m'})$$

Histogram-Free Refactored Distributional Program Putting it all together, we have now rewritten the distributional program $\text{QPLEXMAP}^{(t)}$ as the equivalent distributional program below. Note that there are no longer any sampling statements for histograms on labels, neither in the program itself nor in the distributional programs called within the program.

1: **function** $\text{QPLEXMAP}^{(t)}(\mu)$
2:     $z \sim \mu(Z = \cdot)$
3:     **for** $m = 1, \ldots, M$ **do**
4:        $d_m \sim \text{CATEGORYHISTOGRAM}^{(t)}_{\mu,m}(z)$
5:     $(z', f) \sim \pi^{(t)}(Z' = \cdot, F = \cdot|z, d)$
6:     **for** $m' = 1, \ldots, M$ **do**
7:        $k'_{m'} \sim \text{ENTITYTYPE}^{(t)}_{m'}(f_{m'})$
8:        $\ell'_{m'} \sim \text{NEXTLABEL}^{(t)}_{\mu,m'}(z, z', k'_{m'})$
9:     **return** (z', ℓ')

Algebraic Representation of $\mathcal{Q}^{(t)}$ We can use this program to define several pmfs. Let $q^{(t)}_{\mu,m}(d_m|z)$ be the output of the distributional program CATEGORYHIS-
$\text{TOGRAM}^{(t)}_{\mu,m}$, i.e.,

$$q^{(t)}_{\mu,m}(d_m|z) = \Pr\left[\text{mult}\left(x^{(t)}_m(z), \sum_{\ell_m} \mu(\ell_m|z) \times \pi^{(t),\text{cat}}_m(K_m = \cdot|z, \ell_m)\right) = d_m\right].$$

Let $q^{(t)}_{m'}(k'_{m'}|f_{m'})$ be the output of the distributional program $\text{ENTITYTYPE}^{(t)}_{m'}$, i.e.,

$$q^{(t)}_m(k'_{m'}|f_{m'}) = \begin{cases} \frac{f_{m'}(o_{m'})}{|f_{m'}|} & \text{if } k'_{m'} = o_{m'} \text{ and } |f_{m'}| > 0 \\ \frac{f_{m'}(k'_{m'})}{|f_{m'}|} & \text{if } k'_{m'} = m{:}k_m \text{ and } |f_{m'}| > 0 \\ 1 & \text{if } k'_{m'} = \text{n/a} \text{ and } |f_{m'}| = 0 \\ 0 & \text{otherwise.} \end{cases}$$

Let $q^{(t)}_{\mu,m'}(\ell'_m|z, z', k'_{m'})$ be the output of the distributional program NEXTLA-
$\text{BEL}^{(t)}_{\mu,m'}$, i.e.,

$$q_{\mu,m'}^{(t)}(\ell_{m'}'|z,z',k_{m'}')$$

$$= \begin{cases} \pi_{o_{m'}\to m'}^{(t),\text{join}}(\ell_{m'}'|z,z') & \text{if } k_{m'}' = o_{m'} \\ \sum_{\ell_m} \xi_{\mu,m:k_m}^{(t)}(\ell_m|z) \times \pi_{m:k_m\to m'}^{(t),\text{rel}}(\ell_{m'}'|z,\ell_m,z') & \text{if } k_{m'}' = m{:}k_m \\ \theta_m^0(\ell_{m'}') & \text{if } k_{m'}' = \text{n/a}, \end{cases}$$

where (6.10) defines $\xi_{\mu,m:k_m}^{(t)}$ in terms of μ and $\pi_m^{(t),\text{cat}}(k_m|z,\ell_m)$.

We now define

$$q_\mu^{(t)}(z,d,z',f,k',\ell') = \mu(z) \times \prod_m \left[q_{\mu,m}^{(t)}(d_m|z) \right] \times \pi^{(t)}(z',f|z,d)$$

$$\times \prod_{m'} \left[q_{m'}^{(t)}(k_{m'}'|f_{m'}) \times q_{\mu,m'}^{(t)}(\ell_{m'}'|z,z',k_{m'}') \right]$$

and note that we can express the QPLEX map $\mathcal{Q}^{(t)}$ in terms of this pmf.

Theorem 6.1 *Consider a QPLEX chain with advanced transition dynamics. For every $t \geq 0$, the QPLEX map $\mathcal{Q}^{(t)}$ may be represented via*

$$[\mathcal{Q}^{(t)}(\mu)](z',\ell') = \sum_{z,d,f,k'} q_\mu^{(t)}(z,d,z',f,k',\ell').$$

Chapter 7
Models with Advanced Transition Dynamics

This chapter provides examples of practical models that can be represented as QPLEX chains with advanced transition dynamics via suitable choices of model primitives. None of these examples can be represented using simple transition dynamics model primitives. Each model is a variation of the multiserver queueing model presented in Chap. 2. The last two models—a reentrant flow and service preemption model and a machine repair model—require a highly nontrivial specification of the model primitives.

As in Chap. 5, we do not present models in their fullest generality. For example, all activity duration pmfs in this chapter are taken to be time-homogeneous. We follow our standard convention that both arrivals and activity completions occur at the end of each period. We also assume that an appropriate initial distribution is specified. The size functions do not vary with time for any of the models, so we suppress their superscripts. Each example starts with a model description, after which we define the QPLEX chain and advanced model primitives for a fixed period between time t and $t + 1$, which we call *this period*.

7.1 Multiserver Model with Abandonment

In this example, the customers in the multiserver queueing model from Chap. 2 can *abandon* the buffer. Specifically, upon entering the buffer, each customer obtains a random *patience duration* (i.i.d. and independent of arrivals and services), which is how long the customer is willing to wait in the buffer.

Customers are governed by the following dynamics each period. As in the model from Chap. 2, a customer who completes service leaves the system; otherwise, the customer remains in service. (Customers do not abandon service once service has begun.) A customer in the buffer who runs out of patience leaves the system; otherwise, the customer will either remain in the buffer if no server is available or

© The Author(s) 2025   107
A. B. Dieker, S. T. Hackman, *QPLEX: A Computational Modeling and Analysis Methodology for Stochastic Systems*, Springer Series in Operations Research and Financial Engineering, https://doi.org/10.1007/978-3-031-74870-7_7

start service if a server is available. External arrivals will either start their service if a server is available or join the buffer if no server is available. Customers waiting in the buffer may be viewed as having higher priority for service than external arrivals, a detail that is not relevant if the goal is to calculate the distribution of the number of customers in the system over time. Customers are chosen uniformly at random to begin their service from among those waiting in the buffer; how long customers have been waiting does not play a role (so not first-come, first-serve).

We recall that $\alpha^{(t)}(a)$ denotes the pmf of the number of external arrivals between times t and $t + 1$, n denotes the number of servers in the server pool, and g denotes the service duration pmf. We let ζ denote the pmf of a customer's patience duration. All service and patience durations are assumed to be at least one time period.

Counter and Histograms We define the following random variable:

- $Z^{(t)}$ is the number of customers in the system at time t.
- $H_S^{(t)}$ is the histogram of remaining service durations, ages, or phases of the customers undergoing service at time t.
- $H_B^{(t)}$ is the histogram of remaining patience durations, ages, or phases of the customers in the buffer at time t.

There are two activities, which we denote respectively by S ("service") and B ("buffer"). Completing activity S is synonymous with completing service, whereas completing activity B is synonymous with abandoning the system.

We take the counter set as $\Omega_Z = \mathbb{Z}_+$. For each time $t \geq 0$, the size function for activity S is given by $x_S(z) = \min(z, n)$, and the size function for activity B is given by $x_B(z) = \max(z - n, 0)$. The activity duration pmfs associated with activities S and B are respectively given by the service duration pmf g and the patience duration pmf ζ. We pick an interpretation of the labels (remaining activity duration, age, phase) for each activity and select the label sets accordingly.

Categories The two categories are denoted by "done" and "stay." The feasible category assignments for each activity are listed and interpreted in the table below.

Activity	Category	Interpretation
S	done	Customer completes service
S	stay	Customer does not complete service
B	done	Customer in buffer runs out of patience
B	stay	Customer in buffer does not run out of patience

Entity Types There is a single origin o. The entity types for each activity are listed and described in the table below.

There is a fundamental difference between the feasible entity types displayed in this table and those applicable to the setting of simple transition dynamics recast in the language of advanced transition dynamics as in Example 6.2: there is an additional feasible entity type "B:stay" for activity S. A customer in the buffer

Activity	Entity Type	Description
S	S:stay	Customer continues service
S	B:stay	Customer in buffer moves into service
S	o	New customer begins service
B	B:stay	Customers remains in buffer
B	o	New customer enters buffer

assigned category "stay" (i.e., its patience does not run out) during this period *need not* remain in the buffer at time $t + 1$. Thus, this setting is not compatible with simple transition dynamics. However, as we will see, it *is* compatible with advanced transition dynamics. This is because customers in the buffer who do not leave the system due to impatience can begin service during this period, resulting in an (additional) component $f_S(B$:stay$)$ of the entity-type flow histogram for activity S.

Category, Relabeling, and Join Pmf Primitives The category, relabeling, and join pmfs for activities S and B equal their simple transition dynamics counterparts from Sect. 4.1 (see (6.4), (6.5), (6.8), and (6.9)) and are consistent with the chosen label interpretation (remaining activity duration, age, phase). For instance, if the remaining activity duration interpretation is chosen for both activities, then the join pmfs for activities S and B are given by

$$\pi_{o \to S}^{(t),\text{join}}(\ell_S' | z, z') = g(\ell_S'),$$

$$\pi_{o \to B}^{(t),\text{join}}(\ell_B' | z, z') = \zeta(\ell_B').$$

Routing Pmf Primitive The distributional program below outputs the routing pmf for this model. It takes as input the current counter z and the two category assignment histograms $d_S = (d_S(\text{done}), d_S(\text{stay}))$ and $d_B = (d_B(\text{done}), d_B(\text{stay}))$, and outputs the pmf of the next counter z' and the entity-type flow histograms for activities S and B:

$$f_S = (f_S(S\text{:stay}), f_S(B\text{:stay}), f_S(o))$$

$$f_B = (f_B(B\text{:stay}), f_B(o)).$$

Line 3 uses standard flow balance to assign the value of z'. The $d_S(\text{stay})$ customers that do not complete their service during this period remain in service (Line 4). There are $n - d_S(\text{stay})$ servers available for customers to begin their service, which represents the supply. This supply is first offered to the $d_B(\text{stay})$ customers in the buffer, and any remaining supply is then offered to the a new arrivals. The PRIORITYDEMANDSPLIT program introduced in Example 5.3 is used to partition these two types of demand sequentially into their met and unmet demands and assigns the appropriate components of the entity-type flow histograms (Line 5). For those customers in the buffer, the met demand is $f_S(B$:stay$)$ and the unmet demand is $f_B(B$:stay$)$, and for the system arrivals, the met demand is $f_S(o)$ and the unmet demand is $f_B(o)$.

```
1: function ROUTING⁽ᵗ⁾(z, dₛ, d_B)
2:    a ~ α⁽ᵗ⁾(A = ·)
3:    z' ← z − dₛ(done) − d_B(done) + a
4:    fₛ(S:stay) ← dₛ(stay)
5:    ((fₛ(B:stay), f_B(B:stay)), (fₛ(o), f_B(o)))
         ~ PRIORITYDEMANDSPLIT(d_B(stay), a; n − dₛ(stay))
6:    return (z', fₛ, f_B)
```

7.2 Multiserver Model with Customer Classes

The multiserver queueing model of Chap. 2 is extended to accommodate $C > 1$ classes of customer arrivals, which could be used, for example, to represent patients with different acuity levels arriving to an emergency room. We assume that these classes are ordered so that customers with class 1 have the highest priority for available servers. The number of arrivals A of customers of class c is independent across time periods with pmf $\alpha_c^{(t)}(a)$ for this period. We let $g_c(\ell)$ denote the service duration pmf for a customer of class c, and we assume each service duration is at least one period.

Counters and Histograms We define the following random variables:

- $Z^{(t)}$ is the number of customers in the system at time t.
- $Z_c^{(t)}$ is the number of customers of class $c = 1, \ldots, C$ in the buffer at time t.
- $H_S^{(t)}$ is the histogram of remaining service durations, ages, or phases of the customers undergoing service at time t.

There is one activity corresponding to undergoing service, which we denote by S. (In contrast with simple transition dynamics, this one activity must be "named" so that we may refer to its feasible entity types.)

We take the counter set as $\Omega_Z = \mathbb{Z}_+^{C+1}$. The size function for activity S is given by $x_S(z, z_1, \ldots, z_C) = \min(z, n)$. We pick an interpretation of the labels (remaining activity duration, age, phase) and select the label set accordingly.

Categories The two categories denoted by "done" and "stay" have their usual interpretations. The set of feasible category assignments for activity S is given by {done, stay}.

Entity Types There are C origins $o_1, \ldots, o_C$, where o_c represents the origin for customers of customer class c that begin their activity during this period. The set of feasible entity types for the one activity S is given by {S:stay, $o_1, \ldots, o_C$}.

Category, Relabeling, and Join Pmf Primitives The category, relabeling, and join pmfs for the one activity S equal their simple transition dynamics counterparts from Sect. 4.1 (see (6.4), (6.5), (6.8), and (6.9)) and are consistent with the chosen label interpretation (remaining activity duration, age, phase). For instance, if the remaining activity duration interpretation is chosen, then the join pmfs are

$$\pi_{o_c \to S}^{(t),\text{join}}(\ell_S' | z, z_1, \ldots, z_C, z', z_1', \ldots, z_C') = g_c(\ell_S').$$

Routing Pmf Primitive The distributional program below outputs the routing pmf for this model. It takes as input the current counters $z, z_1, \ldots, z_C$ and the category assignment histogram $d_S = (d_S(\text{done}), d_S(\text{stay}))$, and outputs the pmf of the next counters $z', z_1', \ldots, z_C'$ and the entity-type flow histogram $f_S = (f_S(S:\text{stay}), f_S(o_1), \ldots, f_S(o_C))$ for activity S.

The number of arrivals of each customer class is sampled in Lines 2 and 3. The $d_S(\text{stay})$ customers who do not complete their service during this period will remain in service (Line 4), thereby leaving $n - d_S(\text{stay})$ available servers. The PRIORITYDEMANDSPLIT program introduced in Example 5.3 is used to sequentially partition the demands for available servers by customer class into its met and unmet demands and assign the appropriate components of the entity-type flow histogram f_S and next counters $z_1', \ldots, z_C'$ (Line 5). The first in line are the $z_1 + a_1$ customers in class 1, the next in line are the $z_2 + a_2$ in class 2, etc. Line 6 uses standard flow balance to assign the next counter z'.

1: **function** ROUTING$^{(t)}(z, z_1, \ldots, z_C, d_S)$
2:     **for all** c **do**
3:         $a_c \sim \alpha_c^{(t)}(A_c = \cdot)$
4:     $f_S(S:\text{stay}) \leftarrow d_S(\text{stay})$
5:     $((f_S(o_1), z_1'), \ldots, (f_S(o_C), z_C'))$
            $\sim$ PRIORITYDEMANDSPLIT$(z_1 + a_1, \ldots, z_C + a_C; n - d_S(\text{stay}))$
6:     $z' \leftarrow z - d_S + \sum_c a_c$
7:     **return** $(z', z_1', \ldots, z_C', f_S)$

An alternative QPLEX chain uses an activity for each customer class. For such a model, the counters $Z_{B,c}^{(t)}$ and $Z_{S,c}^{(t)}$ are the numbers of customers of class $c = 1, \ldots, C$ at time t, respectively, in the buffer and in service, and $H_c^{(t)}$ is the histogram of remaining service durations, ages, or phase of the customers *of class c* undergoing service at time t. For this model, the number of counters is $2C$ (instead of $C + 1$), and the number of labels is C (instead of one).

7.3   Multiserver Model with Multiple Activities

In this example, we revisit the multiserver queueing model of Chap. 2, but we define a different QPLEX chain. The QPLEX chain we specify here may provide greater approximation quality, albeit with more computational effort.

Counter and Histograms Fix some pre-specified remaining service duration $\hat{\ell} > 1$. We define the following random variable:

- $Z^{(t)}$ is the number of customers in the system at time t.
- $Z_1^{(t)}$ is the number of customers with remaining service duration less than or equal to $\hat{\ell}$.
- $H_1^{(t)}$ and $H_2^{(t)}$ are, respectively, the histograms of remaining service durations associated with activities 1 and 2 at time t.

There are *two* activities with the following interpretations. Customers undertaking activity 1 have a remaining service duration less than or equal to $\hat{\ell}$, whereas customers undertaking activity 2 have a remaining service duration greater than $\hat{\ell}$.

We take the counter set as $\Omega_Z = \mathbb{Z}_+^2$. The size function for activity 1 is given by $x_1(z, z_1) = z_1$, and the size function for activity 2 is given by $x_2(z, z_1) = \min(z, n) - z_1$.

The label set for activity 1 is given by $\Omega_1 = \{1, 2, \ldots, \hat{\ell}\}$, and the label set for activity 2 is given by $\Omega_2 = \{\hat{\ell} + 1, \hat{\ell} + 2, \ldots\}$.

Categories The two categories are still "done" and "stay." For each activity, the set of feasible categories is given by $\{\text{done}, \text{stay}\}$. For activity 1, categories have their usual interpretations in the sense that customers assigned category "done" leave the system, while those assigned category "stay" continue to undertake activity 1. Customers undertaking activity 2 always *continue* their service, but those assigned category "done" will no longer have a remaining service duration greater than $\hat{\ell}$ at time $t + 1$.

Entity Types There is a single origin o. The set of feasible entity types for activity 1 is given by $\{1:\text{stay}, 2:\text{done}, o\}$, and for activity 2 it is given by $\{2:\text{stay}, o\}$.

Category, Relabeling, and Join Pmf Primitives Customers undertaking activity 1 are assigned categories as before. For activity 2, only customers with label value $\hat{\ell} + 1$ are assigned category "done." Therefore, the category pmfs are given by

$$\pi_1^{(t),\text{cat}}(k_1|z, z_1, \ell_1) = 1(k_1 = \text{done}) \times 1(\ell_1 = 1) + 1(k_1 = \text{stay}) \times 1(\ell_1 > 1)$$

$$\pi_2^{(t),\text{cat}}(k_2|z, z_1, \ell_2) = 1(k_2 = \text{done}) \times 1(\ell_2 = \hat{\ell} + 1)$$

$$+ 1(k_2 = \text{stay}) \times 1(\ell_2 > \hat{\ell} + 1).$$

Customers who stay at their activities are relabeled in the usual way. The relabeling pmfs are given by

$$\pi_{1:\text{stay}\to 1}^{(t),\text{rel}}(\ell_1'|z, z_1, \ell_1, z', z_1') = 1(\ell_1' = \ell_1 - 1),$$

$$\pi_{2:\text{stay}\to 2}^{(t),\text{rel}}(\ell_2'|z, z_1, \ell_2, z', z_1') = 1(\ell_2' = \ell_2 - 1),$$

$$\pi_{2:\text{done}\to 1}^{(t),\text{rel}}(\ell_1'|z, z_1\ell_2, z', z_1') = 1(\ell_1' = \ell_2 - 1).$$

The join pmfs are given by

$$\pi_{o \rightarrow 1}^{(t), \text{join}}(\ell_1' | z, z_1 z', z_1') = \frac{g(\ell_1')}{\sum_{\tilde{\ell} \leq \hat{\ell}} g(\tilde{\ell})},$$

$$\pi_{o \rightarrow 2}^{(t), \text{join}}(\ell_2' | z, z_1 z', z_1') = \frac{g(\ell_2')}{\sum_{\tilde{\ell} > \hat{\ell}} g(\tilde{\ell})}.$$

Routing Pmf Primitive The distributional program below outputs the routing pmf for this model. It takes as input the current counters z, z_1 and the two category assignment histograms $d_1 = (d_1(\text{done}), d_1(\text{stay}))$ and $d_2 = (d_2(\text{done}), d_2(\text{stay}))$, and outputs the pmf of the next counters z', z_1' and the entity-type flow histograms for activities 1 and 2:

$$f_1 = (f_1(1\text{:stay}), f_1(2\text{:done}), f_1(o)),$$

$$f_2 = (f_2(2\text{:stay}), f_2(o)).$$

Line 2 samples the number of arrivals to the system. Line 3 uses standard flow balance to assign the value of z'. Line 4 samples the number of arrivals to activity 1 given the number of arrivals to the system. Line 5 assigns the value of the next counter z_1'. Lines 6–10 calculate the entity-type flow histograms, which use the fact that all entities of type 2:done (with label value of $\hat{\ell}$) are routed to activity 1 from activity 2.

```
 1: function ROUTING(t)(z, z1, d1, d2)
 2:     a ~ α(t)(A = ·)
 3:     z' ← z − d1(done) + a
 4:     a1 ~ bin (a, ∑ℓ̃≤ℓ̂ g(ℓ̃))
 5:     z1' ← d1(stay) + d2(done) + a1
 6:     f1(1:stay) ← d1(stay)
 7:     f1(1:o) ← a1
 8:     f1(2:done) ← d2(done)
 9:     f2(2:stay) ← d2(stay)
10:     f2(2:o) ← a − a1
11:     return (z', z1', f1, f2)
```

7.4 Multiserver Model with Reentrant Flow and Preemption

We consider a multiserver queueing model of Chap. 2 where each customer must be served *twice*. The service duration pmfs for each time through are different. Priority is given to *first-time* customers, i.e., the service of a customer being served for the second time is preempted to allow first-time customers to begin their service. Preempted customers enter a buffer awaiting resumption of their service as soon as there is an available server (not demanded by a first-time customer). The service of each preempted customer is temporarily stopped until it can be resumed where it was left off. The same customer can be preempted multiple times. First-time customers completing their service when there is no server available to begin their second service enter a separate buffer to await their second service. Customers in this buffer have lower priority than those with preempted service. We let g_1 and g_2 denote, respectively, the service duration pmf for the first and second time through service. Service durations are assumed to be at least one time period. We let n stand for the number of servers and let $\alpha^{(t)}(a)$ be the pmf of the number of arrivals A between times t and $t + 1$.

Counter and Histograms We define the following random variables:

- $Z_{S_1}^{(t)}, Z_{S_2}^{(t)}$ are, respectively, the numbers of customers undertaking their first or second service at time t.
- $Z_{B_1}^{(t)}, Z_{B_2}^{(t)}$ are, respectively, the numbers of customers waiting for their first or second service to start at time t.
- $Z_P^{(t)}$ is the number of preempted customers awaiting resumption of their second service at time t.
- $H_{S_1}^{(t)}, H_{S_2}^{(t)}, H_P^{(t)}$ are, respectively, the histograms of remaining service durations, ages, or phases associated with the first, second, and preempted services, respectively, at time t.

There are three activities with the following interpretations. Activities S_1 and S_2 represent, respectively, undergoing their first or second service, and activity P represents awaiting the resumption of the second service while being preempted. We take the counter set as $\Omega_Z = \mathbb{Z}_+^5$, and the size functions for activities S_1, S_2, and P are, respectively, given by

$$
\begin{aligned}
x_{S_1}(z_{S_1}, z_{S_2}, z_{B_1}, z_{B_2}, z_P) &= z_{S_1}, \\
x_{S_2}(z_{S_1}, z_{S_2}, z_{B_1}, z_{B_2}, z_P) &= z_{S_2}, \\
x_P(z_{S_1}, z_{S_2}, z_{B_1}, z_{B_2}, z_P) &= z_P.
\end{aligned}
\tag{7.1}
$$

The label sets for activities S_1 and S_2 are consistent with their chosen interpretations (remaining service duration, age, phase). The label set for activity P is the same as the label set for activity S_2.

Categories There are two categories denoted by "done" and "stay." The feasible category assignments for each activity are listed and interpreted in the table below. Note that for activity P, the category "done" is the only feasible category, as all such customers are always "ready-to-go." (Our choice to use the category "done" for activity I is arbitrary.)

Activity	Category	Interpretation
S_1	done	Customer completes first service
S_1	stay	Customer does not complete first service
S_2	done	Customer completes second service
S_2	stay	Customer does not complete second service
P	done	

Entity Types There is a single origin o. The feasible entity types for each activity are listed and described in the table below.

Activity	Entity Type	Description
S_1	S_1:stay	Customer continues its first service
S_1	o	Customer starts its first service
S_2	S_2:stay	Customer continues its second service
S_2	P:done	Customer resumes second service
S_2	o	Customer starts its second service
P	S_2:stay	Customer undergoing second service is preempted
P	P:done	Customer remains preempted

Category, Relabeling, and Join Pmf Primitives The category, relabeling, and join pmfs for service activities S_1 and S_2 equal their simple transition dynamics counterparts from Sect. 4.1 (see (6.4), (6.5), (6.8), and (6.9)) and are consistent with the chosen label interpretation (remaining service duration, age, phase). For instance, if the remaining service duration interpretation is chosen, then the join pmfs for activities S_i, $i = 1, 2$, are

$$\pi_{o \to S_i}^{(t),\text{join}}(\ell'_{S_i} | z, z') = g_i(\ell'_{S_i}).$$

The category pmf for preemptive activity P is trivial, as there is but one category "done." The relabeling pmf for activity P is given by

$$\pi_{P:\text{done} \to P}^{(t),\text{rel}}(\ell'_P | z, \ell_P, z') = 1(\ell'_P = \ell_P)$$

as no service progress is being made while undertaking activity P. There are no join pmfs for activity P as the one origin is not a feasible entity type for activity P.

Routing Pmf Primitive The distributional program for the routing pmf takes as input the current counters z_{S_1}, z_{B_1}, z_{S_2}, z_{B_2}, z_P and the three category assignment histograms

$$d_{S_1} = (d_{S_1}(\text{done}), d_{S_1}(\text{stay})),$$
$$d_{S_2} = (d_{S_2}(\text{done}), d_{S_2}(\text{stay})),$$
$$d_P = (d_P(\text{done})).$$

It outputs the pmf of the next counters z'_{S_1}, z'_{B_1}, z'_{S_2}, z'_{S_2}, z'_P and the entity-type flow histograms for activities S_1, S_2 and P:

$$fs_1 = (fs_1(S_1\text{:stay}), fs_1(o))$$
$$fs_2 = (fs_2(S_2\text{:stay}), fs_2(P\text{:done}), fs_2(o))$$
$$fp = (fp(S_2\text{:stay}), fp(P\text{:done})).$$

The distributional program below outputs the routing pmf for this model. The d_{S_1}(stay) customers who do not complete their service will remain in service (Line 3). There are $n - d_{S_1}$(stay) servers available for customers to begin their first or second service or resume their second service, and these are allocated to customers according to the following priority rule:

- The $a + z_{B_1}$ customers waiting to begin their first service have the highest priority.
- The d_{S_2}(stay) customers undergoing their second service but not completing it this period have the second highest priority.
- The d_P(done) customers who had been preempted while undergoing their second service have the third highest priority.
- The d_{S_1}(done) $+ z_{B_2}$ customers awaiting the start of their second service have the lowest priority.

The PRIORITYDEMANDSPLIT program introduced in Example 5.3 is used to partition these four customer demands for available servers sequentially into their respective met and unmet demands and to assign the appropriate components of the entity-type flow histograms and next counters (Line 4). Line 6 uses the fact that $|f'_m| = x^{(t+1)}_{m'}(z')$ for each activity m' and the explicit form of the size functions in (7.1).

```
1: function ROUTING^(t)(z_{S_1}, z_{B_1}, z_{S_2}, z_{B_2}, z_P, d_{S_1}, d_{S_2}, d_P)
2:     a ~ α^(t)(A = ·)
3:     f_{S_1}(S_1:stay) ← d_{S_1}(stay)
4:     ((f_{S_1}(o), z'_{B_1}), (f_{S_2}(S_2:stay), f_P(S_2:stay)), (f_{S_2}(P:done), f_P(P:done)), (f_{S_2}(o), z'_{B_2}))
          ~ PRIORITYDEMANDSPLIT(a + z_{B_1}, d_{S_2}(stay), d_P(done), d_{S_1}(done) + z_{B_2}; n - d_{S_1}(stay))
5:     for m' ∈ {S_1, S_2, P} do
6:         z'_{m'} ← |f_{m'}|
7:     return (z'_{S_1}, z'_{B_1}, z'_{S_2}, z'_{B_2}, z'_P, f_{S_1}, f_{S_2}, f_P)
```

7.5 Multiserver Model with Server Repair

We consider a multiserver queueing model of Chap. 2 in which the servers can independently fail, thus requiring repair. In this example, we will see that the servers are the entities undertaking activities, not the customers. We will also see label sets consisting of *pairs* of integers.

We call a server *operational* whenever it is serving a customer (so when it is not idle or being repaired). Servers deteriorate while being operational and fail after being operational for some random amount of time. When a server fails, it immediately undergoes a repair process, and the customer it is serving continues its service only after the server has been repaired. We allow for a customer to complete its service and a server to fail simultaneously, so repairs do not necessarily interrupt the service of customers. After servers are repaired, they are able to serve customers as if they were new.

When a customer enters service and multiple servers are available, the first choice is a newly repaired server, the second choice is an idle server, and the last choice is a server that just completed service and did not fail. (This order is arbitrary and not fundamental for our development.)

We use the symbol L_S for a generic service duration and write $g_S(\ell_S)$ for its pmf. Services can be interrupted by repairs, so we interpret the service duration of a customer as the number of periods it needs to be served. Service durations are independent and identically distributed and are independent of everything else.

We use the symbol L_O for a generic operational duration and write $g_O(\ell_O)$ for its pmf. The operational duration is the number of periods a server can serve customers until it fails, not counting idle time in between operational periods. Operational durations are independent and identically distributed and are independent of everything else.

We use the symbol L_R for a generic repair duration and write $g_R(\ell_R)$ for its pmf. Repairs are not interrupted. Repair durations are independent and identically distributed and are independent of everything else. All service, operational, and repair durations are assumed to be at least one time period.

We let n stand for the number of servers and let $\alpha^{(t)}(a)$ be the pmf of the number of arrivals A between times t and $t + 1$.

Counter and Histograms We define the following random variable:

- $Z_J^{(t)}$ is the number of servers under repair with a customer at time t.
- $Z_W^{(t)}$ is the number of servers under repair without a customer at time t.
- $Z_S^{(t)}$ is the number of servers that serve a customer at time t.
- $Z_B^{(t)}$ is the number of customers in the buffer awaiting service at time t.
- $H_J^{(t)}$ is the histogram of pairs of remaining service durations and repair ages (the number of periods since the repair started) of the servers under repair with a customer at time t.
- $H_W^{(t)}$ is the histogram of repair ages of the servers under repair without a customer at time t.
- $H_S^{(t)}$ is the histogram of pairs of remaining service durations and operational ages of the servers (the number of periods a server has been operational since its last repair) that serve a customer at time t.
- $H_I^{(t)}$ is the histogram of operational ages of the idle servers at time t.

(It has been an arbitrary choice to use remaining service durations, repair ages, and operational ages.) Note that the number of idle servers at time t is $n - (Z_J^{(t)} + Z_W^{(t)} + Z_S^{(t)})$.

Here are the descriptions of the four activities J, W, S, and I, as well as their corresponding labels (and therefore their label sets):

- Activity J represents being repaired with a customer waiting for the repair to complete. A server undertaking this activity has a label represented via a pair (ℓ_S, ℓ_R), where $\ell_S \geq 1$ is the remaining service duration of the customer and $\ell_R \geq 0$ is its repair age.
- Activity W represents being repaired without a customer. A server undertaking this activity has a label $\ell_R \geq 0$ representing its repair age.
- Activity S represents serving a customer. A server undertaking this activity has a label represented via a pair (ℓ_S, ℓ_O), where $\ell_S \geq 1$ is the remaining service duration of the customer and $\ell_O \geq 0$ is the operational age of the server.
- Activity I represents waiting on a customer. A server undertaking this activity has a label $\ell_O \geq 0$ representing the operational age of the server.

We take the counter set as $\Omega_Z = \mathbb{Z}_+^5$, and the size functions for activities J, W, S, and I are, respectively, given by

$$
\begin{aligned}
x_J(z_J, z_W, z_S, z_B) &= z_J, \\
x_W(z_J, z_W, z_S, z_B) &= z_W, \\
x_S(z_J, z_W, z_S, z_B) &= z_S, \\
x_I(z_J, z_W, z_S, z_B) &= n - (z_J + z_W + z_S).
\end{aligned}
\tag{7.2}
$$

Categories The six categories are denoted by "done," "stay," "doneFail," "doneNoFail," "stayFail," and "stayNoFail." The feasible category assignments for each activity are listed and interpreted in the table below. (Our choice to use the category "done" for activity I is arbitrary.)

Activity	Category	Interpretation
J	done	Server repair complete
J	stay	Server repair not complete
W	done	Server repair complete
W	stay	Server repair not complete
S	doneFail	Customer completes service, server fails
S	doneNoFail	Customer completes service, server does not fail
S	stayFail	Customer does not complete service, server fails
S	stayNoFail	Customer does not complete service, server does not fail
I	done	

Entity Types There are no origins. The feasible entity types for each activity are listed and described in the table below.

Activity	Entity Type	Description
J	J:stay	Server does not complete repair
J	S:stayFail	Customer does not complete service, server fails
W	W:stay	Server with no customer does not complete repair
W	S:doneFail	Customer completes service, server fails
S	I:done	Idle server begins new service
S	J:done	Server completes repair, resumes service
S	W:done	Server with no customer completes repair, begins new service
S	S:doneNoFail	Customer completes service, server does not fail, begins new service
S	S:stayNoFail	Customer does not complete service, server does not fail
I	I:done	Idle server remains idle
I	W:done	Server with no customer completes repair, becomes idle
I	S:doneNoFail	Customer completes service, server does not fail, becomes idle

Category Pmf Primitives Many category pmfs use hazard probabilities, so we introduce the following notation:

$$\overline{G}_S(\ell_S) = \sum_{\tilde{\ell}_S > \ell_S} g_S(\tilde{\ell}_S),$$

$$\overline{G}_O(\ell_O) = \sum_{\tilde{\ell}_O > \ell_O} g_O(\tilde{\ell}_O),$$

$$\overline{G}_R(\ell_R) = \sum_{\tilde{\ell}_R > \ell_R} g_R(\tilde{\ell}_R).$$

The category pmfs associated with activity J are

$$\pi_J^{(t),\mathrm{cat}}(k_J | z, (\ell_S, \ell_R)) = \begin{cases} \dfrac{g_R(\ell_R+1)}{\overline{G}_R(\ell_R)} & \text{if } k_J = \text{done} \\ 1 - \dfrac{g_R(\ell_R+1)}{\overline{G}_R(\ell_R)} & \text{if } k_J = \text{stay.} \end{cases}$$

Similarly, the category pmfs associated with activity W are

$$\pi_W^{(t),\mathrm{cat}}(k_W | z, \ell_R) = \begin{cases} \dfrac{g_R(\ell_R+1)}{\overline{G}_R(\ell_R)} & \text{if } k_W = \text{done} \\ 1 - \dfrac{g_R(\ell_R+1)}{\overline{G}_R(\ell_R)} & \text{if } k_W = \text{stay.} \end{cases}$$

The category pmfs associated with activity S are

$$
\pi_S^{(t),\text{cat}}(k_S|z,(\ell_S,\ell_O)) = \begin{cases} 1(\ell_S = 1) \times \frac{g_O(\ell_O+1)}{G_O(\ell_O)} & \text{if } k_S = \text{doneFail} \\ 1(\ell_S = 1) \times 1 - \frac{g_O(\ell_O+1)}{G_O(\ell_O)} & \text{if } k_S = \text{doneNoFail} \\ 1(\ell_S > 1) \times \frac{g_O(\ell_O+1)}{\overline{G}_O(\ell_O)} & \text{if } k_S = \text{stayFail} \\ 1(\ell_S > 1) \times 1 - \frac{g_O(\ell_O+1)}{\overline{G}_O(\ell_O)} & \text{if } k_S = \text{stayNoFail.} \end{cases}
$$

The category pmfs for activity I are trivial, and we do not specify them.

Relabeling Pmf Primitives For each activity, there is one relabeling pmf for each feasible entity type. We describe these model primitives for each activity.

There are two relabeling pmfs for activity J:

- A server under repair with a customer does not complete its repair (entity type "J:stay"). The customer's remaining service duration remains unchanged and the repair age increases by one, so the relabeling pmf for this entity type is

$$
\pi_{J:\text{stay}\to J}^{(t),\text{rel}}((\ell_S', \ell_R')|z, (\ell_S, \ell_R), z') = 1(\ell_S' = \ell_S) \times 1(\ell_R' = \ell_R + 1).
$$

- A server that serves a customer fails and the customer does not complete its service (entity type "S:stayFail"). The customer's remaining service duration decreases by one and the repair age becomes 0, so the relabeling pmf for this entity type is

$$
\pi_{S:\text{stayFail}\to J}^{(t),\text{rel}}((\ell_S', \ell_R')|z, (\ell_S, \ell_O), z') = 1(\ell_S' = \ell_S - 1) \times 1(\ell_R' = 0).
$$

There are two relabeling pmfs for activity W:

- A server under repair without a customer does not complete its repair (entity type "W:stay"). Its age increases by one, so the relabeling pmf for this entity type is

$$
\pi_{W:\text{stay}\to W}^{(t),\text{rel}}(\ell_R'|z, \ell_R, z') = 1(\ell_R' = \ell_R + 1).
$$

- A server that serves a customer completes service and fails (entity type "S:doneFail"). Its age is 0, and the relabeling pmf for this entity type is

$$
\pi_{S:\text{doneFail}\to W}^{(t),\text{rel}}(\ell_R'|z, (\ell_S, \ell_O), z') = 1(\ell_R' = 0).
$$

There are five relabeling pmfs for activity S:

- An idle server now has a customer to service (entity type "I:done"). The remaining service duration is random with pmf g_S, and its age remains unchanged, so the relabeling pmf for this entity type is

$$
\pi_{I:\text{done}\to S}^{(t),\text{rel}}((\ell_S', \ell_O')|z, \ell_O, z') = g_S(\ell_S') \times 1(\ell_O' = \ell_O).
$$

- A server under repair with a customer completes its repair (entity type "J:done"). The remaining service duration remains unchanged and its age is 0, so the relabeling pmf for this entity type is

$$\pi_{J:\text{done}\to S}^{(t),\text{rel}}((\ell_S', \ell_O')|z, (\ell_S, \ell_R), z') = 1(\ell_S' = \ell_S) \times 1(\ell_O' = 0).$$

- A server under repair without a customer completes its repair (entity type "W:done"). The remaining service duration is sampled from the pmf g_S and its operational age becomes 0, so the relabeling pmf for this entity type is

$$\pi_{W:\text{done}\to S}^{(t),\text{rel}}((\ell_S', \ell_O')|z, \ell_R, z') = g_S(\ell_S') \times 1(\ell_O' = 0).$$

- A server that serves a customer completes service and does not fail (entity type "S:doneNoFail"). The remaining service duration is sampled from the pmf g_S and its operational age increases by one, so the relabeling pmf for this entity type is

$$\pi_{S:\text{doneNoFail}\to S}^{(t),\text{rel}}((\ell_S', \ell_O')|z, (\ell_S, \ell_O), z') = g_S(\ell_S') \times 1(\ell_O' = \ell_O + 1).$$

- A server that serves a customer does not complete service and does not fail (entity type "S:stayNoFail"). The remaining service duration decreases by one and the operational age increases by one, so the relabeling pmf for this entity type is

$$\pi_{S:\text{stayNoFail}\to S}^{(t),\text{rel}}((\ell_S', \ell_O')|z, (\ell_S, \ell_O), z') = 1(\ell_S' = \ell_S - 1) \times 1(\ell_O' = \ell_O + 1).$$

There are three relabeling pmfs associated with activity I:

- An idle server remains idle (entity type "I:done"). Its operational age is unchanged, so the relabeling pmf for this entity type is

$$\pi_{I:\text{done}\to I}^{(t),\text{rel}}(\ell_O'|z, \ell_O, z') = 1(\ell_O' = \ell_O).$$

- A server under repair without a customer completes its repair and becomes idle (entity type "W:done"). Its operational age becomes 0, so the relabeling pmf for this entity type is

$$\pi_{W:\text{done}\to I}^{(t),\text{rel}}(\ell_O'|z, \ell_R, z') = 1(\ell_O' = 0).$$

- A server that serves a customer completes service and becomes idle (entity type "S:doneNoFail"). Its operational age increases by one, so the relabeling pmf for this entity type is

$$\pi_{S:\text{doneNoFail}\to I}^{(t),\text{rel}}(\ell_O'|z, (\ell_S, \ell_O), z') = 1(\ell_O' = \ell_O + 1).$$

Join Pmf Primitives Since there are no origins, there are no join pmfs for this model.

Routing Pmf Primitive The distributional program for the routing pmf takes as input the current counters z_J, z_W, z_S, z_B and the four category assignment histograms shown in the table below.

Activity	Category Assignment Histogram
J	$d_J = (d_J(\text{done}), d_J(\text{stay}))$
W	$d_W = (d_W(\text{done}), d_W(\text{stay}))$
S	$d_S = (d_S(\text{doneFail}), d_S(\text{doneNoFail}), d_S(\text{stayFail}), d_S(\text{stayNoFail}))$
I	$d_I = (d_I(\text{done}))$

Its output is the pmf of the next counters z'_J, z'_W, z'_S, and z'_B and the four entity-type flow histograms shown in the table below.

Activity	Entity-type Flow Histogram
J	$f_J = (f_J(J{:}\text{stay}), f_J(S{:}\text{stayFail}))$
W	$f_W = (f_W(W{:}\text{stay}), f_W(S{:}\text{doneFail}))$
S	$f_S = (f_S(I{:}\text{done}), f_S(J{:}\text{done}), f_S(S{:}\text{done}), f_S(S{:}\text{doneNoFail}), f_S(S{:}\text{stayNoFail}))$
I	$f_I = (f_I(I{:}\text{done}), f_I(W{:}\text{done}), f_I(S{:}\text{doneNoFail}))$

The distributional program below outputs the routing pmf for this model. Six components of category assignment histograms identify six components of entity-type flow histograms (Lines 3–8), as follows:

- A server under repair with a customer that does not complete its repair continues to be under repair (Line 3).
- A server under repair with a customer that completes its repair resumes serving this customer (Line 4).
- A server under repair without a customer that does not complete its repair continues to be under repair (Line 5).
- A server that serves a customer that completes service and also fails undergoes repair without a customer (Line 6).
- A server that serves a customer that does not complete service and also fails becomes a server under repair with a customer (Line 7).
- A server that serves a customer that does not complete service and does not fail continues serving this customer (Line 8).

1: **function** ROUTING$^{(t)}(z_I, z_J, z_W, z_S, z_B, d_I, d_J, d_W, d_S)$
2: $\quad a \sim \alpha^{(t)}(A = \cdot)$
3: $\quad f_J(J{:}\text{stay}) \leftarrow d_J(\text{stay})$
4: $\quad f_S(J{:}\text{done}) \leftarrow d_J(\text{done})$
5: $\quad f_W(W{:}\text{stay}) \leftarrow d_W(\text{stay})$
6: $\quad f_W(S{:}\text{doneFail}) \leftarrow d_S(\text{doneFail})$
7: $\quad f_J(S{:}\text{stayFail}) \leftarrow d_S(\text{stayFail})$
8: $\quad f_S(S{:}\text{stayNoFail}) \leftarrow d_S(S{:}\text{stayNoFail})$
9: $\quad ((f_S(W{:}\text{done}), f_I(W{:}\text{done})), (f_S(I{:}\text{done}), f_I(I{:}\text{done})), (f_S(S{:}\text{doneNoFail}), f_I(S{:}\text{doneNoFail})))$
$\quad\quad \sim$ PRIORITYDEMANDSPLIT$(d_W(\text{done}), d_I(\text{done}), d_S(\text{doneNoFail}); z_B + a)$
10: $\quad$ **for** $m' \in \{J, W, S\}$ **do**
11: $\quad\quad z'_{m'} \leftarrow |f_{m'}|$
12: $\quad z'_B \leftarrow z_B + a - z'_S$
13: $\quad$ **return** $(z'_J, z'_W, z'_S, z'_B, f_I, f_J, f_W, f_S)$

Six components of entity-type flow histograms and four counters remain to be specified. Recall that with respect to the assignment of customers to servers, the priority is to first assign a customer to a newly repaired server if one is available; if not, then assign it to an idle server if one is available and, if not, then assign it to a server that has just completed service. These three server demand classes vie to process the supply of available customers. The numbers of these server demand classes are $d_W(\text{done})$, $d_I(\text{done})$, and $d_S(\text{doneNoFail})$, respectively. The PRIORITYDEMANDSPLIT program introduced in Example 5.3 is used to sequentially partition each of the three server demand classes into its met and unmet demands and renames them accordingly (Line 9).

- The three respective met demands, namely, $f_S(W{:}\text{done})$, $f_S(I{:}\text{done})$, and $f_S(S{:}\text{doneNoFail})$, are servers starting service at time $t + 1$.
- The three respective unmet demands, namely, $f_I(W{:}\text{done})$, $f_I(I{:}\text{done})$, and $f_I(S{:}\text{doneNoFail})$, are idle servers at time $t + 1$.
- Line 11 uses the fact that $|f'_m| = x_{m'}^{(t+1)}(z')$ for each activity $m' \neq I$ and the explicit form of the size functions in (7.2).
- The total supply of customers available to be served less the number of customers in service at time $t + 1$ represents the number of customers in the buffer awaiting service at time $t + 1$ (Line 12).

Chapter 8
Conditional and Joint Probabilities

This chapter modifies the QPLEX iterative scheme to reflect information about the counter vector observed at various time epochs and illustrates its use to perform Bayesian updates and to make prediction and dynamic adjustments. We define *conditional QPLEX iterates* and use them to define the joint QPLEX pmf of the counter vectors at different time epochs. We show via counterexample that the corresponding finite-dimensional joint pmfs of the counter vectors at different time epochs are *not*, in general, consistent. Consequently, these pmfs cannot be associated with a stochastic process (such as a Markov chain or a hidden Markov model). We use the original "unconditional" QPLEX iterates to define finite-dimensional joint *ex-post pmfs* of the counter vectors at different time epochs. These pmfs are consistent, which leads to a so-called nonlinear Markov chain of the counter vectors. However, there is a conceptual flaw underlying these *ex-post* pmfs.

Throughout this chapter, we fix some QPLEX chain and take a collection of QPLEX maps $\{\mathcal{Q}^{(t)} : t \geq 0\}$ as given. The results in this chapter do not require any structure of the transition probabilities of the QPLEX chain; in particular, we do not impose simple or advanced transition dynamics. On the other hand, our counterexample uses simple transition dynamics.

8.1 Observation Sets and Conditioning Maps

This section introduces the notions of observation sets and conditioning maps and shows the first examples of their use. These concepts form the foundation of the rest of this chapter.

Observation Sets Suppose that the counter vector is observed to lie in some set $O \subseteq \Omega_Z$. We call such an O an *observation set*. Typically, these sets are defined in

© The Author(s) 2025
A. B. Dieker, S. T. Hackman, *QPLEX: A Computational Modeling and Analysis Methodology for Stochastic Systems*, Springer Series in Operations Research and Financial Engineering, https://doi.org/10.1007/978-3-031-74870-7_8

terms of data collected on one or more of the components of the counter vector. If the data collected is the full counter vector, say z_o, then there is complete information and $O = \{z_o\}$ is a singleton. Observation sets can also represent partial information, for instance, when data is collected on a single component of the counter vector or on some function of the counter vector (e.g., an indicator that some counter exceeds a threshold). No information corresponds to $O = \Omega_Z$.

Conditioning Maps The *conditioning map* $\mathcal{O}$ associated with an observation set $O \subseteq \Omega_Z$ is defined via the distributional program below. Its input is a pmf μ of (Z, L) with $\mu(Z \in O) > 0$.

```
1: function CONDITIONINGMAP_O(μ)
2:     (z, ℓ) ~ μ(Z = ·, L = ·)
3:     if z ∉ O then
4:         skip
5:     return (z, ℓ)
```

In algebraic form, the conditioning map is given by

$$[\mathcal{O}(\mu)](z, \ell) = \frac{\mu(z, \ell) \times 1(z \in O)}{\mu(Z \in O)}. \tag{8.1}$$

For notational convenience, we suppress the functional dependence of O on $\mathcal{O}$. Note that $\mathcal{O}$ is the identity map if $\mu(Z \in O) = 1$, which is evidently the case if $O = \Omega_Z$. When the observation set is a singleton $\{z_o\}$, we write $\mathcal{O}_{z_o}$ for the corresponding conditioning map.

QPLEX Iterative Scheme with Observation Sets Modifying the original QPLEX iterative scheme to incorporate observation sets is straightforward. Fix some sequence $\{O^{(t)} : t \geq 0\}$ of observation sets. Such a sequence of observations can be interpreted as an "information scenario," the likelihood of which a decision-maker may wish to assess at time zero. Let $\{\mathcal{O}^{(t)} : t \geq 0\}$ denote the sequence of conditioning maps corresponding to $\{O^{(t)} : t \geq 0\}$, respectively. Given an initial pmf $\mu_C^{(0)}$ of (Z, L), we modify the basic QPLEX iterative scheme, for $t \geq 0$, via

$$\mu_C^{(t+1)} = (Q^{(t)} \circ \mathcal{O}^{(t)})(\mu_C^{(t)}).$$

Let τ be the first time t such that $\mu_C^{(t)}(Z \in O^{(t)}) < 1$. The modified QPLEX iterates $\mu_C^{(0)}, \ldots, \mu_C^{(\tau)}$ are identical to the QPLEX iterates $\mu_Q^{(0)}, \ldots, \mu_Q^{(\tau)}$ generated *without* observation sets, but the subsequent iterates at times $\tau + 1, \tau + 2, \ldots$ are different.

Here is an application of this modified QPLEX iterative scheme.

Example 8.1 (Bayesian Model Selection) The premise is that there is some prior distribution over several models and that data about the system are revealed over time. As the data are observed, the prior distribution needs to be updated. At each time epoch, the data revealed can correspond to no information, partial information, or complete information.

Specifically, suppose we have several QPLEX models indexed by i with the same counter set. For instance, we could have multiserver queueing models from Chap. 2 with different Poisson arrival rate functions in each model, say, corresponding to "quiet," "normal," and "busy" service demand environments. Another example is to have models with different possible service duration pmfs in applications where service duration data are not directly available.

Suppose each model i has an associated initial pmf $\mu_{C,i}^{(0)}$ and QPLEX maps $\{Q_i^{(t)} : t \geq 0\}$. Given a sequence of observation sets $\{O^{(t)} : t \geq 0\}$, we define the pmfs $\mu_{C,i}^{(t)}$ of (Z, L) for $t \geq 0$ and model i via

$$\mu_{C,i}^{(t+1)} = (Q_i^{(t)} \circ O^{(t)})(\mu_{C,i}^{(t)})$$

as outlined above, thus taking these observation sets into account.

Suppose we have a *prior* pmf $\zeta^{(t)} = (\zeta^{(t)}(i))_i$ on models *before* the observation set $O^{(t)}$ (e.g., data on components of the counter vector) is observed. Thus, $\zeta^{(t)}(i)$ is the belief at time t that model i is the "true" model. The *posterior* pmf $\zeta^{(t+1)} = (\zeta^{(t+1)}(i))_i$, i.e., the beliefs on which model is the "true" model *after* $O^{(t)}$ is observed, is found via Bayes' rule:

$$\zeta^{(t+1)}(i) = \frac{\zeta^{(t)}(i) \times \mu_{C,i}^{(t)}(O^{(t)})}{\sum_j \zeta^{(t)}(j) \times \mu_{C,j}^{(t)}(O^{(t)})}.$$

The *posterior* pmf can then be used to calculate the posterior pmf of the counter vector at some future time epoch, which may be used for operational decision-making. △

8.2 Conditional QPLEX Iterates

In this section, we work with a specific sequence of observation sets. Given some fixed pmf $\mu_C^{(0)}$ of (Z, L), arbitrary $R \geq 1, 0 \leq t_1 < \cdots < t_R$, and counter vectors $z^{(t_1)}, \ldots, z^{(t_R)} \in \Omega_Z$, we define the associated sequence of observation sets $\{O^{(t)} : t \geq 0\}$ via

$$O^{(t)} = \begin{cases} O_{z^{(t_r)}} & \text{if } t = t_r \text{ for some } r = 1, \ldots, R \\ \Omega_Z & \text{otherwise.} \end{cases}$$

The *conditional QPLEX iterates* $\mu_C^{(t)}$ are the pmfs defined, for $t \geq 0$, via

$$\mu_C^{(t+1)} = (Q^{(t+1)} \circ O^{(t)})(\mu_C^{(t)}).$$

Keep in mind that when $O^{(t)} = \Omega_z$, the conditional QPLEX iterate $\mu_C^{(t+1)}$ can be obtained from the previous conditional QPLEX iterate $\mu_C^{(t)}$ by only applying the QPLEX map $\mathcal{Q}^{(t)}$. To emphasize the values of the counter vectors that have been incorporated in these pmfs, for $1 \leq r \leq R$, we also denote the conditional QPLEX iterate $\mu_C^{(t)}(z, \ell)$ for time t with $t_r < t \leq t_{r+1}$ by

$$\mu_C^{(t)}(z, \ell \| z^{(t_1)}, \ldots, z^{(t_r)}).$$

For $r = R$, this is to be interpreted as being defined for all $t > t_r$.

We use a double bar as a visual reminder that a conditional QPLEX iterate is *not* a conditional pmf in the sense that there is no underlying joint pmf $\mu_C^{(t)}(z^{(t_1)}, \ldots, z^{(t_r)}, z, \ell)$. Instead, conditional QPLEX iterates should be viewed as unconditional pmfs that are parameterized by past observed counter vector values. This distinction is important in this chapter because we will discover a lack of consistency of appropriately defined joint pmfs in Sect. 8.4, a subtlety that is not encountered when working with stochastic processes.

Note that, by convention, $\mu_C^{(0)}$ is a conditional QPLEX iterate. Also note that we suppress the functional dependence of the conditional QPLEX iterates on $\mu_C^{(0)}$, and for brevity, we also suppress $t_1, \ldots, t_R$ as these time epochs can be inferred from the notation used to define the counter vector values. Finally, note that the counter vector(s) in the condition of a conditional QPLEX iterate cannot necessarily be arbitrary elements of the counter set, only "feasible" values.

Example 8.2 Consider the diagram below.

The first four conditional QPLEX iterates for $R = 2$ and $(t_1, t_2) = (0, 2)$ and arbitrary feasible $z^{(0)}, z^{(2)}$ are shown in the boxes. (We use the argument "." instead of "z, ℓ" to save space.) $\triangle$

Example 8.3 (Prediction and Dynamic Adjustment) We now consider the QPLEX maps as being parameterized by decision variables. For example, the variables could represent the number of servers ("capacity") at each station in a queueing network. The values of these parameters are initially set at time zero, and we start from some initial distribution. We then apply the QPLEX maps until time t_1, at which time we observe the counter vector. This need not be the full counter vector, but for simplicity, we assume it is. In response to the data we observe at time t_1, we may wish to explore options to change the values of the parameters (e.g., increase, retain, or decrease capacities). Given some performance metric at time T, we can calculate the QPLEX iterate at time T associated with each option and make the best decision at time t_1. We could then choose the initial values of the parameters, knowing that we have options to change the parameter values in response to the observation at time t_1.

This example is illustrated in the diagram below. (Once again, we use the argument "·" instead of "z, ℓ" to save space.)

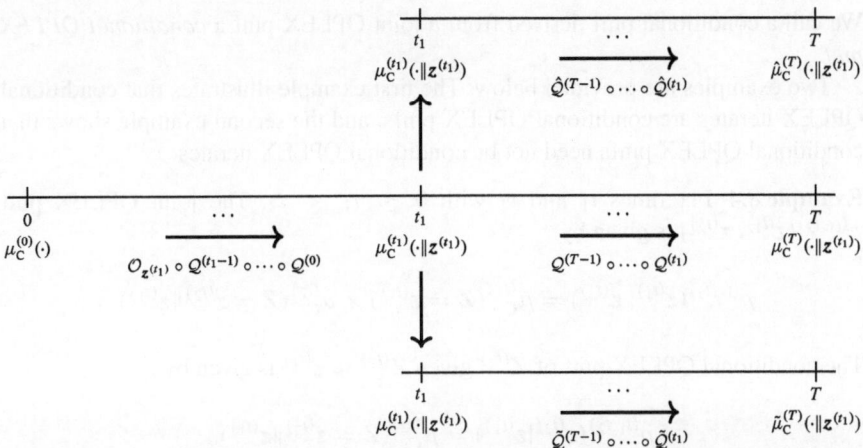

This diagram depicts three options, one of which is to make no changes to parameter values. The two other options result in different QPLEX maps $\check{Q}^{(t)}$ and $\hat{Q}^{(t)}$ for $t \geq t_1$. This example can be extended to multiple observation time epochs and more than three options.   △

8.3 Joint QPLEX Pmfs

We are now in a position to define joint QPLEX pmfs. We assume that $R \geq 1$ and $0 \leq t_1 < \cdots < t_R$ are fixed but arbitrary. Calculating a joint QPLEX pmf in full requires calculation for all possible choices of the arguments, so it can be computationally expensive even for moderate values of R.

Definition 8.1 Given an initial pmf $\mu_C^{(0)}$ of $(\mathbf{Z}, \mathbf{L})$, the *joint QPLEX pmf* is the pmf $\rho^{(t_1,\dots,t_R)}$ of the counter vectors $(\mathbf{Z}^{(t_1)}, \dots, \mathbf{Z}^{(t_R)})$ at times $t_1, \dots, t_R$ defined via

$$\rho^{(t_1,\dots,t_R)}(\mathbf{z}^{(t_1)}, \dots, \mathbf{z}^{(t_R)}) = \prod_{r=1}^{R} \mu_C^{(t_r)}(\mathbf{Z} = \mathbf{z}^{(t_r)} \| \mathbf{z}^{(t_1)}, \dots, \mathbf{z}^{(t_{r-1})}),$$

where the term for $r = 1$ should be interpreted as $\mu_C^{(t_1)}(\mathbf{Z} = \mathbf{z}^{(t_1)})$. If any of the terms on the right-hand side are undefined, the right-hand side should be interpreted as 0.

As a consequence of this definition, the marginal pmf of the counter vector of each conditional QPLEX iterate can be written in terms of the joint QPLEX pmf:

$$\mu_C^{(t_r)}(\mathbf{Z} = z^{(t_r)} \| z^{(t_1)}, \ldots, z^{(t_{r-1})}) = \rho^{(t_1, \ldots, t_R)}(z^{(t_r)} | z^{(t_1)}, \ldots, z^{(t_{r-1})}).$$

We call a conditional pmf derived from a joint QPLEX pmf a *conditional QPLEX pmf*.

Two examples are provided below. The first example illustrates that conditional QPLEX iterates are conditional QPLEX pmfs, and the second example shows that conditional QPLEX pmfs need not be conditional QPLEX iterates.

Example 8.4 Fix times t_1 and t_2 with $0 \leq t_1 < t_2$. The joint QPLEX pmf $\rho^{(t_1, t_2)}(z^{(t_1)}, z^{(t_2)})$ is given by

$$\rho^{(t_1, t_2)}(z^{(t_1)}, z^{(t_2)}) = \mu_C^{(t_1)}(\mathbf{Z} = z^{(t_1)}) \times \mu_C^{(t_2)}(\mathbf{Z} = z^{(t_2)} \| z^{(t_1)}).$$

The conditional QPLEX pmf of $\mathbf{Z}^{(t_2)}$ given $\mathbf{Z}^{(t_1)} = z^{(t_1)}$ is given by

$$\rho^{(t_1, t_2)}(z^{(t_2)} | z^{(t_1)}) = \mu_C^{(t_2)}(\mathbf{Z} = z^{(t_2)} \| z^{(t_1)}).$$

This is a conditional QPLEX iterate.   △

Example 8.5 Fix times t_1, t_2, and t_3 with $0 \leq t_1 < t_2 < t_3$. The conditional QPLEX pmf of $\mathbf{Z}^{(t_2)}$ given $\mathbf{Z}^{(t_1)} = z^{(t_1)}$ and $\mathbf{Z}^{(t_3)} = z^{(t_3)}$ is given by

$$\rho^{(t_1, t_2, t_3)}(z^{(t_2)} | z^{(t_1)}, z^{(t_3)}) = \frac{\rho^{(t_1, t_2, t_3)}(z^{(t_1)}, z^{(t_2)}, z^{(t_3)})}{\sum_{\tilde{z}^{(t_2)}} \rho^{(t_1, t_2, t_3)}(z^{(t_1)}, \mathbf{Z}^{(t_2)} = \tilde{z}^{(t_2)}, z^{(t_3)})}. \tag{8.2}$$

By definition of the joint QPLEX pmf $\rho^{(t_1, t_2, t_3)}$, we have, for all $\tilde{z}^{(t_2)}$, that

$$\rho^{(t_1, t_2, t_3)}(z^{(t_1)}, \mathbf{Z}^{(t_2)} = \tilde{z}^{(t_2)}, z^{(t_3)}) =$$
$$\mu_C^{(t_1)}(\mathbf{Z} = z^{(t_1)}) \times \mu_C^{(t_2)}(\mathbf{Z} = \tilde{z}^{(t_2)} \| z^{(t_1)}) \times \mu_C^{(t_3)}(\mathbf{Z} = z^{(t_3)} \| z^{(t_1)}, \tilde{z}^{(t_2)}),$$

and so the conditional QPLEX pmf can also be written as

$$\rho^{(t_1, t_2, t_3)}(z^{(t_2)} | z^{(t_1)}, z^{(t_3)}) =$$

$$\frac{\mu_C^{(t_2)}(\mathbf{Z} = z^{(t_2)} \| z^{(t_1)}) \times \mu_C^{(t_3)}(\mathbf{Z} = z^{(t_3)} \| z^{(t_1)}, z^{(t_2)})}{\sum_{\tilde{z}^{(t_2)}} \mu_C^{(t_2)}(\mathbf{Z} = \tilde{z}^{(t_2)} \| z^{(t_1)}) \times \mu_C^{(t_3)}(\mathbf{Z} = z^{(t_3)} \| z^{(t_1)}, \tilde{z}^{(t_2)})}.$$

This is *not* a conditional QPLEX iterate.   △

8.4   Violation of Consistency

Having defined the QPLEX pmf $\rho^{(t_1,\ldots,t_R)}$ for fixed $R \geq 1$ and $0 \leq t_1 < \cdots < t_R \leq T$, the collection of *all* such joint QPLEX pmfs is well-defined. In this section, we show that this collection is not consistent in general, and therefore it is not possible to define an associated stochastic process. In particular, one cannot associate a Markov chain or a hidden Markov model with the collection of joint QPLEX pmfs. Lack of consistency is closely tied to the nonlinearity of QPLEX maps. For instance, this lack of consistency implies that the denominator of the conditional QPLEX pmf in (8.2) of Example 8.5 does *not* equal $\rho^{(t_1,t_3)}(z^{(t_1)}, z^{(t_3)})$ in general.

Theorem 8.2 *Fix some time* $T > 0$. *The collection*

$$\{\rho^{(t_1,\ldots,t_R)} : R \geq 1, 0 \leq t_1 < \cdots < t_R \leq T\}.$$

of joint QPLEX pmfs is not *consistent in general.*

We establish this claim via a counterexample that is a variation of the numerical illustration in Sect. 4.4. Instead of working with labels that represent remaining service durations, we let the labels represent the number of periods since the start of service ("service age"). This offers a way to compare the numerical results with the two different label interpretations, a topic we revisit in more detail in Sect. 18.5. We consider the following example of a multiserver queueing model from Chap. 2 for three periods:

- The number of servers is $n = 2$. Therefore, the size function $x^{(t)}(z)$ equals $\min(z, 2)$ as in Sect. 4.4.
- Service is completed in three periods or less. The probabilities of a service completion given service age $\ell = 0, 1, 2$ are constant in time and independent of the number of customers in the system z and are given by $\gamma^{(t)}(z, 0) = 0.6$, $\gamma^{(t)}(z, 1) = 0.75$, and $\gamma^{(t)}(z, 2) = 1$. These values are consistent with the service duration pmf given in Sect. 4.4.
- There are either two or three arrivals at the end of the first period and no arrivals in subsequent periods. The probability of two arrivals at the end of the first period is 0.3, independent of the number of customers in the system at time 0.
- The system is initially empty.
- Labels represent service ages, so the label set is given by $\{0, 1, 2\}$.

We show below that

$$\rho^{(2,3)}(2, 2) = 0.03825375$$

$$\neq \rho^{(1,2,3)}(2, 2, 2) + \rho^{(1,2,3)}(3, 2, 2) = 0.003 + 0.03549 = 0.03849.$$

Here and throughout this section, we write $\rho^{(1,2,3)}(2, 2, 2)$ as an abbreviation of a probability such as $\rho^{(1,2,3)}(Z^{(1)} = 2, Z^{(2)} = 2, Z^{(3)} = 2)$. The two probabilities on the right-hand side can be thought of as "path probabilities" of the two (linearly interpolated) paths shown in the diagram below. Since 2 and 3 are the only possible values of $Z^{(1)}$, this counterexample establishes the theorem.

For convenience, we interpret the expression bin(n, p) as the pmf of a binomial random variable with number of trials parameter n and success probability p in what follows (as opposed to the random variable itself), as we do in distributional programs.

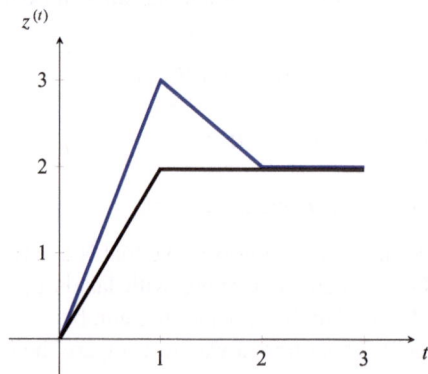

Calculation of $\rho^{(1,2,3)}(2, 2, 2)$ From Definition 8.1, the desired probability is given by

$$\rho^{(1,2,3)}(2, 2, 2) = \mu_C^{(1)}(Z = 2) \times \mu_C^{(2)}(Z = 2 \| Z^{(1)} = 2)$$
$$\times \mu_C^{(3)}(Z = 2 \| Z^{(1)} = 2, Z^{(2)} = 2). \qquad (8.3)$$

We calculate the terms on the right-hand side of (8.3) in turn. As we subsequently need the results, we will calculate more for the first and second terms, namely, $\mu_C^{(1)}(z, \ell)$ and $\mu_C^{(2)}(Z = 2, \ell \| Z^{(1)} = 2)$.

Calculation of $\mu_C^{(1)}(Z = 2)$ All customers entering service during the first period have age 0 at time 1. From the input data, it immediately follows that

$$\mu_C^{(1)}(Z = 2, L = \cdot) = 0.3 \times (1, 0, 0) \qquad (8.4)$$
$$\mu_C^{(1)}(Z = 3, L = \cdot) = 0.7 \times (1, 0, 0). \qquad (8.5)$$

In particular, $\mu_C^{(1)}(Z = 2) = 0.3$.

Calculation of $\mu_C^{(2)}(Z = 2 \| Z^{(1)} = 2)$ By definition of conditional QPLEX iterate, the pmf $\mu_C^{(2)}(z, \ell \| Z^{(1)} = 2)$ is obtained by applying the QPLEX map $\mathcal{Q}^{(1)}$ to the pmf $\mu_o^{(1)} = \mathcal{O}_2(\mu^{(1)})$, which by (8.1) and (8.4) can be represented via

$$\mu_o^{(1)}(Z = 2, L = \cdot) = (1, 0, 0).$$

We now apply the QPLEX map $\mathcal{Q}^{(1)}$ to $\mu_o^{(1)}$. Under (the multinomial lift map applied to) $\mu_o^{(1)}$, there are two customers in the system and each customer has a service age of 0. Therefore, in view of $\gamma^{(1)}(2, 0) = 0.6$, the number of service completions has pmf

$$\text{NUMBEROFCOMPLETIONS}_{\mu_o^{(1)}}^{(1)}(2) = \text{bin}(2, 0.6).$$

Since there are no arrivals in the second period, the only way for the counter to remain at 2 is for the number of service completions to equal 0. A uniformly chosen at random customer at time 2 will then have entity type "old" as both customers remain to complete their respective service, so the entity type pmf is given by

$$[\text{ENTITYTYPE}^{(1)}(2, 0, 2)](k') = 1(k' = \text{old}),$$

and since the ages of each of these customers will increase to 1 at time 2, the next label pmf is given by

$$\text{NEXTLABEL}_{\mu_o^{(1)}}^{(1)}(2, 2, \text{old}) = (0, 1, 0).$$

We conclude that

$$\mu_C^{(2)}(Z = 2, L = \cdot \| Z^{(1)} = 2) =$$
$$\Pr[\text{bin}(2, 0.6) = 0] \times (0, 1, 0) = 0.4^2 \times (0, 1, 0), \qquad (8.6)$$

so in particular $\mu_C^{(2)}(Z = 2 \| Z^{(1)} = 2) = 0.4^2$.

Calculation of $\mu_C^{(3)}(Z = 2 \| Z^{(1)} = 2, Z^{(2)} = 2)$ By definition of conditional QPLEX iterate, the pmf $\mu_C^{(3)}(z, \ell \| Z^{(1)} = 2, Z^{(2)} = 2)$ is obtained by applying the QPLEX map $\mathcal{Q}^{(2)}$ to the pmf

$$\mu_o^{(2)} = \mathcal{O}_2(\mu_C^{(2)}(Z = \cdot, L = \cdot \| Z^{(1)} = 2)).$$

Thus, from (8.1) and (8.6), we find that

$$\mu_o^{(2)}(Z = 2, L = \cdot) = (0, 1, 0).$$

We now apply the QPLEX map $\mathcal{Q}^{(2)}$ to $\mu_o^{(2)}$. Under (the multinomial lift map applied to) $\mu_o^{(2)}$, there are two customers in the system and each customer has a service age of 1. Therefore, in view of $\gamma^{(2)}(2, 1) = 0.75$, the number of service completions has pmf

$$\text{NUMBEROFCOMPLETIONS}_{\mu_o^{(2)}}^{(2)}(2) = \text{bin}(2, 0.75).$$

Since there are no arrivals in the third period, the only way for the counter to remain at 2 is for the number of service completions to equal 0, and so

$$\mu_C^{(3)}(Z = 2 \| Z^{(1)} = 2, Z^{(2)} = 2) = \Pr[\text{bin}(2, 0.75) = 0] = 0.25^2.$$

(We do not need to calculate the output of other utility programs as this is the sought-after third probability.)

Putting it all together, we conclude that

$$\rho^{(1,2,3)}(2, 2, 2) = 0.3 \times 0.4^2 \times 0.25^2 = 0.003.$$

Calculation of $\rho^{(1,2,3)}(3, 2, 2)$ From Definition 8.1, the desired probability is given by

$$\rho^{(1,2,3)}(3, 2, 2) = \mu_C^{(1)}(Z = 3) \times \mu_C^{(2)}(Z = 2 \| Z^{(1)} = 3)$$
$$\times \mu_C^{(3)}(Z = 2 \| Z^{(1)} = 3, Z^{(2)} = 2). \qquad (8.7)$$

We know that $\mu_C^{(1)}(Z = 3) = 0.7$. We proceed to calculate each of the remaining two terms on the right-hand side of (8.7). As before, we need to calculate more for the second term, namely, $\mu_C^{(2)}(Z = 2, L = \cdot \| Z^{(1)} = 3)$. We reuse the notation $\mu_o^{(1)}$ and $\mu_o^{(2)}$, although these pmfs are different here.

Calculation of $\mu_C^{(2)}(Z = 2 \| Z^{(1)} = 3)$ By definition of conditional QPLEX iterate, the pmf $\mu_C^{(2)}(z, \ell \| Z^{(1)} = 3)$ is obtained by applying the QPLEX map $\mathcal{Q}^{(1)}$ to the pmf $\mu_o^{(1)} = \mathcal{O}_3(\mu_C^{(1)})$, which by (8.1) and (8.5) can be represented via

$$\mu_o^{(1)}(Z = 3, L = \cdot) = (1, 0, 0).$$

We now apply the QPLEX map $\mathcal{Q}^{(1)}$ to $\mu_o^{(1)}$. Under (the multinomial lift map applied to) $\mu_o^{(1)}$, there are three customers in the system, two of which are in service and have a service age of 0. In view of $\gamma^{(1)}(3, 0) = 0.6$, the number of service completions has pmf

$$\text{NUMBEROFCOMPLETIONS}_{\mu_o^{(1)}}^{(1)}(3) = \text{bin}(2, 0.6).$$

Since there are no arrivals in the second period, for the number of customers in the system to decrease from 3 to 2, the number of service completions must be 1. In that case, of the two customers in service at time 2, one customer has entity type "old" and the other has entity type "new." When choosing one customer in service uniformly at random, we obtain the entity type pmf via

$$[\text{ENTITYTYPE}^{(1)}(3, 1, 2)](k') = 0.5 \times 1(k' = \text{old}) + 0.5 \times 1(k' = \text{new}).$$

Under (the multinomial lift map of) $\mu_o^{(1)}$, the conditional pmf of the age of a uniformly chosen at random customer with entity type "old" is $(0, 1, 0)$. The conditional pmf of the age of a uniformly chosen at random customer with entity type "new" is always $(1, 0, 0)$, so the next label pmfs are given by

$$\text{NEXTLABEL}_{\mu_o^{(1)}}^{(1)}(3, 2, \text{old}) = (0, 1, 0),$$

$$\text{NEXTLABEL}_{\mu_o^{(1)}}^{(1)}(3, 2, \text{new}) = (1, 0, 0).$$

Therefore, we have that

$$\mu_C^{(2)}(Z = 2, L = \cdot \| Z^{(1)} = 3)$$

$$= \Pr[\text{bin}(2, 0.6) = 1] \times [0.5 \times (0, 1, 0) + 0.5 \times (1, 0, 0)]$$

$$= 2 \times 0.6 \times 0.4 \times (0.5, 0.5, 0), \qquad (8.8)$$

and thus $\mu_C^{(2)}(Z = 2 | Z^{(1)} = 3) = 2 \times 0.6 \times 0.4$.

Calculation of $\mu_C^{(3)}(Z = 2 \| Z^{(1)} = 3, Z^{(2)} = 2)$ By definition of conditional QPLEX iterate, the pmf $\mu_C^{(3)}(z, \ell \| Z^{(1)} = 3, Z^{(2)} = 2)$ is obtained by applying the QPLEX map $\mathcal{Q}^{(2)}$ to the pmf

$$\mu_o^{(2)} = \mathcal{O}_2(\mu_C^{(2)}(Z = \cdot, L = \cdot \| Z^{(1)} = 3)).$$

Thus, from (8.1) and (8.8), we find that

$$\mu_o^{(2)}(Z = 2, L = \cdot) = (0.5, 0.5, 0).$$

We now apply the QPLEX map $\mathcal{Q}^{(2)}$ to $\mu_o^{(2)}$. Under (the multinomial lift map applied to) $\mu_o^{(2)}$, there are two customers in the system, and each customer is equally likely to have an age of 0 or 1. Therefore, either customer completes its service in the third period independently with probability

$$0.5 \times \gamma^{(2)}(2, 0) + 0.5 \times \gamma^{(2)}(2, 1) = 0.5 \times 0.6 + 0.5 \times 0.75 = 0.675.$$

As a result, we have that

$$\text{NUMBEROFCOMPLETIONS}^{(2)}_{\mu_o^{(2)}}(2) = \text{bin}(2, 0.675).$$

Since there are no arrivals in the third period, the only way for the number of customers in the system to remain at 2 is for the number of service completions to equal 0. Therefore, we have that

$$\mu_C^{(3)}(Z = 2\|Z^{(1)} = 3, Z^{(2)} = 2) = \Pr[\text{bin}(2, 0.675) = 0] = 0.325^2.$$

Putting it all together, we conclude that

$$\rho^{(1,2,3)}(3, 2, 2) = 0.7 \times (2 \times 0.6 \times 0.4) \times 0.325^2 = 0.03549.$$

Calculation of $\rho^{(2,3)}(2, 2)$ From Definition 8.1, the desired probability is given by

$$\rho^{(2,3)}(2, 2) = \mu_C^{(2)}(Z = 2) \times \mu_C^{(3)}(Z = 2\|Z^{(2)} = 2). \tag{8.9}$$

We proceed to calculate each of the two terms on the right-hand side of (8.9). For the first term, we calculate more, namely, $\mu_C^{(2)}(Z = 2, L = \cdot)$.

Calculation of $\mu_C^{(2)}(Z = 2)$ The pmf $\mu_C^{(1)}$ is given in (8.4) and (8.5), and the pmf $\mu_C^{(2)}$ is found via $\mu_C^{(2)} = Q^{(1)}(\mu_C^{(1)})$. All customers in service at time 1 have a service age of 0, and since $\gamma^{(1)}(z, 0) = 0.6$ for all z, we thus have for $z \geq 2$ that

$$\text{NUMBEROFCOMPLETIONS}^{(1)}_{\mu_C^{(1)}}(z) = \text{bin}(2, 0.6).$$

The pmfs $\text{ENTITYTYPE}^{(1)}(2, 0, 2)$ and $\text{ENTITYTYPE}^{(1)}(3, 1, 2)$ are the same as before since they do not depend on the QPLEX iterate. Similarly, the next label pmfs are, as before, given by, for $z = 2, 3$,

$$\text{NEXTLABEL}^{(1)}_{\mu_C^{(1)}}(z, 2, \text{old}) = (0, 1, 0),$$

$$\text{NEXTLABEL}^{(1)}_{\mu_C^{(1)}}(z, 2, \text{new}) = (1, 0, 0).$$

Thus, we have that

$$\mu_C^{(2)}(Z = 2, L = \cdot)$$
$$= \mu_C^{(1)}(Z = 2) \times \Pr[\text{bin}(2, 0.6) = 0] \times (0, 1, 0)$$
$$+ \mu_C^{(1)}(Z = 3) \times \Pr[\text{bin}(2, 0.6) = 1] \times [0.5 \times (0, 1, 0) + 0.5 \times (1, 0, 0)].$$

Since $\mu_C^{(1)}(Z = 2) = 0.3$, we have that

$$\mu_C^{(2)}(Z = 2, L = \cdot)$$

$$= 0.3 \times 0.4^2 \times (0, 1, 0) + 0.7 \times (2 \times 0.6 \times 0.4) \times (0.5, 0.5, 0)$$

$$= (0.168, 0.216, 0), \tag{8.10}$$

and so $\mu_C^{(2)}(Z = 2) = 0.168 + 0.216 = 0.384$.

Calculation of $\mu_C^{(3)}(Z = 2 \| Z^{(2)} = 2)$ By definition of conditional QPLEX iterate, the pmf $\mu_C^{(3)}(z, \ell \| Z^{(2)} = 2)$ is obtained by applying the QPLEX map $\mathcal{Q}^{(2)}$ to the pmf $\mu_o^{(2)} = \mathcal{O}_2(\mu_C^{(2)})$, where we again reuse the symbol $\mu_o^{(2)}$ with a different meaning. By (8.1), we need to renormalize (8.10), so we find that

$$\mu_o^{(2)}(Z = 2, L = \cdot) = \left(\frac{0.168}{0.384}, \frac{0.216}{0.384}, 0\right) = \left(\frac{7}{16}, \frac{9}{16}, 0\right).$$

We now apply the QPLEX map $\mathcal{Q}^{(2)}$ to $\mu_o^{(2)}$. Under (the multinomial lift map applied to) $\mu_o^{(2)}$, there are two customers in the system and the ages of the two customers in service are independent with pmf $(\frac{7}{16}, \frac{9}{16}, 0)$. Using $\gamma^{(2)}(2, 0) = 0.6$ and $\gamma^{(2)}(2, 1) = 0.75$, the probability that either customer completes their service in the third period equals

$$\frac{7}{16} \times \gamma^{(2)}(2, 0) + \frac{9}{16} \times \gamma^{(2)}(2, 1) = \frac{7}{16} \times 0.6 + \frac{9}{16} \times 0.75 = 0.684375.$$

Therefore, the number of service completions has pmf

$$\text{NUMBEROFCOMPLETIONS}_{\mu_o^{(2)}}^{(2)}(2) = \text{bin}(2, 0.684375).$$

Since there are no arrivals in the third period, the only way for the counter to remain at 2 is for the number of service completions to equal 0, and so

$$\mu_C^{(3)}(Z = 2 \| Z^{(2)} = 2) = \Pr[\text{bin}(2, 0.684375) = 0] = 0.315625^2.$$

Putting it all together, we conclude that

$$\rho^{(2,3)}(2, 2) = 0.384 \times 0.315625^2 = 0.03825375.$$

8.5 QPLEX Trajectory Probabilities

This section discusses the joint QPLEX pmf $\rho^{(0,\dots,T)}(z^{(0)}, \dots, z^{(T)})$ defined in Definition 8.1 for arbitrary $T > 0$ and fixed $z^{(0)}, \dots, z^{(T)}$. We call the associated probabilities *QPLEX trajectory probabilities* as the time epochs are adjacent. We show that this probability can be obtained by iterating with lower-dimensional pmfs than the conditional QPLEX iterates, exploiting the fact that the counter vectors are fixed.

Another Iterative Scheme The diagram below, which corresponds to $T = 2$, plays a central role in this section. We assume that we are given QPLEX maps $Q^{(0)}, \ldots, Q^{(T)}$ and some initial pmf $\mu_C^{(0)}$ of (Z, L). For $t \geq 0$, we define $\mu_o^{(t)} = O_{z^{(t)}}(\mu_C^{(t)})$ and $\mu_C^{(t+1)} = Q^{(t)}(\mu_o^{(t)})$. In addition to the iterative scheme

$$\mu_C^{(t+1)} = (Q^{(t)} \circ O_{z^{(t)}})(\mu_C^{(t)})$$

for conditional QPLEX iterates on the top row, the diagram depicts the scheme

$$\mu_o^{(t+1)} = (O_{z^{(t+1)}} \circ Q^{(t)})(\mu_o^{(t)})$$

on the bottom row. Note that a QPLEX map and a conditioning map are applied in reverse order on the two levels.

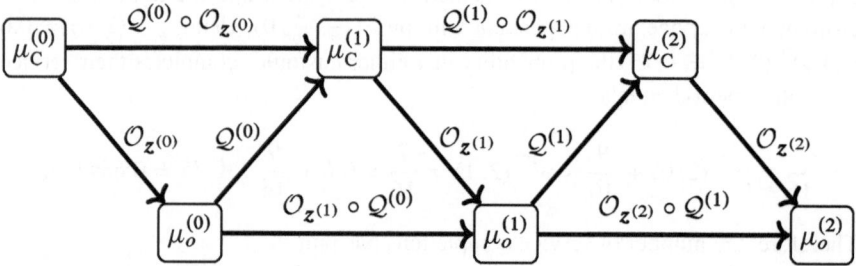

It is readily seen from (8.1) that $O_{z^{(t)}} \circ O_{z^{(t)}} = O_{z^{(t)}}$, so we have, for all $t \geq 0$, that

$$\mu_o^{(t)} = O_{z^{(t)}}(\mu_o^{(t)})$$

or equivalently

$$\mu_o^{(t)}(z, \ell) = 1(z = z^{(t)}) \times \mu_o^{(t)}(\ell).$$

Thus, the $\mu_o^{(t)}$ can be viewed as a pmf on labels *only*. To avoid introducing further notation, we continue to work with them as pmfs of (Z, L).

The pmfs on the top level of this diagram are conditional QPLEX iterates, but we have suppressed their dependence on previous counter vectors. The pmf $\mu_C^{(t)}(z, \ell)$ stands for $\mu_C^{(t)}(z, \ell \| z^{(0)}, \ldots, z^{(t-1)})$ for $t \geq 0$. Similarly, to stress that the $\mu_o^{(t)}$ depend on $z^{(0)}, \ldots, z^{(t)}$, in the rest of this section, we write

$$\mu_o^{(t)}(z, \ell \| z^{(0)}, \ldots, z^{(t)}) \tag{8.11}$$

instead of $\mu_o^{(t)}(z, \ell)$, where we highlight the additional $z^{(t)}$ in this convention relative to the corresponding conditional QPLEX iterate.

The notation $\mu_o^{(t)}$ used in the above diagram is directly related to the notation used in the counterexample of the previous section. In our calculation of $\rho^{(1,2,3)}(2,2,2)$, for instance, we have (partially) calculated the pmfs shown in the diagram below. Note that we have shifted the time indices of the pmfs.

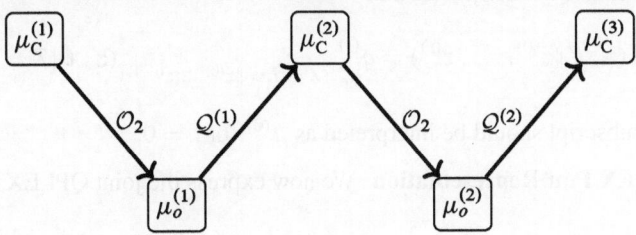

One-Step QPLEX Probabilities Given any pmf μ of (Z, L), for $z \in \Omega_Z$ with $\mu(z) > 0$, we define the *one-step QPLEX probabilities* using the conditioning map $\mathcal{O}_z$ via

$$q_\mu^{(t)}(z', \ell'|z) = [(\mathcal{Q}^{(t)} \circ \mathcal{O}_z)(\mu)](z', \ell'). \tag{8.12}$$

The one-step QPLEX probabilities $q_\mu^{(t)}(z', \ell'|z)$ are the analogs of the transition probabilities $p^{(t)}(z', h'|z, h)$ of the QPLEX chain. Just as (3.2) expresses the kernel map $\mathcal{K}^{(t)}$ in terms of these transition probabilities, the QPLEX map $\mathcal{Q}^{(t)}$ can be expressed in terms of one-step QPLEX probabilities via

$$[\mathcal{Q}^{(t)}(\mu)](z', \ell') = \sum_z \mu(z) \times q_\mu^{(t)}(z', \ell'|z). \tag{8.13}$$

To see this, we note that the distributional program for the QPLEX map (see Sect. 3.2) has the following form:

```
1: function QPLEXMAP(t)(μ)
2:     z ~ μ(Z = ·)
3:     for m = 1, ..., M do
4:         hm ~ mult (x_m^(t)(z), μ(Lm = ·|z))
        ⋮
11:    return (z′, ℓ′)
```

The instructions corresponding to Lines 5–10 are not shown but do not involve the input μ. It immediately follows that

$$Q^{(t)}(\mu) = \sum_z \mu(z) \times (Q^{(t)} \circ \mathcal{O}_z)(\mu),$$

as required in view of $q_\mu^{(t)}(z', \ell'|z)$ in (8.12).

Since the conditional QPLEX iterate at time $t + 1$ is found from the conditional QPLEX iterate at time t by applying the map $Q^{(t)} \circ \mathcal{O}_{z^{(t)}}$, it follows from (8.12) and the definition of QPLEX iterates that

$$\mu_C^{(t+1)}(z', \ell'\|z^{(0)}, \ldots, z^{(t)}) = q_{\mu_C^{(t)}(Z=\cdot, L=\cdot\|z^{(0)}, \ldots, z^{(t-1)})}^{(t)}(z', \ell'|Z = z^{(t)}),$$
(8.14)

where the subscript should be interpreted as $\mu^{(0)}$ for $t = 0$.

Joint QPLEX Pmf Representation We now express the joint QPLEX pmf

$$\rho^{(0,\ldots,T)}(z^{(0)}, \ldots, z^{(T)})$$

$$= \mu_C^{(0)}(Z = z^{(0)}) \times \prod_{t=0}^{T-1} \mu_C^{(t+1)}(Z = z^{(t+1)}\|z^{(0)}, \ldots, z^{(t)}) \qquad (8.15)$$

in terms of the one-step QPLEX probabilities and the iterates in (8.11). Using $\mathcal{O}_z \circ \mathcal{O}_z = \mathcal{O}_z$, we have that

$$q_{\mathcal{O}_z(\mu)}^{(t)}(z', \ell'|z) = q_\mu^{(t)}(z', \ell'|z).$$

Therefore, we can change the subscript on the right-hand side of (8.14) and sum over ℓ to obtain that

$$\mu_C^{(t+1)}(z\|z^{(0)}, \ldots, z^{(t)}) = q_{\mu_o^{(t)}(Z=\cdot, L=\cdot\|z^{(0)}, \ldots, z^{(t)})}^{(t)}(Z' = z|Z = z^{(t)}).$$

We have thus shown that

$$\rho^{(0,\ldots,T)}(z^{(0)}, \ldots, z^{(T)})$$

$$= \mu_C^{(0)}(Z = z^{(0)}) \times \prod_{t=0}^{T-1} q_{\mu_o^{(t)}(Z=\cdot, L=\cdot\|z^{(0)}, \ldots, z^{(t)})}^{(t)}(Z' = z^{(t+1)}|Z = z^{(t)}). \qquad (8.16)$$

Let us now compare (8.15) and (8.16). The pmf in the subscript in (8.16) can be viewed as a pmf on labels *only*. As such, it is intrinsically lower-dimensional than the pmf in the subscript in (8.15). It is possible to further iterate only over the set of one-dimensional pmfs $\{\mu_o^{(t)}(\ell_m|z^{(0)}, \ldots, z^{(t)}) : 1 \leq m \leq M\}$, a point we will encounter again in Sect. 17.6.

8.6 Joint Ex-Post Probabilities

We have seen that the collection of joint QPLEX pmfs $\{\rho^{(t_1,\ldots,t_R)}\}$, with each pmf calculated using conditional QPLEX iterates, cannot (in general) be associated with a stochastic process. In this section, we show that it is possible to define a pmf using the (unconditional) QPLEX iterates and use it to define a so-called nonlinear Markov chain.

Ex-post Pmfs This pmf uses the QPLEX iterates $\{\mu_Q^{(t)} : t \geq 0\}$ iteratively defined given $\mu_Q^{(0)}$ via $\mu_Q^{(t+1)} = Q^{(t)}(\mu_Q^{(t)})$ for $t \geq 0$. To distinguish them from conditional QPLEX iterates, we refer to these iterates as *unconditional QPLEX iterates*. Here is the formal definition of the *ex-post* pmf ψ.

Definition 8.3 The *ex-post* pmf associated with the unconditional QPLEX iterates $\mu^{(0)}, \ldots, \mu^{(T-1)}$ is the pmf ψ of $(\mathbf{Z}^{(0)}, \mathbf{L}^{(0)}, \ldots, \mathbf{Z}^{(T)}, \mathbf{L}^{(T)})$ given by

$$
\psi(z^{(0)}, \ell^{(0)}, \ldots, z^{(T)}, \ell^{(T)})
$$

$$
= \mu_Q^{(0)}(\mathbf{Z} = z^{(0)}, \mathbf{L} = \ell^{(0)}) \times \prod_{t=0}^{T-1} q_{\mu_Q^{(t)}}^{(t)}(\mathbf{Z}' = z^{(t+1)}, \mathbf{L}' = \ell^{(t+1)}|\mathbf{Z} = z^{(t)}).
$$

(8.17)

We use the adjective *ex-post* because the unconditional QPLEX iterates appear in this definition, so they need to have been calculated first.

The conditional *ex-post* probabilities

$$
\psi(z^{(t+1)}, \ell^{(t+1)}|z^{(0)}, \ell^{(0)}, \ldots, z^{(t)}, \ell^{(t)})
$$

$$
= q_{\mu_Q^{(t)}}^{(t)}(\mathbf{Z}' = z^{(t+1)}, \mathbf{L}' = \ell^{(t+1)}|\mathbf{Z} = z^{(t)})
$$

are constant in $z^{(0)}, \ell^{(0)}, \ldots, z^{(t-1)}, \ell^{(t-1)}$ yet do *not* satisfy the Markov property. Unlike for a Markov chain with *exogenously* specified transition probabilities, the right-hand side is *endogenous* as these probabilities are not known a *priori*: they depend on unconditional QPLEX iterates. Still, *ex-post* pmfs satisfy a suitably modified Markov property, as we discuss next.

Nonlinear Markov Chains We now show that the stochastic process associated with the *ex-post* pmf is a *nonlinear Markov chain*. This notion was originally introduced by McKean [18] and plays an important role in mean-field stochastic models. We define a discrete-time, finite-dimensional, and time-inhomogeneous version. In the definition below, we implicitly assume there is an underlying set Ω_X in which each of the $X^{(t)}$ takes its values.

Definition 8.4 A pmf ξ of the random vector $(X^{(0)}, \ldots, X^{(T)})$ has the *nonlinear Markov chain property* if, for each t, there exists a conditional pmf $K_\nu^{(t)}(x'|x)$ parameterized by pmfs ν of $X^{(t)}$ such that, for all $0 \leq t < T$,

$$\xi(x^{(t+1)}|x^{(0)}, \ldots, x^{(t)}) = K_{\xi(X^{(t)}=\cdot)}^{(t)}(X' = x^{(t+1)}|X = x^{(t)}). \tag{8.18}$$

Upon conditioning on $(X^{(0)}, \ldots, X^{(t)})$ under ξ, we find with (8.18) that the nonlinear Markov chain property yields the following iterative scheme for the time-t marginal pmfs:

$$\xi(X^{(t+1)} = x') = \sum_{x \in \Omega_X} \xi(X^{(t)} = x) \times K_{\xi(X^{(t)}=\cdot)}^{(t)}(x'|x).$$

This scheme should be compared with that of the QPLEX iterates $\mu_Q^{(t+1)} = Q^{(t)}(\mu_Q^{(t)})$, which can be written with (8.13) as

$$\mu_Q^{(t+1)}(z', \ell') = \sum_{z \in \Omega_Z} \mu_Q^{(t)}(z) \times q_{\mu_Q^{(t+1)}}^{(t)}(z', \ell'|z). \tag{8.19}$$

It is not by accident that these two equations are similar, as the following proposition formalizes.

Proposition 8.5 *The ex-post pmf ψ associated with the unconditional QPLEX iterates $\mu_Q^{(0)}, \ldots, \mu_Q^{(T-1)}$ has the nonlinear Markov chain property with*

$$K_\mu^{(t)}\left(z', \ell' \,|\, z, \ell\right) = q_\mu^{(t)}(z', \ell'|z).$$

The proof relies on (8.17), which implies that, for $t \geq 0$,

$$\psi(z^{(0)}, \ell^{(0)}, \ldots, z^{(t+1)}, \ell^{(t+1)}) = \psi(z^{(0)}, \ell^{(0)}, \ldots, z^{(t)}, \ell^{(t)})$$

$$\times q_{\mu_Q^{(t)}}^{(t)}(Z' = z^{(t+1)}, L' = \ell^{(t+1)}|Z = z^{(t)}). \tag{8.20}$$

To establish the claim, we need to show that the unconditional QPLEX iterate $\mu_Q^{(t)}$ on the right-hand side equals the time-t marginal pmf of the ex-*post* pmf ψ, i.e.,

$$\mu_Q^{(t)}(Z = z^{(t)}, L = \ell^{(t)}) = \psi(z^{(t)}, \ell^{(t)}). \tag{8.21}$$

This identity immediately holds for $t = 0$. It holds for $t \geq 1$ because summing (8.20) over $z^{(0)}, \ell^{(0)}, \ldots, z^{(t)}, \ell^{(t)}$ yields

$$\psi(z^{(t+1)}, \ell^{(t+1)}) = \sum_{z^{(t)}} \psi(z^{(t)}) \times q_{\mu_Q^{(t)}}^{(t)}(Z' = z^{(t+1)}, L' = \ell^{(t+1)} | Z = z^{(t)}).$$

$$(8.22)$$

Comparing (8.22) with (8.19) establishes (8.21) via an inductive argument.

Comparison of *Ex-post* and Joint QPLEX Pmfs We now illustrate the key difference between the ex-post pmf and the joint QPLEX pmf. We do so in the following specific setting. Fix some $\mu_Q^{(0)}$ of (Z, L) and counter vector values $z^{(0)}$, $z^{(1)}$, and $z^{(2)}$. We compare the probability $\psi(z^{(2)}|z^{(0)}, z^{(1)})$ based on the *ex-post* pmf with the probability $\rho^{(0,1,2)}(z^{(2)}|z^{(0)}, z^{(1)})$ based on the joint QPLEX pmf.

We start with the probability $\psi(z^{(2)}|z^{(0)}, z^{(1)})$ based on the *ex-post* pmf. Summing (8.20) for $t = 1$ over $\ell^{(0)}, \ell^{(1)}, \ell^{(2)}$ and subsequently dividing through by $\psi(z^{(0)}, z^{(1)})$ yields

$$\psi(z^{(2)}|z^{(0)}, z^{(1)}) = q_{\mu_Q^{(1)}}^{(1)}(Z' = z^{(2)}|Z = z^{(1)}),$$

$$(8.23)$$

where $\mu_Q^{(1)} = Q^{(0)}(\mu_Q^{(0)})$ on the right-hand side is calculated via (8.19) as

$$\mu_Q^{(1)}(z', \ell') = \sum_z \mu_Q^{(0)}(z) \times q_{\mu_Q^{(0)}}^{(0)}(z', \ell'|z).$$

$$(8.24)$$

Note that $\psi(z^{(2)}|z^{(0)}, z^{(1)})$ does not depend on $z^{(0)}$, which is a manifestation of the nonlinear Markov chain property described above.

Now consider the probability $\rho^{(0,1,2)}(z^{(2)}|z^{(0)}, z^{(1)})$ based on the joint QPLEX pmf, which equals $\mu_C^{(2)}(Z = z^{(2)}\|z^{(0)}, z^{(1)})$. To calculate it, we use (8.14) first for $t = 0$:

$$\mu_C^{(1)}(z', \ell'\|z^{(0)}) = q_{\mu_Q^{(0)}}^{(0)}(z', \ell'|Z = z^{(0)})$$

$$(8.25)$$

and then for $t = 1$ while summing over ℓ:

$$\mu_C^{(2)}(Z = z^{(2)}\|z^{(0)}, z^{(1)}) = q_{\mu_C^{(1)}(Z=\cdot, L=\cdot\|z^{(0)})}^{(1)}(Z' = z^{(2)}|Z = z^{(1)}).$$

$$(8.26)$$

We can now compare the *ex-post* pmf $\psi(z^{(2)}|z^{(0)}, z^{(1)})$ calculated via (8.23) with the QPLEX pmf $\mu_C^{(2)}(Z = z^{(2)}\|z^{(0)}, z^{(1)})$ calculated via (8.26). We see that the right-hand sides are identical except for the subscripts. For the ex-post pmf $\psi(z^{(2)}|z^{(0)}, z^{(1)})$, the subscript is the unconditionalQPLEX iterate $\mu_Q^{(1)}$, which is a weighted average over *all* possible values z for $\mathbf{Z}^{(0)}$; see (8.24). For $\mu_C^{(1)}(\mathbf{Z} = z^{(2)}\|z^{(0)}, z^{(1)})$, on the other hand, the subscript is the conditional QPLEX iterate $\mu_C^{(1)}(\mathbf{Z} = \cdot, \mathbf{L} = \cdot\|z^{(0)})$, which (correctly) only involves a single term for $\mathbf{Z}^{(0)}$, namely, $z^{(0)}$; see (8.25).

Part II
Graphical QPLEX Calculus

Chapter 9
Introduction to Graphical QPLEX Calculus

In the worst case, the computational effort to evaluate a QPLEX map grows at least linearly in the cardinality of the support of the marginal pmf of the counter vector. Thus, for models involving many counters, such as (large) network models, evaluating the QPLEX map (and calculating QPLEX iterates) can require an effort that is exponential in the number of counters, resulting in a second curse of dimensionality. We call this the *curse of dimensionality for counters*.

Instead of working with pmfs μ of a counter vector Z and a label vector L, we overcome this curse of dimensionality for counters in Part II by working with *collections* $\iota = \{\iota_C\}_{C \in \mathcal{C}}$ of pmfs of (generally) fewer variables. Here, $\mathcal{C}$ is a set consisting of sets of components of Z and L, and each ι_C is a pmf of the variables in C. We call the components of Z *counter variables* and the components of L *label variables*. Label variables play a central role in Part II, while histograms play no role at all. In fact, the QPLEX chain concept is no longer relevant for the key ideas in this part, but we use it as a source of examples that give rise to counter and label variables.

9.1 Subsystem QPLEX Calculus

As a first step towards overcoming this curse of dimensionality for counters, in Chap. 10, the index set $\mathcal{C}$ is a *partition* of the set of all counter and label variables.

Subsystems To stress that $\mathcal{C}$ is a partition of the set of counter and label variables in the model, we denote a generic element of $\mathcal{C}$ by σ instead of C and refer to σ as a *subsystem*. We use this terminology since groups of counter and label variables are often associated with parts of a system. We write $\iota = \{\iota_\sigma\}_{\sigma \in \mathcal{C}}$, so there is one pmf in the collection ι for each subsystem.

© The Author(s) 2025
A. B. Dieker, S. T. Hackman, *QPLEX: A Computational Modeling and Analysis Methodology for Stochastic Systems*, Springer Series in Operations Research and Financial Engineering, https://doi.org/10.1007/978-3-031-74870-7_9

Although each subsystem σ is defined as a set of counter and label variables, we also use σ as an index, as with ι_σ above. For this purpose, it is useful to identify each subsystem σ with some symbol. When used as an index set, $\mathcal{C}$ can then be taken as the set of such symbols and σ can be replaced with its corresponding index. The following examples illustrate how parts of a queueing network can be grouped into subsystems, how elements of $\mathcal{C}$ can be identified with symbols, and how these symbols are used as indices.

Example 9.1 Let $\mathcal{C}$ be the trivial partition consisting of a single element, namely, the set of all counter and label variables in the model. The collection ι contains a single pmf of all these variables. The QPLEX calculus developed in Part I corresponds to this single subsystem case. $\triangle$

Example 9.2 Consider a model of three multiserver stations with counter variables Z_1, Z_2, and Z_3 and label variables L_1, L_2, and L_3 as in Sects. 5.4 or 5.5. Let $\{Z_m, L_m\}$ for $m = 1, 2, 3$ be subsystems so that each subsystem corresponds to a station. We may identify the set $\mathcal{C}$ of these subsystems with $\{1, 2, 3\}$ so that the collection $\iota = \{\iota_1, \iota_2, \iota_3\}$ contains one pmf for each station. Each ι_m is a pmf of (Z_m, L_m) for $m = 1, 2, 3$. $\triangle$

Example 9.3 Consider the setting of the previous example, but now with the subsystems $\{Z_1, Z_2, L_1, L_2\}$ and $\{Z_3, L_3\}$. We may identify the set of subsystems $\mathcal{C}$ with $\{A, B\}$, with subsystem A corresponding to stations 1 and 2 and subsystem B corresponding to station 3. The collection $\iota = \{\iota_A, \iota_B\}$ consists of a pmf ι_A of (Z_1, Z_2, L_1, L_2) and a pmf ι_B of (Z_3, L_3). $\triangle$

Subsystem QPLEX Maps and Iterates We add another level to our diagram; see below. The top level of this diagram plays no significant role in Part II but serves as a reminder that the QPLEX map is a composition itself. For each $\sigma \in \mathcal{C}$, we introduce a *subsystem QPLEX map* $\mathcal{Q}_\sigma^{(t)}$, which takes the collection $\iota = \{\iota_{\tilde{\sigma}}\}_{\tilde{\sigma} \in \mathcal{C}}$ as input and produces a *single* pmf ι_σ' pertaining to subsystem σ. The collection of these maps can be viewed as taking the collection ι as input and producing another collection $\iota' = \{\iota_\sigma'\}_{\sigma \in \mathcal{C}}$ (highlighted in blue).

Fix some partition $\mathcal{C}$. The lower left corner of the diagram starts with some collection $\iota = \{\iota_\sigma\}_{\sigma \in \mathcal{C}}$. Following along the edge labeled "independence," we produce the pmf μ on all counter and label variables by imposing independence of the counter and label variables across subsystems, i.e., $\mu = \prod_{\sigma \in \mathcal{C}} \iota_\sigma$. Such an assumption is often referred to as a *mean field* assumption in the literature; see Sect. 10.5 for more details. The QPLEX map $\mathcal{Q}^{(t)}$ takes the pmf μ as input and generates the pmf $\hat{\mu}'$. For each σ, we then let ι_σ' be the marginal pmf of $\hat{\mu}'$ of the counter and label variables in subsystem σ.

Our objective is to "go directly across" the bottom level without having to pass through the middle level. To achieve this objective, we assume that the QPLEX map uses simple or advanced transition dynamics and that all pmf primitives except the routing pmf are "local" to each subsystem. This means that such primitives only use counter variables in a single subsystem.

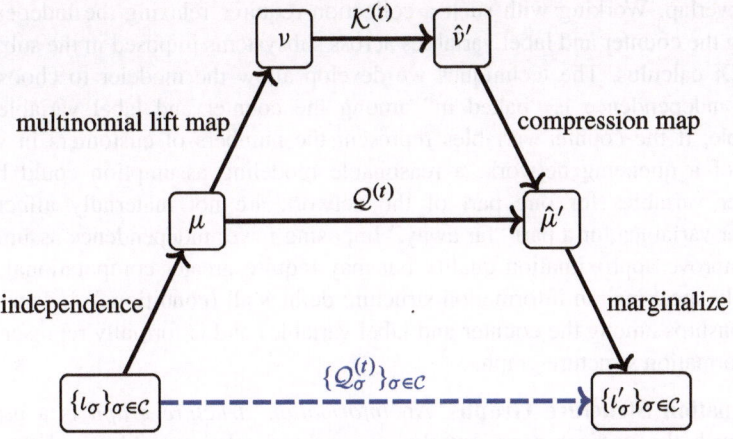

We develop an iterative scheme that generates the *subsystem iterates* $\iota_{\text{sub}}^{(t)} = \{\iota_{\sigma,\text{sub}}^{(t)}\}_{\sigma \in \mathcal{C}}$, which are calculated by iteratively applying subsystem QPLEX maps $\mathcal{Q}_{\sigma}^{(t)}$ via

$$\iota_{\sigma,\text{sub}}^{(t+1)} = \mathcal{Q}_{\sigma}^{(t)}(\iota_{\text{sub}}^{(t)}).$$

Note that the input to $\mathcal{Q}_{\sigma}^{(t)}$ is a collection of pmfs that includes pmfs for subsystems other than σ, while its output is a pmf of the random variables in σ only. To find the next iterate, instead of applying a single (full) QPLEX map as in Part I, we apply a subsystem QPLEX map for each subsystem. This can be viewed as a *decomposition approach*.

The premise of this decomposition approach is that it can be computationally much less demanding to calculate subsystem iterates than to calculate QPLEX iterates, which are pmfs of *all* counter and label variables in the model. For example, consider a network of M multiserver stations with probabilistic routing as described in Sect. 5.5, where each station is identified with a subsystem. We will see that calculating a subsystem iterate is essentially equivalent to *separately* calculating M QPLEX iterates for a multiserver single station model, with the important caveat that each station's routing pmf uses distributional summaries of the other stations. Numerical results for a challenging nine-station network with reentrant flow show that the approximation quality using this decomposition approach can be excellent.

9.2 Information Structure

In the subsequent chapters of Part II, we will generalize the subsystem QPLEX calculus by working with collections $\iota = \{\iota_C\}_{C \in \mathcal{C}}$ for which the sets $C \in \mathcal{C}$ need no longer be mutually exclusive. Thus, the scopes of the pmfs in the collection ι

may overlap. Working with such a collection requires relaxing the independence among the counter and label variables across subsystems imposed in the subsystem QPLEX calculus. The techniques we develop allow the modeler to choose how much independence is "baked in" among the counter and label variables. For example, if the counter variables represent the numbers of customers in various parts of a queueing network, a reasonable modeling assumption could be that counter variables for one part of the network are not materially affected by counter variables for a part "far away." Imposing fewer independence assumptions can improve approximation quality but may require greater computational effort. Loosely speaking, an information structure defines all (conditional) independence relationships among the counter and label variables and is formally represented by an information structure graph.

Information Structure Graphs An *information structure graph* is a graph G for which the vertices represent the counter and label variables and the edges (implicitly) define the scopes of each pmf ι_C in the collection $\iota = \{\iota_C\}_{C \in \mathcal{C}}$. (Additional technical conditions are also required for a graph to be an information structure graph; see Sect. 12.1.) Specifically, the set $\mathcal{C}$ is taken to be the set $\mathcal{C}(G)$ of maximal cliques in G so that the maximal clique C is a scope of each pmf ι_C. (Recall that a clique is a subset of vertices such that distinct vertices are adjacent, and a clique is maximal if it is not strictly contained in another clique.)

We develop an iterative scheme working with such collections $\iota = \{\iota_C\}_{C \in \mathcal{C}}$, and it is an important requirement that these pmfs be consistent. Recall that this means that the marginal pmfs corresponding to any intersection of scopes are equal for each pmf in the collection that contains this intersection in its scope. Given an information structure graph G, we say that a collection $\iota = \{\iota_C\}_{C \in \mathcal{C}(G)}$ is a *G-information collection* if it satisfies this consistency requirement. Note that when $\mathcal{C}$ is a partition, the consistency requirement for a collection $\iota = \{\iota_\sigma\}_{\sigma \in \mathcal{C}}$ associated with subsystems is automatically satisfied since the sets of counter and label variables associated with each subsystem do not intersect.

Example 9.4 Depicted below is an information structure graph with vertices corresponding to the counter and label variables associated with a network of three multiserver stations. This graph has two maximal cliques $C_1 = \{Z_1, Z_2, L_1, L_2\}$ and $C_2 = \{Z_3, L_3\}$, and the subgraphs $G[C_1]$ and $G[C_2]$ induced by these two maximal cliques are each components of the graph.

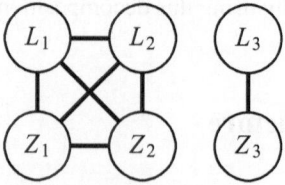

The set of maximal cliques partitions the set of all counter and label variables into two subsystems, so the consistency requirement is automatically satisfied.   △

Example 9.5 Consider a model with two counter variables, Z_1 and Z_2, and one label variable L. Two information structure graphs are depicted below.

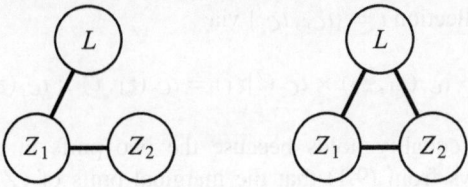

The leftmost graph G_1 has two maximal cliques $C_1 = \{Z_1, Z_2\}$ and $C_2 = \{Z_1, L\}$, which *do* overlap. A collection $\iota = \{\iota_{C_1}, \iota_{C_2}\}$ is a G_1-information collection if it satisfies the consistency requirement:

$$\sum_{z_2} \iota_{C_1}(z_1, z_2) = \sum_{\ell} \iota_{C_2}(z_1, \ell).$$

This is the only consistency requirement. For the rightmost information graph G_2, any G_2-information collection is a singleton set consisting of a pmf $\mu(z_1, z_2, \ell)$ on all variables in the model, and there are no consistency requirements. △

Lift and Marginalization Maps Given an information structure graph G, we introduce maps that "translate" between G-information collections $\iota = \{\iota_C\}_{C \in \mathcal{C}(G)}$ on the one hand and pmfs μ of all counter and label variables on the other hand, similarly to what we did in the subsystem QPLEX calculus. Given a pmf μ of all counter and label variables in the model, it is most natural to simply *marginalize* in order to obtain a G-information collection, in which case consistency is evidently guaranteed. We let I_G denote the map that takes an input pmf μ and outputs the collection of marginal pmfs of the maximal cliques in G via marginalization. We call this map the G-*marginalization map*.

We also define a map that *lifts* a G-information collection $\iota = \{\iota_C\}_{C \in \mathcal{C}(G)}$ to a pmf μ of all counter and label variables by imposing (conditional) independence relationships encoded in G. The resulting pmf is consistent with each of the pmfs in the G-information collection. We denote this map by L_G and call it the G-*lift map*. (The map L_G is not to be confused with the random variables L_1, L_2, etc.) We illustrate the map L_G for two simple examples. Its general construction is described in Chap. 11.

Example 9.6 For the information structure graph of Example 9.4, we can define a pmf μ of $(Z_1, Z_2, Z_3, L_1, L_2, L_3)$ from any G-information collection $\iota = \{\iota_{C_1}, \iota_{C_2}\}$ via

$$\mu(z_1, z_2, z_3, \ell_1, \ell_2, \ell_3) = \iota_{C_1}(z_1, z_2, \ell_1, \ell_2) \times \iota_{C_2}(z_3, \ell_3).$$

We will see that this pmf μ is $L_G(\iota)$. For this example, the map L_G simply imposes independence. △

Example 9.7 Consider a model with two counter variables, one label variable, and the leftmost information graph G_1 of Example 9.5. By imposing that Z_2 and L are conditionally independent given Z_1, we can define a pmf μ of (Z_1, Z_2, L) from any G_1-information collection $\iota = \{\iota_{C_1}, \iota_{C_2}\}$ via

$$\mu(z_1, z_2, \ell) = \iota_{C_1}(z_1, z_2) \times \iota_{C_2}(\ell|z_1) = \iota_{C_2}(z_1, \ell) \times \iota_{C_1}(z_2|z_1), \qquad (9.1)$$

where the second equality holds because the two pmfs in ι are consistent. It immediately follows from (9.1) that the marginal pmfs of (Z_1, Z_2) and (Z_1, L) under μ agree with ι_{C_1} and ι_{C_2}, respectively. We will see that this pmfy μ is $L_G(\iota)$ for this example.   △

A key property of the G-lift map is that its output has the so-called global Markov property with respect to G. This property can be formulated in terms of the graph G and generalizes the classical Markov property. In fact, the pmf μ defined in (9.1) of Example 9.7 shows that (Z_1, Z_2, L) is a Markov chain under this lifted pmf. Furthermore, as illustrated in this example, the resulting joint pmf $\mu = L_G(\iota)$ can always be represented as a product of "factors," each of which can explicitly be expressed in terms of a pmf in the G-information collection ι. The details are provided in Chap. 11. It is this explicit factorization that we exploit to achieve computational efficiencies.

Graphical QPLEX Maps and Iterates We create a similar diagram as the one for the subsystem QPLEX calculus; see below. In this diagram, we have fixed some t and two arbitrary information structure graphs G and G'. The *graphical QPLEX map* $Q^{(t)}_{G \to G'}$ (highlighted in blue) is defined as the composition of three maps. The input of the graphical QPLEX map is a G-information collection $\iota = \{\iota_C\}_{C \in \mathcal{C}(G)}$, and the output is a G'-information collection $\iota' = \{\iota'_{C'}\}_{C' \in \mathcal{C}(G')}$. Starting from the G-information collection ι in the lower left corner, we obtain a pmf μ on counters and labels by applying the G-lift map. We then apply the QPLEX map $Q^{(t)}$ to μ and obtain $\hat{\mu}'$, a new pmf of all counters and label variables. The G'-information collection ι' is obtained from $\hat{\mu}'$ by applying the G'-marginalization map. The details are provided in Chap. 13.

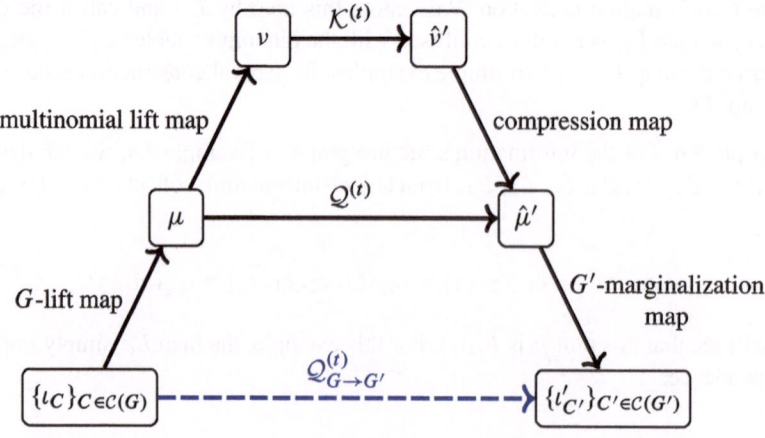

Fix some sequence of information structure graphs $\{G^{(t)} : t \geq 0\}$. Given some initial $G^{(0)}$-information collection $\iota_{\text{IS}}^{(0)}$, the iterative scheme $\iota_{\text{IS}}^{(t+1)} = \mathcal{Q}_{G^{(t)} \to G^{(t+1)}}^{(t)}(\iota_{\text{IS}}^{(t)})$ for $t \geq 0$ generates *information iterates*

$$\iota_{\text{IS}}^{(0)}, \iota_{\text{IS}}^{(1)}, \iota_{\text{IS}}^{(2)}, \ldots.$$

Each information iterate $\iota_{\text{IS}}^{(t)}$ is a $G^{(t)}$-information collection.

The collection of information structure graphs $\{G^{(t)} : t \geq 0\}$ is the one additional model primitive in Part II. Changing information structure graphs, possibly only for certain time epochs, results in different QPLEX output. A trade-off exists between approximation quality and computational efficiency in choosing an information structure graph. An information structure graph characterizes the conditional independence relationships between these variables. As we will see, adding edges to an information graph imposes *fewer* conditional independence relationships, which can improve approximation quality, albeit at potentially greater computational expense.

In the special case where there are no label variables, the top level in this diagram disappears, and the QPLEX map $\mathcal{Q}^{(t)}$ can be *any* kernel map $\mathcal{K}^{(t)}$ of a Markov chain on some Cartesian product. Although the techniques we present in this part have been known in the literature for this special case, they are not commonly used for stochastic modeling in operations research.

9.3 Circumventing the Curse of Dimensionality for Counters

The QPLEX calculus developed in Part I circumvented the curse of dimensionality for histograms by imposing structure on the transition dynamics, so that a QPLEX map can be evaluated by "going directly across" the middle level of the diagrams from this chapter without having to pass through the top level. In a similar vein, the decomposition approach circumvents the curse of dimensionality for counters by imposing independence to "go directly across" the bottom level without having to pass through the middle or top levels. In the much more general information structure setting, Chaps. 13 and 14 will similarly show how to efficiently evaluate graphical QPLEX maps (e.g., for the purpose of calculating information iterates). We call this the *graphical QPLEX calculus*. This calculus builds on the QPLEX calculus of Part I to go directly across the middle level, in combination with properties of the G-lift and G'-marginalization maps. We use the perspective of *probabilistic graphical models* [14, 19] to leverage a sizeable (and largely complete) body of existing work in the areas of algorithmic graph theory and machine learning to obtain efficient algorithms that can dramatically lower computational burdens.

Here is a brief overview of the graphical QPLEX calculus. A graphical QPLEX map can be represented by a collection of distributional programs, which are constructed as a combination of the distributional programs of the three maps, namely, L_G, $\mathcal{Q}^{(t)}$, and I_G, that define the graphical QPLEX map as a composition. Each program in this collection requires the *full* pmf μ to be calculated from ι. In Chap. 13, we refactor these programs to represent the graphical QPLEX map efficiently, and the resulting programs can be viewed as "going directly across" the bottom level of our diagram. Efficiency is understood to mean that each program: (1) No longer requires μ to be calculated but instead uses ι directly (2) Does not introduce variables that are subsequently unused (3) Suppresses unused arguments of what we call utility programs

Chapter 14 discusses techniques to calculate the output pmf of a distributional program efficiently. This calculation is a so-called exact computational inference task at its core that involves a (large) sum of products, each of which only depends on a subset of the variables. We use the so-called sum-product message passing algorithm to calculate the sums and products in a judicious order. Here is a simple illustration of the basic idea. Suppose we seek to calculate the following sum of products:

$$\sum_{x_1, x_3, x_5} f_1(x_1, x_2) \times f_2(x_2, x_3) \times f_3(x_3, x_4) \times f_4(x_4, x_5) \times f_5(x_5, x_6),$$

which is a function of x_2, x_4 and x_6. If each variable takes B possible values, then a naive calculation takes order B^6 time. However, by rewriting the above expression as

$$\left(\sum_{x_1} f_1(x_1, x_2) \right) \times \left(\sum_{x_3} f_2(x_2, x_3) \times f_3(x_3, x_4) \right) \times \left(\sum_{x_5} f_4(x_4, x_5) \times f_5(x_5, x_6) \right),$$

the three functions in parentheses can be calculated first, which takes order B^3 time, and then they can be used to calculate the sought-after sum. The result is a calculation of order B^3 instead of B^6. This example is simple enough to "eyeball" how to obtain the best organization of the sums and products. The standard machinery for probabilistic graphical models defines a general step-by-step procedure for calculating the sums and products in the best order. The computational savings can be significant with many variables and a pmf that can be written as the product of many factors.

Chapter 10
Subsystem QPLEX Calculus

This chapter develops a decomposition technique that can be used to circumvent the curse of dimensionality for counters. It can be viewed as a step towards more general techniques to be presented in subsequent chapters in Part II. The basic idea underlying the decomposition technique applies to QPLEX maps using either simple or advanced transition dynamics. To ease notational burdens, we work with QPLEX maps defined by simple transition dynamics.

We fix time $t \geq 0$ and work with a fixed QPLEX map $\mathcal{Q}^{(t)}$ on the set of pmfs of some pair (Z, L) of counter and label vectors. Recall from the previous chapter that we define a *subsystem* as a subset of the counter and label variables such that the collection of subsystems $\mathcal{C}$ partitions the set of components of Z and L. Also recall that we can think of σ as a symbol identifying part of a system when it is used as an index. We let

$$M_\sigma = \{m = 1, \dots, M : L_m \in \sigma\}$$

denote the set of activities corresponding to subsystem σ.

We write Z_σ and L_σ, respectively, for the subvectors of Z and L corresponding to subsystem σ, so that we may write $Z = (Z_\sigma)_{\sigma \in \mathcal{C}}$ and $L = (L_\sigma)_{\sigma \in \mathcal{C}}$. Note that L_σ can also be written as $(L_m)_{m \in M_\sigma}$. We use the notation Z'_σ and L'_σ for the corresponding vectors at time $t + 1$, and we furthermore use the same notational conventions for the possible values z and ℓ of Z and L.

10.1 Subsystem QPLEX Maps

In this chapter, we assume that all model primitives underlying the simple transition dynamics except the routing pmf primitive are "local" to each subsystem, formalized in Assumption 10.1 below using the notion of scope. Since a function's scope

© The Author(s) 2025
A. B. Dieker, S. T. Hackman, *QPLEX: A Computational Modeling and Analysis Methodology for Stochastic Systems*, Springer Series in Operations Research and Financial Engineering, https://doi.org/10.1007/978-3-031-74870-7_10

identifies the minimal set of "relevant" input variables and we use the vertical bar in the pmf primitives merely for notational purposes, the scope may include some of the variables to the right of the vertical bar; for instance, the scope of $\pi_m^{(t),\text{rel}}(\ell_m'|z_\sigma, \ell_m, z_\sigma')$ can include ℓ_m as well as components of the vectors z_σ and z_σ'.

Assumption 10.1 *For each subsystem $\sigma \in C$ and each of its activities $m \in M_\sigma$, the current (and next, if applicable) counters in the scopes of the size function $x_m^{(t)}(z)$, completion probability function $\gamma_m^{(t)}(z, \ell_m)$, relabeling pmf $\pi_m^{(t),\text{rel}}(\ell_m'|z, \ell_m, z')$, and join pmf $\pi_m^{(t),\text{join}}(\ell_m'|z, z')$ are associated with subsystem σ.*

In view of this assumption, we write, for $m \in M_\sigma$, $x_m^{(t)}(z_\sigma)$, $\gamma_m^{(t)}(z_\sigma, \ell_m)$, $\pi_m^{(t),\text{rel}}(\ell_m'|z_\sigma, \ell_m, z_\sigma')$, and $\pi_m^{(t),\text{join}}(\ell_m'|z_\sigma, z_\sigma')$.

Below, we reproduce the bottom part of the diagram from Sect. 9.1, where we introduce subsystem QPLEX maps. The lower left corner of the diagram starts with some collection $\iota = \{\iota_\sigma\}_\sigma$, which is input to each of the subsystem QPLEX maps. Following along the edge labeled "independence," we create the pmf μ on all counter and label variables from ι by simply setting, for all z and ℓ,

$$\mu(z, \ell) = \prod_\sigma \iota_\sigma(z_\sigma, \ell_\sigma). \tag{10.1}$$

We abbreviate this operation by $\mu = \prod_\sigma \iota_\sigma$. Under the pmf μ, the counter and label variables are independent across subsystems. The QPLEX map $\mathcal{Q}^{(t)}$ takes the pmf μ as input and generates the pmf $\hat{\mu}'$. For each σ, we then set $\iota_\sigma'(z_\sigma', \ell_\sigma') = \hat{\mu}'(z_\sigma', \ell_\sigma')$, i.e., ι_σ' is the marginal pmf of $\hat{\mu}'$ of the counter and label variables associated with subsystem σ.

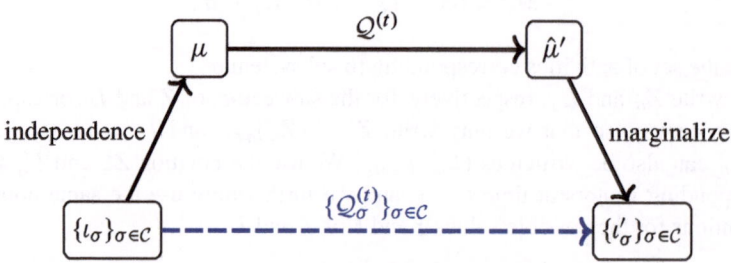

The distributional program below represents the *subsystem QPLEX map* $\mathcal{Q}_\sigma^{(t)}$ for subsystem σ. It takes a collection $\iota = \{\iota_{\tilde{\sigma}}\}_{\tilde{\sigma} \in C}$ as input and outputs a pmf $\iota_\sigma'(z_\sigma', \ell_\sigma')$ obtained via marginalization.

```
1: function SUBSYSTEMQPLEXMAP_σ^(t)(ι)
2:     (z', ℓ') ~ QPLEXMAP^(t) (∏_σ̃ ι_σ̃)
3:     return (z'_σ, ℓ'_σ)
```

We stress that under $\hat{\mu}' = \mathcal{Q}^{(t)}(\prod_\sigma \iota_\sigma)$ in the diagram, the counter and label variables are *not* necessarily independent across subsystems. The pmf $\hat{\mu}'$ in our diagram *cannot* be guaranteed to admit the product form $\prod_\sigma \iota'_\sigma$. The hat serves as a visual reminder that $\hat{\mu}'$ does not have this structure.

Evaluating the subsystem QPLEX maps produces *no* computational benefits over using the (regular) QPLEX map because a subsystem QPLEX map, by its definition, still requires working with pmfs on *all* counter and label variables. To overcome the curse of dimensionality for counters, it is necessary to evaluate the subsystem QPLEX maps *without* having to work with such pmfs. The next two sections show how to achieve this objective.

10.2 Subsystem Routing Pmfs

This section shows that the distributional program representing the subsystem QPLEX map $\mathcal{Q}_\sigma^{(t)}$ for each subsystem σ can be refactored as a distributional program that is identical in form to the distributional program that represents the QPLEX map $\mathcal{Q}^{(t)}$. Doing so exposes a *subsystem routing pmf*, which is a function of the pmfs $\iota_{\tilde{\sigma}}$ for $\tilde{\sigma} \neq \sigma$. Thus, the subsystem routing pmf for subsystem σ uses distributional summaries of the counter and label variables of the *other* subsystems. Although the program for $\mathcal{Q}_\sigma^{(t)}$ contains (much) fewer counter and label variables, these variables *do* appear in the subsystem routing pmf. As such, isolating the subsystem routing pmfs by itself does not necessarily achieve significant computational efficiencies. The efficient calculation of subsystem routing pmfs is the topic of the next section.

The distributional program for $\mathcal{Q}^{(t)}$ is reproduced below. The programs NUMBEROFCOMPLETIONS$_{\mu,m}^{(t)}$, ENTITYTYPE$_m^{(t)}$, and NEXTLABEL$_{\mu,m}^{(t)}$ used are the "global" distributional programs defined in Chap. 4. This program serves as a starting point for refactoring the distributional program representation for the subsystem QPLEX maps $\mathcal{Q}_\sigma^{(t)}$, but it also serves as a reminder of the form we seek for the programs representing these maps.

1: **function** QPLEXMAP$^{(t)}(\mu)$
2:     $z \sim \mu(\mathbf{Z} = \cdot)$
3:     **for** $m = 1, \ldots, M$ **do**
4:         $d_m \sim$ NUMBEROFCOMPLETIONS$_{\mu,m}^{(t)}(z)$
5:     $z' \sim \pi^{(t)}(\mathbf{Z}' = \cdot | z, d)$
6:     **for** $m = 1, \ldots, M$ **do**
7:         $k'_m \sim$ ENTITYTYPE$_m^{(t)}(z, d_m, z')$
8:         $\ell'_m \sim$ NEXTLABEL$_{\mu,m}^{(t)}(z, z', k'_m)$
9:     **return** (z', ℓ')

A pmf μ of the form $\mu = \prod_\sigma \iota_\sigma$ has an important property that we fully exploit. Fix some activity m and counter vector z such that $\mu(z) > 0$, and hence $\iota_\sigma(z_\sigma) > 0$

for all σ, too. We then have, for $m \in M_\sigma$, that

$$\mu(\ell_m | z) = \iota_\sigma(\ell_m | z_\sigma). \tag{10.2}$$

This is true because (10.1) implies

$$\mu(z) = \prod_\sigma \iota_\sigma(z_\sigma)$$

and because

$$\frac{\mu(z, \ell_m)}{\mu(z)} = \frac{\sum_{\{\ell_{\tilde m}\}_{\tilde m \neq m}} \prod_{\tilde\sigma} \iota_{\tilde\sigma}(z_{\tilde\sigma}, \ell_{\tilde\sigma})}{\prod_{\tilde\sigma} \iota_{\tilde\sigma}(z_{\tilde\sigma})}$$

$$= \frac{\left(\sum_{\{\ell_{\tilde m}\}_{\tilde m \in M_\sigma : \tilde m \neq m}} \iota_\sigma(z_\sigma, \ell_\sigma)\right) \times \left(\prod_{\tilde\sigma \neq \sigma} \sum_{\ell_{\tilde\sigma}} \iota_{\tilde\sigma}(z_{\tilde\sigma}, \ell_{\tilde\sigma})\right)}{\prod_{\tilde\sigma} \iota_{\tilde\sigma}(z_{\tilde\sigma})}$$

$$= \frac{\iota_\sigma(z_\sigma, \ell_m) \times \prod_{\tilde\sigma \neq \sigma} \iota_{\tilde\sigma}(z_{\tilde\sigma})}{\prod_{\tilde\sigma} \iota_{\tilde\sigma}(z_{\tilde\sigma})}$$

$$= \frac{\iota_\sigma(z_\sigma, \ell_m)}{\iota_\sigma(z_\sigma)}.$$

Under Assumption 10.1, if $\mu = \prod_\sigma \iota_\sigma$, then the distributional programs NUMBEROFCOMPLETIONS$_{\mu,m}^{(t)}(z)$, ENTITYTYPE$_m^{(t)}(z, d_m, z')$, and NEXTLABEL$_{\mu,m}^{(t)}(z, z', k'_m)$ introduced in Sect. 4.3 only use z and z' through the subvectors $z_{\sigma(m)}$ and $z'_{\sigma(m)}$, where $\sigma(m)$ is the unique subsystem containing L_m. Indeed, each of the primitives appearing in these programs is specific to the subsystem σ. Moreover, since $\mu(\ell_m | z) = \iota_{\sigma(m)}(\ell_m | z_{\sigma(m)})$ in view of (10.2), these programs only use μ through $\iota_{\sigma(m)}$. As a result, these programs have subsystem counterparts defined below for each subsystem σ and $m \in M_\sigma$.

```
1: function SUBSYSTEMNUMBEROFCOMPLETIONS_{ι_σ,m}^{(t)}(z_σ)
2:     d_m ~ bin (x_m^{(t)}(z_σ), Σ_{ℓ_m} ι_σ(ℓ_m|z_σ) × γ_m^{(t)}(z_σ, ℓ_m))
3:     return d_m
```

```
1: function SUBSYSTEMENTITYTYPE_m^{(t)}(z_σ, d_m, z'_σ)
2:     if x_m^{(t+1)}(z'_σ) = 0 then
3:         return n/a
4:     probabilityOld ← (x_m^{(t)}(z_σ) − d_m)/x_m^{(t+1)}(z'_σ)
5:     entityIsOld ~ ber(probabilityOld)
6:     if entityIsOld then
7:         return old
8:     else
9:         return new
```

1: **function** SUBSYSTEMNEXTLABEL$_{\iota_\sigma,m}^{(t)}(z_\sigma, z'_\sigma, k'_m)$
2:     **switch** k'_m **do**
3:         **case** old
4:             $\ell_m \propto \iota_\sigma(L_m = \cdot|z_\sigma) \times (1 - \gamma_m^{(t)}(z_\sigma, \cdot))$
5:             $\ell'_m \sim \pi_m^{(t),\text{rel}}(L'_m = \cdot|z_\sigma, \ell_m, z'_\sigma)$
6:         **case** new
7:             $\ell'_m \sim \pi_m^{(t),\text{join}}(L'_m = \cdot|z_\sigma, z'_\sigma)$
8:         **case** n/a
9:             $\ell'_m \sim \theta_m^0(L_m = \cdot)$
10:     **return** ℓ'_m

Fix some subsystem σ. We proceed by refactoring the distributional program representing the subsystem map $\mathcal{Q}_\sigma^{(t)}$ by substituting and appropriately adjusting the instructions in the distributional program representing the QPLEX map $\mathcal{Q}^{(t)}$. The distributional program for $\mathcal{Q}_\sigma^{(t)}$ takes a collection $\iota = \{\iota_{\tilde{\sigma}}\}_{\tilde{\sigma}}$ as input and outputs a pmf $\iota'_\sigma(z'_\sigma, \ell'_\sigma)$. It calls the program for the QPLEX map $\mathcal{Q}^{(t)}$ with input $\prod_{\tilde{\sigma}} \iota_{\tilde{\sigma}}$, so the first instruction in the program of $\mathcal{Q}^{(t)}$ can be replaced with a set of sampling instructions of the form $z_{\tilde{\sigma}} \sim \iota_{\tilde{\sigma}}(Z_{\tilde{\sigma}} = \cdot)$, as independence is implicit in such sampling instructions. Moreover, only the counter vector z'_σ and the label vector ℓ'_σ associated with subsystem σ need to be returned; all other counter and label variables need not be sampled. This corresponds to the marginalization step. Furthermore, calling the appropriate subsystem programs, we thus obtain the following distributional program representing $\mathcal{Q}_\sigma^{(t)}$.

1: **function** SUBSYSTEMQPLEXMAP$_\sigma^{(t)}(\iota)$
2:     **for all** $\tilde{\sigma}$ **do**
3:         $z_{\tilde{\sigma}} \sim \iota_{\tilde{\sigma}}(Z_{\tilde{\sigma}} = \cdot)$
4:         **for** $m \in M_{\tilde{\sigma}}$ **do**
5:             $d_m \sim$ SUBSYSTEMNUMBEROFCOMPLETIONS$_{\iota_{\tilde{\sigma}},m}^{(t)}(z_{\tilde{\sigma}})$
6:     $z'_\sigma \sim \pi^{(t)}(Z'_\sigma = \cdot|z, d)$
7:     **for** $m \in M_\sigma$ **do**
8:         $k'_m \sim$ SUBSYSTEMENTITYTYPE$_m^{(t)}(z_\sigma, d_m, z'_\sigma)$
9:         $\ell'_m \sim$ SUBSYSTEMNEXTLABEL$_{\iota_\sigma,m}^{(t)}(z_\sigma, z'_\sigma, k'_m)$
10:     **return** $(z'_\sigma, \ell'_\sigma)$

We now refactor this program so that its form is closer to that of the (global) QPLEX map $\mathcal{Q}^{(t)}$. We first introduce $d_\sigma = d_{M_\sigma}$, where we recall that M_σ is the set of activities corresponding to subsystem σ. The loop over all subsystems $\tilde{\sigma}$ in Line 2 includes the subsystem σ. For subsystems $\tilde{\sigma} \neq \sigma$, the vectors $z_{\tilde{\sigma}}$ and $d_{\tilde{\sigma}}$ generated in Lines 3–6 are only used for sampling from the routing pmf in Line 6. From the perspective of subsystem σ, Lines 2–6 generate z_σ, d_σ, and z'_σ in analogy to Lines 2–5 in the program for the QPLEX map $\mathcal{Q}^{(t)}$. We may now move all other samples into a separate distributional program SUBSYSTEMROUTING$_{\iota,\sigma}^{(t)}$, shown below, that outputs the *subsystem routing pmf* for subsystem σ. This pmf uses information about the other subsystems via $\{\iota_{\tilde{\sigma}}\}_{\tilde{\sigma} \neq \sigma}$ and (an appropriate marginal of) the global routing pmf $\pi^{(t)}(z' | z, d)$.

1: **function** SUBSYSTEMROUTING$^{(t)}_{\iota,\sigma}(z_\sigma, d_\sigma)$
2:     **for all** $\tilde{\sigma} \neq \sigma$ **do**
3:         $z_{\tilde{\sigma}} \sim \iota_{\tilde{\sigma}}(\mathbf{Z}_{\tilde{\sigma}} = \cdot)$
4:         **for** $m \in M_{\tilde{\sigma}}$ **do**
5:             $d_m \sim$ SUBSYSTEMNUMBEROFCOMPLETIONS$^{(t)}_{\iota_{\tilde{\sigma}},m}(z_{\tilde{\sigma}})$
6:     $z'_\sigma \sim \pi^{(t)}(\mathbf{Z}'_\sigma = \cdot | z, d)$
7:     **return** z'_σ

Incorporating the SUBSYSTEMROUTING$^{(t)}_{\iota,\sigma}$ program, we obtain the following refactored distributional program representing $\mathcal{Q}^{(t)}_\sigma$.

1: **function** SUBSYSTEMQPLEXMAP$^{(t)}_\sigma(\iota)$
2:     $z_\sigma \sim \iota_\sigma(\mathbf{Z}_\sigma = \cdot)$
3:     **for** $m \in M_\sigma$ **do**
4:         $d_m \sim$ SUBSYSTEMNUMBEROFCOMPLETIONS$^{(t)}_{\iota_\sigma,m}(z_\sigma)$
5:     $z'_\sigma \sim$ SUBSYSTEMROUTING$^{(t)}_{\iota,\sigma}(z_\sigma, d_\sigma)$
6:     **for** $m \in M_\sigma$ **do**
7:         $k'_m \sim$ SUBSYSTEMENTITYTYPE$^{(t)}_m(z_\sigma, d_m, z'_\sigma)$
8:         $\ell'_m \sim$ SUBSYSTEMNEXTLABEL$^{(t)}_{\iota_\sigma,m}(z_\sigma, z'_\sigma, k'_m)$
9:     **return** $(z'_\sigma, \ell'_\sigma)$

10.3   Efficient Representation via Decomposition

The subsystem routing pmf for subsystem σ introduced in the previous section contains sampling instructions for variables pertaining to the other subsystems $\tilde{\sigma} \neq \sigma$. This subsystem routing pmf needs to be calculated in order to evaluate a subsystem QPLEX map. In its current form, it is generally infeasible to do so because each sampling instruction in the program for the subsystem routing pmf requires the enumeration of the possible values of the variable being sampled. In this section, we rewrite the subsystem routing pmf with computational efficiency in mind, thus enabling efficient evaluation of subsystem QPLEX maps and therefore subsystem iterates. This is best done through examples since the specific form of the (global) routing pmf must be exploited to achieve efficiency.

We rewrite a subsystem routing pmf to correspond to a (regular) routing pmf for a model where the subsystem corresponds to the full system, with a single call to a distributional program that provides the requisite information on the other subsystems. (This distributional program can call further distributional programs, etc.) For example, the subsystem routing pmf may use the pmf of the number of arrivals the subsystem receives from the other subsystems. Such information is *endogenous* in the sense that it cannot be calculated prior to knowing the input collection $\iota = \{\iota_\sigma\}_\sigma$. Combined with the previous section, this means that a subsystem QPLEX map is a (regular) QPLEX map for just that subsystem, albeit with a routing pmf that uses an additional endogenous pmf, unlike a (regular) routing pmf. The collection of subsystem QPLEX maps "decomposes" the QPLEX map for the full model. Evaluating a subsystem QPLEX map amounts to evaluating a QPLEX map for each individual subsystem plus calculating these endogenous pmfs. We show that this decomposition approach is, in general, computationally (much) more efficient than evaluating the QPLEX map for the full model. (Chap. 14 develops a more general message passing algorithmic approach.)

The examples we work out are the flow line and the queueing network with probabilistic routing. We let all subsystems consist of one or more multiserver stations. If each station corresponds to a subsystem, the resulting distributional programs for the subsystem routing pmfs can be viewed as (minor variations of) the distributional program for the routing pmf primitive associated with the multiserver queueing model with reentrant flow, as described in Sect. 5.1. We also consider subsystems consisting of two stations.

Flow Line We revisit the flow line model from Sect. 5.4. Recall that the pmf of the number of arrivals to the first station during this period is denoted by $\alpha_1^{(t)}(a)$. Reproduced below is the distributional program for the routing pmf primitive.

1: **function** ROUTING$^{(t)}(z_1, \ldots, z_M, d_1, \ldots, d_M)$
2:     $d_0 \sim \alpha_1^{(t)}(A = \cdot)$
3:     **for** $m = 1, \ldots, M$ **do**
4:         $z_m' \leftarrow z_m - d_m + d_{m-1}$
5:     **return** $(z_1', \ldots, z_M')$

Stations as Subsystems We first consider the case when each station is considered as a subsystem, i.e., each $\sigma = \{Z_m, L_m\}$ corresponds to exactly one station m. Thus, the collection $\iota = \{\iota_1, \ldots, \iota_M\}$ consists of pmfs ι_m of (Z_m, L_m). We see from the (global) routing pmf that, if $m > 1$, given z_m, d_m, the sample of z_m' only requires a sample of d_{m-1}, the number of arrivals to station m. Thus, for $m > 1$, we define the program NUMBEROFARRIVALS$^{(t)}_{\iota_{m-1},m}$ to correspond to the pmf of the number of arrivals to station m. This pmf is endogenous because it uses ι_{m-1}.

1: **function** NUMBEROFARRIVALS$^{(t)}_{\iota_{m-1},m}()$
2:     $z_{m-1} \sim \iota_{m-1}(Z_{m-1} = \cdot)$
3:     $d_{m-1} \sim$ SUBSYSTEMNUMBEROFCOMPLETIONS$^{(t)}_{\iota_{m-1},m-1}(z_{m-1})$
4:     **return** d_{m-1}

Using this program, we can express the subsystem routing pmf as we did for the routing pmf primitive associated with the multiserver queueing model of Chap. 2 but with the number of arrivals being sampled from the aforementioned endogenous pmf. This is done below.

1: **function** SUBSYSTEMROUTING$^{(t)}_{\iota,m}(z_m, d_m)$
2:     **if** $m = 1$ **then**

3:         $d_0 \sim \alpha_1^{(t)}(A = \cdot)$
4:     **else**
5:         $d_{m-1} \sim \text{NUMBEROFARRIVALS}_{\iota_{m-1},m}^{(t)}()$
6:     $z_m' \leftarrow z_m - d_m + d_{m-1}$
7:     **return** z_m'

The diagram below visualizes the procedure for evaluating the subsystem QPLEX maps given the input collection $\iota = \{\iota_m\}$ as follows. For each $m = 1, \ldots, M$, let $\mathcal{Q}_m^{(t)}$ denote the QPLEX map associated with the multiserver queueing model from Chap. 2 with arrival pmf input $\alpha_m^{(t)}$, service duration pmf input g_m, and number of servers input n_m. In the diagram, $\alpha_1^{(t)}$ is the exogenous input, but each $\alpha_m^{(t)}$ for $m > 1$ is *endogenous* and is the output of the distributional program $\text{NUMBEROFARRIVALS}_{\iota_{m-1},m}^{(t)}$. The diagram highlights that calculating $\alpha_m^{(t)}$ for $m > 1$ uses ι_{m-1}, as indicated by the blue arrows. Note that the order in which the ι_m' are calculated is arbitrary, as the $\alpha_m^{(t)}$ are calculated using the preceding subsystem iterate.

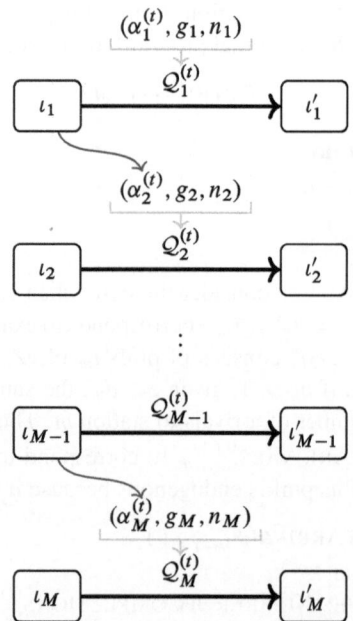

We now explain why it is generally (much) more efficient to evaluate the subsystem QPLEX maps than the (full) QPLEX map for the flow line. Evaluating the full QPLEX map requires summing over all possible values of z, d, and k' (see Sect. 4.3), so the number of summands is generally exponential in the number of stations M. The decomposition approach instead requires calculating the $M - 1$ endogenous pmfs represented by the programs $\text{NUMBEROFARRIVALS}_{\iota_{m-1},m}^{(t)}$, the workload of which is linear in the number of stations M, and then requires evaluating M QPLEX maps for a *single station* multiserver queueing model. Therefore, the subsystem QPLEX maps can be evaluated in time that is linear in M, whereas the full QPLEX map can only be calculated in time that is exponential in M.

Pairs of Stations as Subsystems Next, we consider a flow line with $M = 4$ stations, where the first two and the last two stations constitute a subsystem. To prevent a notational clash between subsystems and stations, we call the subsystem $\{Z_1, Z_2, L_1, L_2\}$ corresponding to the first two stations subsystem A and the subsystem $\{Z_3, Z_4, L_3, L_4\}$ corresponding to the last two stations subsystem B. The collection $\iota = \{\iota_A, \iota_B\}$ consists of a pmf ι_A of (Z_1, Z_2, L_1, L_2) and a pmf ι_B of (Z_3, Z_4, L_3, L_4).

Below is the distributional program representing the subsystem routing pmf of subsystem A. This program also represents the (regular) routing pmf primitive of a flow line with two stations and input to the first station, which would be used to evaluate the QPLEX map for such a system. Note that it does not use ι.

1: **function** SUBSYSTEMROUTING$_{\iota,A}^{(t)}(z_1, z_2, d_1, d_2)$
2:     $d_0 \sim \alpha_1^{(t)}(A = \cdot)$
3:     $z_1' \leftarrow z_1 - d_1 + d_0$
4:     $z_2' \leftarrow z_2 - d_2 + d_1$
5:     **return** (z_1', z_2')

The number of departures d_2 at station 2 of subsystem A is the number of "external arrivals" to station 3 from the perspective of subsystem B. The distributional program NUMBEROFARRIVALSTOB$_{\iota_A}^{(t)}$ below represents the pmf of the number of such arrivals. Note that calculating the (unconditional) pmf of d_2 requires sampling *both* z_1 and z_2 but *not* z_3 and z_4. As such, the program only uses ι via the pmf ι_A.

1: **function** NUMBEROFARRIVALSTOB$_{\iota_A}^{(t)}()$
2:     $(z_1, z_2) \sim \iota_A(Z_1 = \cdot, Z_2 = \cdot)$
3:     $d_2 \sim$ SUBSYSTEMNUMBEROFCOMPLETIONS$_{\iota_A,2}^{(t)}(z_1, z_2)$
4:     **return** d_2

Below is the distributional program that outputs the subsystem routing pmf for subsystem B. This program again represents the (regular) routing pmf primitive of a flow line system with two stations and input only to the first station, where now the pmf of the number of arrivals to the first station is given by the endogenous pmf represented by the above distributional program.

1: **function** SUBSYSTEMROUTING$_{\iota,B}^{(t)}(z_3, z_4, d_3, d_4)$
2:     $d_2 \sim$ NUMBEROFARRIVALSTOB$_{\iota_A}^{(t)}()$
3:     $z_3' \leftarrow z_3 - d_3 + d_2$
4:     $z_4' \leftarrow z_4 - d_4 + d_3$
5:     **return** (z_1', z_2')

The independence assumption imposed here is evidently less strong than the independence assumption imposed when each station comprises a subsystem. We should thus expect better approximation quality. However, it is computationally more demanding, as we now have to evaluate two QPLEX maps for flow lines with two stations, with one of these having an endogenous pmf for the number of arrivals to the first station. Even so, this decomposition approach presents significant computational savings relative to evaluating a QPLEX map for a flow line with four stations.

Networks with Probabilistic Routing We revisit the queueing network with probabilistic routing system example from Sect. 5.5. Recall that $\beta_{m\to\tilde{m}}^{(t)}$ denotes the probability that an entity at station $m \neq 0$ is routed to station $\tilde{m}$ after service completion during this period and that $\tilde{m} = 0$ serves as the source of new arrivals as well as the sink of entities exiting the system. Reproduced below is the distributional program for the routing pmf primitive.

1: **function** ROUTING$^{(t)}(z_1, \ldots, z_M, d_1, \ldots, d_M)$
2:     **for** $m = 1, \ldots, M$ **do**
3:         $\phi_{0\to m} \sim \alpha_m^{(t)}(A = \cdot)$
4:         $(\phi_{m\to 0}, \ldots, \phi_{m\to M}) \sim \text{mult}\left(d_m, \beta_{m\to\cdot}^{(t)}\right)$
5:     **for** $m = 1, \ldots, M$ **do**
6:         $a_m \leftarrow \phi_{0\to m} + \sum_{\tilde{m}\notin\{0,m\}} \phi_{\tilde{m}\to m}$
7:         $z_m' \leftarrow z_m - d_m + \phi_{m\to m} + a_m$
8:     **return** $(z_1', \ldots, z_M')$

Stations as Subsystems We first let each subsystem correspond to a station. Thus, as for the first flow line example, we let $\{Z_m, L_m\}$ be subsystems for $m = 1, \ldots, M$ and write $\iota = \{\iota_1, \ldots, \iota_M\}$ where each ι_m is a pmf of (Z_m, L_m). Since the analysis for each station is the same due to symmetry, we focus the discussion on the first station only.

The first step is to calculate the pmfs of the number of arrivals to station 1 from every other station. This is accomplished via the distributional program below by conditioning on the number of departures at the origin station.

1: **function** FLOW$_{\iota_m, m\to 1}^{(t)}()$
2:     $z_m \sim \iota_m(Z_m = \cdot)$
3:     $d_m \sim \text{SUBSYSTEMNUMBEROFCOMPLETIONS}_{\iota_m, m}^{(t)}(z_m)$
4:     $\phi_{m\to 1} \sim \text{bin}\left(d_m, \beta_{m\to 1}^{(t)}\right)$
5:     **return** $\phi_{m\to 1}$

Note that the pmf represented by this program is endogenous as it uses ι_m. Also note that Lines 5 and 6 can be replaced with the following single binomial sampling instruction:

$$\phi_{m\to 1} \sim \text{bin}\left(x_m^{(t)}(z), \sum_{\ell_m} \iota_m(\ell_m|z) \times \gamma_m^{(t)}(z, \ell_m) \times \beta_{m\to 1}^{(t)}\right)$$

The second step is to calculate the pmf of the total number of arrivals to station 1 from every other station and from the source, as shown in the distributional program below. Recall that each sampling statement is implicitly assumed to be independent. As shown in this program, the flows from the different stations to station 1 are assumed to be independent. This is consistent with the independence of the count and label variables assumed across the subsystems. In fact, in this "decomposition" approach, the internal flows to station 1 from all other stations in the network and its external arrivals are considered independent and also independent of the service durations.

1: **function** TOTALNUMBEROFARRIVALS$_{\iota,1}^{(t)}()$

2: $\quad \phi_{0 \to 1} \sim \alpha_1^{(t)}(A = \cdot)$

3: $\quad$ **for** $m = 2, \ldots, M$ **do**

4: $\quad\quad \phi_{m \to 1} \sim \text{FLOW}_{\iota_m, m \to 1}^{(t)}()$

5: $\quad a_1 \leftarrow \phi_{0 \to 1} + \sum_{m > 1} \phi_{m \to 1}$

6: $\quad$ **return** a_1

Below is the distributional program that represents the subsystem routing pmf for the first station. It matches the routing pmf primitive for a single multiserver station with reentrant flow (see Sect. 5.1), except that here, the number of external arrivals to the station is sampled from an *endogenous* pmf that depends on ι.

1: **function** SUBSYSTEMROUTING$_{\iota,1}^{(t)}(z_1, d_1)$

2: $\quad a_1 \sim \text{TOTALNUMBEROFARRIVALS}_{\iota,1}^{(t)}()$

3: $\quad \phi_{1 \to 1} \sim \text{bin}\left(d_1, \beta_{1 \to 1}^{(t)}\right)$

4: $\quad z_1' \leftarrow z_1 - d_1 + \phi_{1 \to 1} + a_1$

5: $\quad$ **return** z_1'

The decomposition procedure is visualized below for the case of three stations. The order in which the subsystem QPLEX maps are evaluated again does not matter.

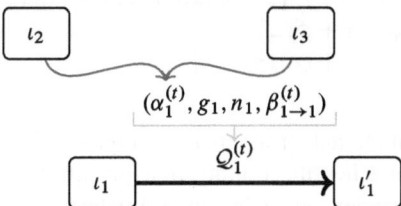

We now explain why the subsystem QPLEX maps can be evaluated much more efficiently than the (full) QPLEX maps, especially if the number of stations M is large. Evaluating the full QPLEX map requires summing over all possible values of z and d, as well as the flow variables $\phi_{m \to \tilde{m}}$. This computational effort is exponential in the number of stations M. The decomposition approach instead requires calculating the flow pmfs; this needs to be done for every pair of stations.

The distributional program $\text{TOTALNUMBEROFARRIVALS}_{\iota,1}^{(t)}$ represents the pmf that is a convolution of M pmfs, so it can be calculated efficiently for each of the stations in the network. Finally, we need to evaluate M subsystem QPLEX maps for multiserver stations with reentrant flow. For moderate values of M, the calculation of these subsystem QPLEX maps is often the computational bottleneck. Therefore, the overall computational effort is linear in M.

Pairs of Stations as Subsystems   Next, we consider the case of four stations denoted by 1, 2, 3, and 4. Here, stations 1 and 2 constitute a subsystem, which we call subsystem A, and stations 3 and 4 constitute another subsystem, which we call subsystem B. Some of the forthcoming programs can be simplified if a specific network structure is assumed, as we did with the flow line system, but we consider a general case here. In view of this generality, by symmetry, it suffices to focus on subsystem A.

We start with a distributional program that outputs the endogenous pmf of the number of arrivals to each of the stations of subsystem A. After generating the numbers of departures from stations 3 and 4 (Lines 2–4), the program generates the flow into stations 1 and 2, both external (Lines 5 and 6) and from subsystem B (Lines 7 and 8). The sum of these is the total number of arrivals to subsystem A (Lines 9 and 10). Note that the pmf output by this program is endogenous as it uses ι_B.

1: **function** $\text{NUMBEROFARRIVALSTOA}_{\iota_B}^{(t)}()$

2:    $(z_3, z_4) \sim \iota_B(Z_3 = \cdot, Z_4 = \cdot)$

3:    $d_3 \sim \text{SUBSYSTEMNUMBEROFCOMPLETIONS}_{\iota_B,3}^{(t)}(z_3, z_4)$

4:    $d_4 \sim \text{SUBSYSTEMNUMBEROFCOMPLETIONS}_{\iota_B,4}^{(t)}(z_3, z_4)$

5:    $\phi_{0\rightarrow 1} \sim \alpha_1^{(t)}(A = \cdot)$

6:    $\phi_{0\rightarrow 2} \sim \alpha_2^{(t)}(A = \cdot)$

7:    $(\phi_{3\rightarrow 1}, \phi_{3\rightarrow 2}, -) \sim \text{mult}\left(d_3, (\beta_{3\rightarrow 1}^{(t)}, \beta_{3\rightarrow 2}^{(t)}, 1 - \beta_{3\rightarrow 1}^{(t)} - \beta_{3\rightarrow 2}^{(t)})\right)$

8:    $(\phi_{4\rightarrow 1}, \phi_{4\rightarrow 2}, -) \sim \text{mult}\left(d_4, (\beta_{4\rightarrow 1}^{(t)}, \beta_{4\rightarrow 2}^{(t)}, 1 - \beta_{4\rightarrow 1}^{(t)} - \beta_{4\rightarrow 2}^{(t)})\right)$

9:    $a_1 \leftarrow \phi_{0\rightarrow 1} + \phi_{3\rightarrow 1} + \phi_{4\rightarrow 1}$

10:   $a_2 \leftarrow \phi_{0\rightarrow 2} + \phi_{3\rightarrow 2} + \phi_{4\rightarrow 2}$

11:   **return** (a_1, a_2)

Below is the distributional program that outputs the subsystem routing pmf for subsystem A. It matches the routing pmf primitive for a network with two stations and arbitrary (independent increment) external arrivals. After generating the number of arrivals from outside subsystem A (Line 2), we calculate the flows within subsystem A (Lines 3 and 4). Lines 5–6 use standard inventory balance to determine z_1' and z_2'.

1: **function** $\text{SUBSYSTEMROUTING}_{\iota,A}^{(t)}(z_1, z_2, d_1, d_2)$

2:    $(a_1, a_2) \sim \text{NUMBEROFARRIVALSTOA}_{\iota_B}^{(t)}()$

3:    $(\phi_{1\rightarrow 1}, \phi_{1\rightarrow 2}, -) \sim \text{mult}\left(d_1, (\beta_{1\rightarrow 1}^{(t)}, \beta_{1\rightarrow 2}^{(t)}, 1 - \beta_{1\rightarrow 1}^{(t)} - \beta_{1\rightarrow 2}^{(t)})\right)$

4:    $(\phi_{2\to1}, \phi_{2\to2}, -) \sim \text{mult}\left(d_2, (\beta^{(t)}_{2\to1}, \beta^{(t)}_{2\to2}, 1 - \beta^{(t)}_{2\to1} - \beta^{(t)}_{2\to2})\right)$

5:    $z'_1 \leftarrow z_1 - d_1 + a_1 + \phi_{1\to1} + \phi_{2\to1}$

6:    $z'_2 \leftarrow z_2 - d_2 + a_2 + \phi_{1\to2} + \phi_{2\to2}$

7:    **return** (z'_1, z'_2)

As for the flow line, the weaker independence assumption imposed here relative to the case where each station comprises a subsystem should result in better approximation quality. However, it is computationally more demanding, as we now have to evaluate two QPLEX maps for two-station networks with probabilistic routing, and each map uses an endogenous (bivariate) pmf for the number of arrivals to each of the stations.

10.4 Numerical Results

We show some numerical experiments using an instance of a multiserver queueing network with probabilistic routing, as shown below. A single server version of this network appears in a paper by Kuehn [15], one of the first investigations into the performance of stochastic networks via decomposition into smaller networks. It is challenging to analyze this network because of the reentrant flow from station E to itself, from station H to station F, and from station I back to station D. The routing probabilities are displayed in the matrix shown below the network. Note that not all row sums are 1; the complementary probability is the probability of leaving the system. External arrivals are split equally over stations A, B, and C.

The time horizon is 48 hours. Each time period represents 0.05 hours (or 3 minutes). The QPLEX output consists of 8640 distributions (9 stations, $48 \times 20 = 960$ distributions per station).

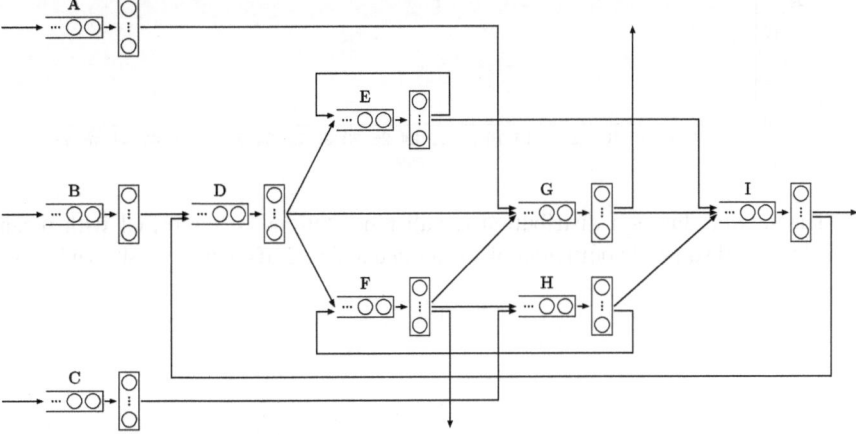

	A	B	C	D	E	F	G	H	I
A	0	0	0	0	0	0	1	0	0
B	0	0	0	1	0	0	0	0	0
C	0	0	0	0	0	0	0	1	0
D	0	0	0	0	0.3	0.6	0.1	0	0
E	0	0	0	0	0.5	0	0	0	0.5
F	0	0	0	0	0	0	0.2	0.1	0
G	0	0	0	0	0	0	0	0	0.3
H	0	0	0	0	0	0.2	0	0	0.8
I	0	0	0	0.3	0	0	0	0	0

We may work with a single *total* arrival process that is split uniformly across stations A, B, and C due to the arrivals being Poisson. Shown below is the total arrival rate function. The average total arrival rate is $\lambda_{avg} = 27\frac{1}{3}$/hr, and the maximum total arrival rate is $\lambda_{max} = 40$/hr. The peak over average is 1.46.

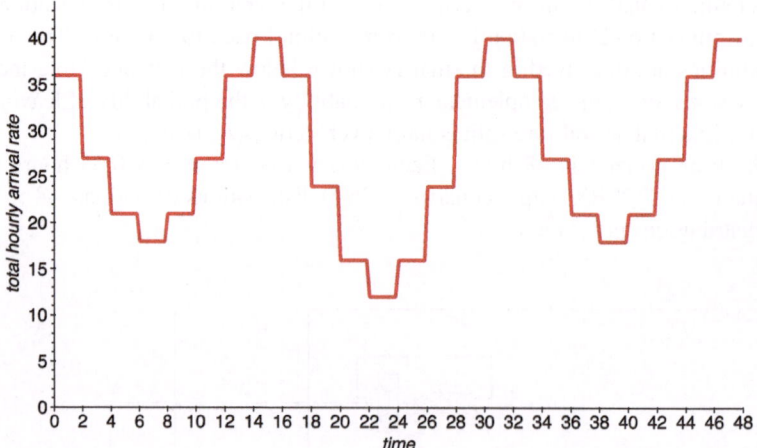

The service duration distribution for all nine stations is lognormal with mean equal to 1 and squared coefficient of variation equal to 2. Its density is shown below.

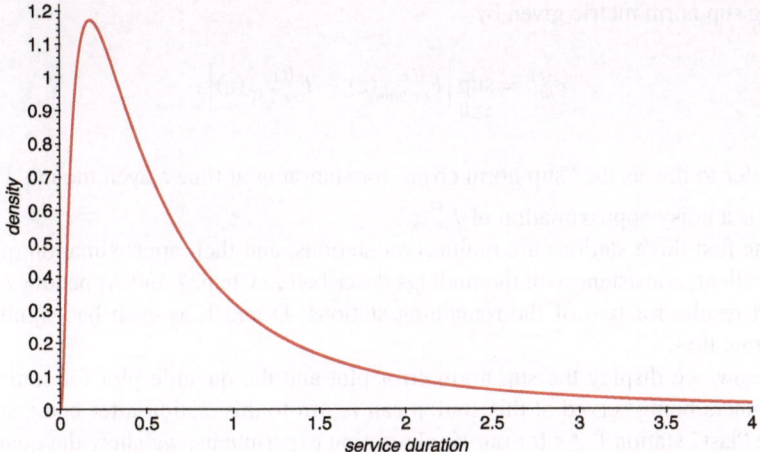

To set the number of servers at each station, we first solve the traffic equations using the average station arrival rates to find the average effective load $\rho_{avg, eff}$ for each station. We then set the number of servers at each station to the largest integer less than or equal to

$$\left(\rho_{avg, eff} + 1.96 \times \sqrt{\rho_{avg, eff}}\right)$$

using a stationary version of the familiar square-root staffing formula; see, for instance, [9, 31] for background. We have chosen this allocation to prevent excessive delays *on average* so that the time-varying arrival rate results in temporary overload at each station (possibly at different times due to lag effects). The table below shows the average effective loads and the number of servers for each station.

	Station						
	A–C	D	E	F	G	H	I
Average effective load	9.44	14.44	8.66	10.77	13.04	10.52	16.66
Number of servers	15	21	14	17	20	17	24

Although the service duration pmf is identical across the stations, the effective loads are not identical, which explains why the numbers of servers at the stations are not the same. The methods for discretization and truncation are the same as for the single-station experiments described in Appendix A.

Approximation Quality Let $F_{m,\text{sub}}^{(t)}$ and $F_{m,1M}^{(t)}$ denote, respectively, the cumulative distribution functions of the number of customers at station m at time t generated from the subsystem iterate and by simulation using one million replications, the latter of which we treat as the true cumulative distribution function $F_{m,\infty}^{(t)}$. One measure of the difference between these two distributions is the exceptionally strong sup norm metric given by

$$\epsilon_m^{(t)} = \sup_{z \geq 0} \left| F_{m,\text{sub}}^{(t)}(z) - F_{m,1M}^{(t)}(z) \right|.$$

We refer to this as the "sup norm error" for station m at time t even though $F_{m,1M}^{(t)}$ itself is a noisy approximation of $F_{m,\infty}^{(t)}$.

The first three stations are multiserver stations, and their approximation quality is excellent, consistent with the findings described in Chap. 2 and Appendix A. We report results for two of the remaining stations, D and I, as each has significant reentrant flow.

Below, we display the sup norm error plot and the quantile plot for station D. Customers being served at this station can return to this station after being served at the "last" station I. As for our single-station experiments, we show the quantiles 0.5, 0.95, 0.99, and 0.995 as a function of time. The quality is excellent.

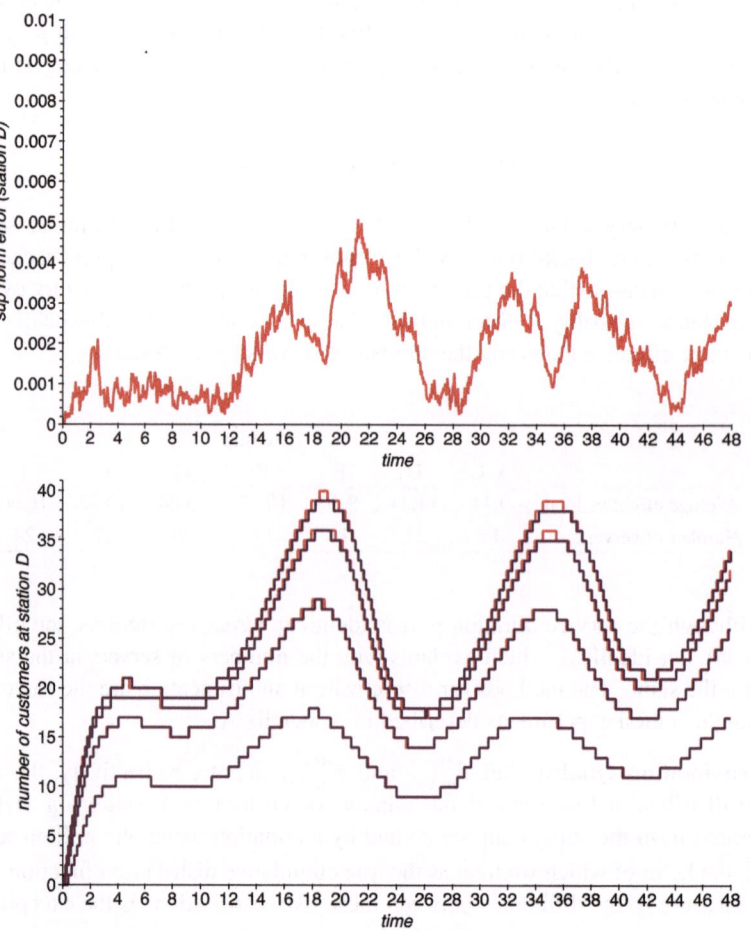

We show analogous plots for station I below. Once again, the quality is excellent.

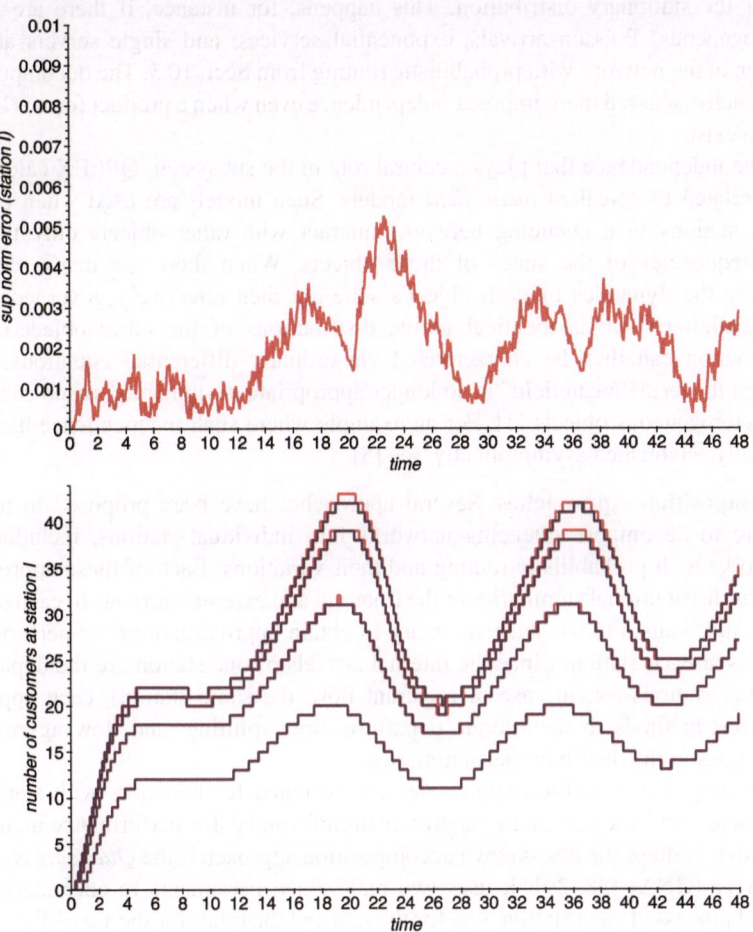

10.5 Bibliographical Notes

There are many excellent texts on queueing networks, each with its own focus. We mention [23] with a point process perspective, [6] with a scaling perspective, [7] with a fluid model and stability perspective, and [27] with a scheduling and control perspective. In this section, we focus on aspects that have a direct bearing on the examples we worked out in Sect. 10.3.

Product Forms and Independence Some Markov models have a stationary distribution that factorizes as a product over state coordinates. Such stationary distributions are known as *product forms* and are closely related to time-reversibility; see the latest edition of Kelly's 1979 book [13] for details. If the state space is a

Cartesian product, a product form implies that the state coordinates are independent under the stationary distribution. This happens, for instance, if there are (time-homogeneous) Poisson arrivals, exponential services, and single servers at each station of the network with probabilistic routing from Sect. 10.3. The decomposition approach discussed there imposes independence even when a product form is known not to exist.

The independence that plays a central role in the subsystem QPLEX calculus is also related to so-called *mean field* models. Such models are used when objects (e.g., stations in a queueing network) interact with other objects only through the frequencies of the states of those objects. When there are many identical objects, the dynamics of each object's state are then effectively governed by its distribution (which is identical to the distributions of the other objects). This distribution can then be characterized via ordinary differential equations. Even though the term "mean field" is no longer appropriate, a similar approach can work for heterogeneous objects [1]. For an example where such independence has been formally established asymptotically, see [5].

Decomposition Approaches Several approaches have been proposed in the literature to decompose queueing networks into individual stations, including for networks with probabilistic routing and their variations. Each of these approaches creates distributional summaries of the (internal and external) arrivals to each station and a mechanism to use each summary to obtain approximations for performance metrics at each station. Since the internal arrivals of one station are the departures at other stations (or, in case of reentrant flow, the same station), each approach develops methods to incorporate departure, flow splitting, and flow aggregation operations in the distributional summaries.

Existing decomposition approaches are designed for networks with stationary parameters and use parametric approximation formulas for performance metrics at a station. Perhaps the best-known decomposition approach is the *Queueing Network Analyzer* (QNA) [28, 29]. It uses two real-valued parameters to characterize the arrival process at each station, one for the rate and the other for the variability. (This is also done in the paper [15] we encountered in the previous section, which is a precursor to this work.) A different recently introduced decomposition approach is the *Robust Queueing Network Analyzer* (RQNA) [32], which is designed for networks of single server stations. (RQNA is so named because it combines a robust optimization perspective with scaling results.) It uses a real-valued function of time instead of a single number to represent the variability of the arrival process at each station.

The subsystem QPLEX calculus results in a different decomposition approach. It uses the distributional programs SUBSYSTEMNUMBEROFCOMPLETIONS$_{l_m,m}^{(t)}$, FLOW$_{l_m,m\to 1}^{(t)}$, and TOTALNUMBEROFARRIVALS$_{l,1}^{(t)}$ for the departure, flow splitting, and flow aggregation operations to generate *distributions* of the total number of arrivals to each station for each time period. It then applies the QPLEX calculus for a single station to obtain the desired distributions of the number of customers at each station over time. The distributions produced by the calculus are not exogenously specified but are obtained iteratively upon substituting the subsystem iterates $\iota_{sub}^{(t)}$ for ι in these distributional programs. As such, it captures "temporal dependencies" in the arrival process to each station endogenously. This decomposition approach is not limited to subsystems corresponding to single stations; see [8] for a different approach with subsystems that form bottlenecks.

Chapter 11
Conditional Independence

This chapter provides the requisite foundations for the remaining chapters in Part II and for some of the results established in Part III. We use graphs to represent conditional independence among counter and label variables. This leads to a *probabilistic graphical model*, defined via a graph $G = (V, E)$ in which the vertex set V is the set of counter and label variables, and the edge set E effectively determines the degree of conditional independence among these variables.

We use the *global Markov property* to connect the graph-theoretic notion of separation with the probabilistic notion of conditional independence. There are two key reasons why we use the global Markov property. The first reason is that pmfs with this property can be characterized with a very convenient structural representation. The second reason is that we can leverage the existing literature on efficient algorithms to dramatically lower the computational burdens of calculating information iterates. The graphical QPLEX calculus that underlies the iterative scheme for these information iterates is the subject of Chaps. 13 and 14.

This chapter considers conditional independence for arbitrary random variables, not necessarily counter and label variables. It is self-contained (apart from mathematical preliminaries), and no QPLEX-specific terminology plays a role. The appendix to this chapter provides proofs of several propositions that require more technical arguments.

We consider graphs for which the vertex set consists of random variables, but these are *not* random labeled graphs in a graph-theoretic sense. Indeed, we will not evaluate these random variables at outcomes in some underlying sample space. As such, the elements of the vertex set can be viewed as tokens representing the random variables. For each of the random variables, we fix some set in which it takes values, which need not be the same for each random variable and can be infinite. We assume that these random variables are ordered, so that (say) marginal pmfs can be indexed by subsets of the vertex set. This ordering can be arbitrary.

Although it is customary in graph theory to use lower case letters for vertices and upper case letters for sets of vertices, we do not adopt this convention since,

© The Author(s) 2025

A. B. Dieker, S. T. Hackman, *QPLEX: A Computational Modeling and Analysis Methodology for Stochastic Systems*, Springer Series in Operations Research and Financial Engineering, https://doi.org/10.1007/978-3-031-74870-7_11

the vertices we work with are random variables and we use upper case letters to represent them. We alert the reader to infer the meaning of a symbol from the lower or upper case letter we use.

We use the following notation. Given a set V of random variables, let $\mathcal{P}_V$ denote the set of all possible (joint) pmfs of the random variables represented in V. For a nonempty set $A \subseteq V$ and any $B \subseteq V$, $p_{A|B}$ denotes the pmf of the variables represented in A given the variables represented in B under $p \in \mathcal{P}_V$. For such a set A, we write p_A as shorthand for $p_{A|\emptyset}$. Throughout, it is understood that identities involving conditional pmfs hold only when they are well-defined, with the caveat that the multiplication of undefined probabilities by zero should be interpreted as zero. A statement such as $p_{A|B} = p_A$ should be interpreted as an identity for all possible values of the variables in B. We also write $\sum_A p$ for $p_{V \setminus A}$.

11.1 Global Markov Property

We begin by defining the probabilistic notion of conditional independence associated with a collection of pairwise disjoint sets in V.

Definition 11.1 For any pairwise disjoint sets $B_1, \ldots, B_K, S \subseteq V$ with $K \geq 2$, where S may be empty but $B_1, \ldots, B_K$ must be nonempty, we say that $B_1, \ldots, B_K$ are *conditionally independent given S under $p \in \mathcal{P}_V$* if

$$p_{B_1 \cup \cdots \cup B_K | S} = p_{B_1 | S} \times \cdots \times p_{B_K | S}. \tag{11.1}$$

As an immediate consequence of this definition, which we use repeatedly in what follows, if B_1 and B_2 are conditionally independent given S under $p \in \mathcal{P}_V$, then we have that

$$p_{B_1 | B_2 \cup S} = p_{B_1 | S}. \tag{11.2}$$

This is readily seen from the definition of conditional probability and (11.1) for $K = 2$:

$$p_{B_1 | B_2 \cup S} = \frac{p_{B_1 \cup B_2 \cup S}}{p_{B_2 \cup S}} = \frac{p_{B_1 \cup B_2 | S}}{p_{B_2 | S}} = p_{B_1 | S}.$$

Next, we define the graph-theoretic notion of separation for an undirected graph $G = (V, E)$ with vertex set V.

Definition 11.2 For any pairwise disjoint sets $B_1, B_2, S \subseteq V$, where S may be empty but B_1 and B_2 must be nonempty, S *separates B_1 and B_2 in G* if there is no path from B_1 to B_2 in $G - S$.

Note that the empty set separates the vertex sets of any two components of G. If $(v_1, v_2) \notin E$ for two distinct vertices v_1 and v_2, then $V \setminus \{v_1, v_2\}$ separates $\{v_1\}$ and

$\{v_2\}$. Moreover, for any vertex v, the set of neighbors of v separates $\{v\}$ and the set of the other vertices. However, sets satisfying this definition need not be of these forms, as seen from the examples given later in this section.

The global Markov property concept connects the probabilistic notion of conditional independence with the graph-theoretic notion of separation.

Definition 11.3 A pmf $p \in \mathcal{P}_V$ has the *global Markov property* with respect to the undirected graph $G = (V, E)$ if, for any pairwise disjoint sets $B_1, \ldots, B_K, S \subseteq V$ with $K \geq 2$, where S may be empty but $B_1, \ldots, B_K$ must be nonempty, $B_1, \ldots, B_K$ are conditionally independent given S whenever S separates each pair of $B_1, \ldots, B_K$.

We let $\mathcal{M}_G$ denote the set of pmfs with the global Markov property with respect to the graph G. If G is the complete graph, then there are no conditional independence relationships and $\mathcal{M}_G = \mathcal{P}_V$. If the graph G has no edges, then $S = \emptyset$ separates every pair of vertices and the random variables represented by the vertices are (unconditionally) independent under any $p \in \mathcal{M}_G$.

Example 11.1 Consider the set of pmfs $\mathcal{M}_G$ for random variables X_1, X_2, and X_3, where G is depicted below.

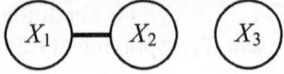

The set $S = \emptyset$ separates $B_1 = \{X_1, X_2\}$ and $B_2 = \{X_3\}$ in G, so (X_1, X_2) is independent of X_3 under any pmf $p \in \mathcal{M}_G$. It follows from (11.1) that, for $p \in \mathcal{M}_G$,

$$p(x_1, x_2, x_3) = p(x_1, x_2) \times p(x_3).$$

Since no other sets satisfy Definition 11.2, this is the only conditional independence relationship that holds for every $p \in \mathcal{M}_G$.   △

Example 11.2 Consider the set of pmfs $\mathcal{M}_G$ for random variables X_1, X_2, X_3, and X_4, where G is depicted below.

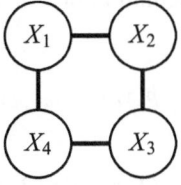

The set $S = \{X_1, X_3\}$ separates $B_1 = \{X_2\}$ and $B_2 = \{X_4\}$ in G, and so X_2 and X_4 are conditionally independent given (X_1, X_3) under any pmf $p \in \mathcal{M}_G$. It follows from (11.1) that, for $p \in \mathcal{M}_G$,

$$p(x_2, x_4 | x_1, x_3) = p(x_2 | x_1, x_3) \times p(x_4 | x_1, x_3). \qquad (11.3)$$

By symmetry, X_2 and X_4 are conditionally independent given (X_1, X_3) under any pmf $p \in \mathcal{M}_G$, and so we also have

$$p(x_1, x_3 | x_2, x_4) = p(x_1 | x_2, x_4) \times p(x_3 | x_2, x_4). \tag{11.4}$$

Since no other sets satisfy Definition 11.2, these are the only conditional independence relationships that hold for every $p \in \mathcal{M}_G$.   △

The next two examples show how the global Markov property on graphs relates to Markov chains.

Example 11.3 Consider the set of pmfs $\mathcal{M}_G$ for random variables $X_1, \ldots, X_N$, where G is a so-called *path graph*, an instance of which is depicted below with $N = 5$ random variables.

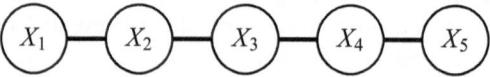

For each $n = 2, \ldots, N - 1$, the set $S = \{X_n\}$ separates $B_1 = \{X_1, \ldots, X_{n-1}\}$ and $B_2 = \{X_{n+1}, \ldots, X_N\}$. Interpreting n as a time epoch, we see that, under each $p \in \mathcal{M}_G$, the future $(X_{n+1}, \ldots, X_N)$ is conditionally independent of the past $(X_1, \ldots, X_{n-1})$ given the present X_n. Applying (11.2) yields

$$p(x_{n+1}, \ldots, x_N | x_1, \ldots, x_n) = p(x_{n+1}, \ldots, x_N | x_n).$$

Thus, $X_1, \ldots, X_N$ is a Markov chain.

The Chapman-Kolmogorov equations for the transition probabilities can also be derived from the global Markov property. Indeed, pick $n, m, k \geq 1$ with $n + m + k \leq N$. The set $S = \{X_{n+m}\}$ separates the sets $B_1 = \{X_n\}$ and $B_2 = \{X_{n+m+k}\}$, so X_n and X_{n+m+k} are conditionally independent given X_{n+m} under any $p \in \mathcal{M}_G$. By (11.2), we thus obtain that

$$p(x_{n+m+k} | x_n, x_{n+m}) = p(x_{n+m+k} | x_{n+m}).$$

Upon combining this with the definition of conditional probability and the chain rule, we obtain

$$p(x_{n+m+k} | x_n) = \sum_{x_{n+m}} p(x_{n+m}, x_{n+m+k} | x_n)$$

$$= \sum_{x_{n+m}} p(x_{n+m} | x_n) \times p(x_{n+m+k} | x_n, x_{n+m})$$

$$= \sum_{x_{n+m}} p(x_{n+m} | x_n) \times p(x_{n+m+k} | x_{n+m}),$$

as required.   △

Example 11.4 Consider the "strong product" of a complete graph with two vertices and a path graph with N vertices, which is depicted below with $N = 3$. (G is constructed by replacing each vertex in the path graph with copies of the complete graph and then inserting all possible edges between the complete graphs that are adjacent in the path graph.) This example studies the set $\mathcal{M}_G$ of pmfs for random variables $X_{1,1}, X_{1,2}, \ldots, X_{N,1}, X_{N,2}$.

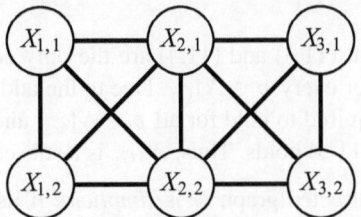

For each $n = 2, \ldots, N - 1$, the set $S = \{X_{n,1}, X_{n,2}\}$ separates the sets

$$B_1 = \{X_{1,1}, X_{1,2}, \ldots, X_{n-1,1}, X_{n-1,2}\}$$

and

$$B_2 = \{X_{n+1,1}, X_{n+1,2}, \ldots, X_{N,1}, X_{N,2}\}.$$

For $p \in \mathcal{M}_G$, the vectors

$$(X_{1,1}, X_{1,2}, \ldots, X_{n-1,1}, X_{n-1,2}), \; (X_{n+1,1}, X_{n+1,2}, \ldots, X_{N,1}, X_{N,2})$$

are conditionally independent given $(X_{n,1}, X_{n,2})$. In other words,

$$(X_{1,1}, X_{1,2}), \ldots, (X_{N,1}, X_{N,2})$$

is a Markov chain.   △

The next lemma shows that adding edges enlarges the set of pmfs with the global Markov property with respect to the resulting graph.

Lemma 11.4 *If $G_1 = (V_1, E_1)$ and $G_2 = (V_2, E_2)$ satisfy $V_1 = V_2$ and $E_1 \subseteq E_2$, then we have that $\mathcal{M}_{G_1} \subseteq \mathcal{M}_{G_2}$.*

To see why this is true, consider a graph G_2 constructed from another graph G_1 by adding one or more edges. Since *fewer* sets satisfy Definition 11.2 for G_2 than for G_1, there are *fewer* conditional independence requirements for $p \in \mathcal{M}_{G_2}$ than for $p \in \mathcal{M}_{G_1}$. Therefore, it is weaker to require that $p \in \mathcal{M}_{G_2}$ than it is to require that $p \in \mathcal{M}_{G_1}$.

Example 11.5 Consider the sets of pmfs $\mathcal{M}_{G_1}$ and $\mathcal{M}_{G_2}$ for random variables X_1, X_2, X_3, and X_4, where G_1 and G_2 are depicted below.

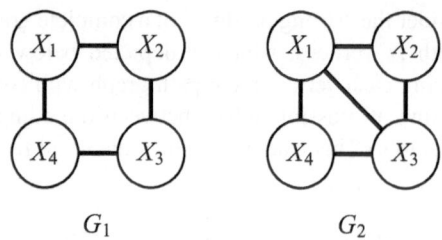

$$G_1 \qquad\qquad G_2$$

Example 11.2 shows that (11.3) and (11.4) are the only conditional independence relationships that hold for every $p \in \mathcal{M}_{G_1}$. Due to the additional edge between X_1 and X_3, (11.4) is *not* required to hold for all $p \in \mathcal{M}_{G_2}$, and so $\mathcal{M}_{G_2}$ consists of all pmfs p for which only (11.3) holds. Thus, $\mathcal{M}_{G_1}$ is a subset of $\mathcal{M}_{G_2}$.   △

A vertex v in an undirected graph G is *simplicial* if its neighbors in G form a clique. Removing a simplicial vertex v from G preserves the global Markov property of the corresponding marginal pmf.

Lemma 11.5 *If v is a simplicial vertex of an undirected graph G, then $p \in \mathcal{M}_G$ implies that $p_{V \setminus \{v\}} \in \mathcal{M}_{G - \{v\}}$.*

To see why this is true, suppose that $K \geq 2$ and $B_1, \ldots, B_K, S$ satisfy the conditions in Definition 11.3 for the graph $G - \{v\}$, i.e., S separates each pair of $B_1, \ldots, B_K$ in $G - \{v\}$. We must show that $B_1, \ldots, B_K$ are conditionally independent given S under $p_{V \setminus \{v\}}$. This fact immediately follows if we show that S separates each pair of $B_1, \ldots, B_K$ in G, too, since $p \in \mathcal{M}_G$ then implies that $B_1, \ldots, B_K$ are conditionally independent given S under p, hence $p_{V \setminus \{v\}}$. If, to the contrary, S did not separate each pair of $B_1, \ldots, B_K$ in G, then v must be adjacent to both a vertex in B_{k_1} and a vertex in B_{k_2} for some $k_1 \neq k_2$. Since v is simplicial, there must be an edge between these vertices, which contradicts the assumption that S separates B_{k_1} and B_{k_2} in $G - \{v\}$.

11.2   Triangulated Graphs

A graph is called *triangulated* (or *chordal*) if every cycle of length four or more possesses a *chord*, namely, an edge that is not part of the cycle but connects two vertices of the cycle.

Example 11.6 Consider the graph below.

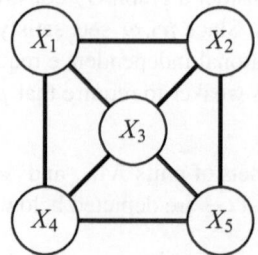

This graph is not triangulated, because the cycle $X_1 - X_2 - X_5 - X_4 - X_1$ has no chord.   △

We next discuss how to test whether a graph is triangulated. A graph is *recursively simplicial with respect to an ordering of the vertices* if each subsequent vertex is simplicial in the subgraphs that remain after eliminating the vertices in the given order. A graph is *recursively simplicial* if there exists an ordering of the vertices such that it is recursively simplicial with respect to that ordering.

Tarjan and Yannakakis [25] show that it can be verified in linear time in $|V|+|E|$ whether G is recursively simplicial. The following proposition is well-known; see, for example, [4, Thm. 2.3].

Proposition 11.6 *A graph G is triangulated if and only if it is recursively simplicial.*

Example 11.7 Consider the graph below.

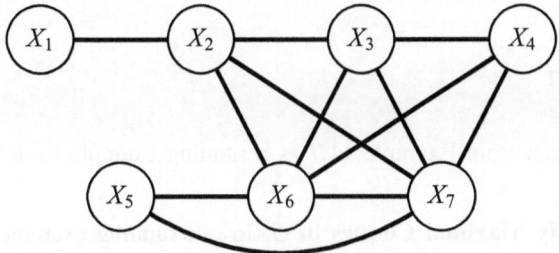

This graph is triangulated, since it is recursively simplicial (for instance) with respect to the ordering $X_1, X_5, X_4, X_6, X_7, X_3, X_2$.   △

Example 11.8 Consider the graph below.

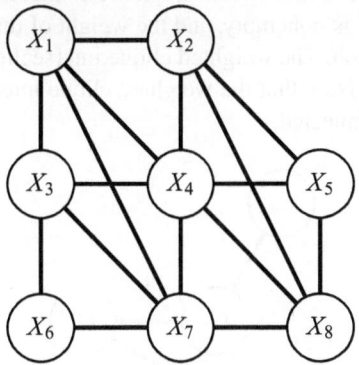

This graph is not triangulated. Vertices X_5 and X_6 are each simplicial, and vertex X_3 becomes simplicial in the subgraph that remains after removing each of these vertices. The subgraph that remains after removing these three vertices is isomorphic to the graph of Example 11.6, i.e., the cycle $X_1 - X_2 - X_8 - X_7 - X_1$ does not have a chord, and none of these vertices is simplicial.   △

11.3 Junction In-Trees

The next section provides a representation of pmfs satisfying the global Markov property with respect to a triangulated graph G. This representation relies on a so-called *junction in-tree* for G. The vertices of a junction in-tree are the maximal cliques in G. The latter are sets of vertices in G, and we will exploit relationships between these sets. From the perspective of a junction in-tree, however, they can be viewed as tokens.

In this section, we show how to construct a junction in-tree given a triangulated graph G. For ease of exposition, we present this construction as a four-step procedure without regard to considerations of computational efficiency. It is possible to execute all four steps simultaneously and efficiently; see [4, Fig. 11]. The construction needs to be applied to each component of G; so in this section, we assume that G is connected and triangulated.

Construction

We use the graph from Example 11.7 as a running example to describe the four steps.

Step 1: Identify Maximal Cliques in G In our running example, the maximal cliques are $C_1 = \{X_5, X_6, X_7\}$, $C_2 = \{X_3, X_4, X_6, X_7\}$, $C_3 = \{X_2, X_3, X_6, X_7\}$, and $C_4 = \{X_1, X_2\}$.

Step 2: Construct the Weighted Clique Intersection Graph The vertex set of the *weighted clique intersection graph* associated with G is the set of maximal cliques in G. Two distinct vertices C and $\tilde{C}$ in this graph are connected by an edge if and only if the intersection $C \cap \tilde{C}$ is nonempty, and the weight of this edge is the cardinality $|C \cap \tilde{C}|$ of this intersection. The weighted clique intersection graph for our running example is given below. Note that the weighted clique intersection graph is always connected, since G is connected.

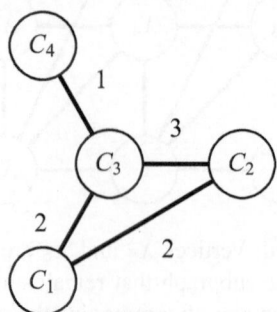

Step 3: Find a Maximum Weight Spanning Tree A maximum weight spanning tree of the weighted clique intersection graph is called a *junction tree*. It does not need to be unique, and any choice will work. Shown below are the two possible junction trees in our running example.

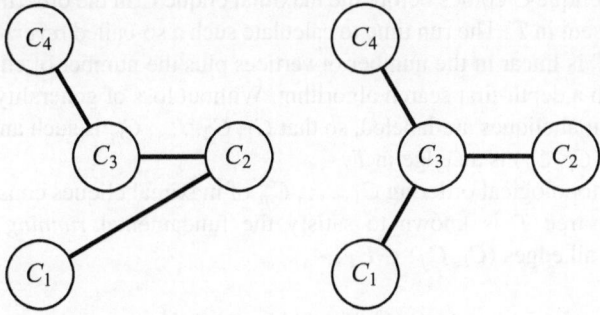

Step 4: Create a Junction In-Tree This step creates a directed version of the junction tree from the previous step. One vertex is designated as the root, and each edge is then toward this root, i.e., we construct an *in-tree* from the junction tree. We refer to the resulting directed graph as a *junction in-tree*.

We write $\mathcal{T}(G)$ for the set of all junction in-trees that can be constructed from a given graph G via these four steps. We denote a generic element of $\mathcal{T}(G)$ by $T = (V_T, E_T)$. We use the symbol V_T for the vertex set of the junction in-tree T to distinguish it from the vertex set V of the graph G. The vertex set V_T is the set of all maximal cliques of G, so it is unaffected by the choice of $T \in \mathcal{T}(G)$. We write R_T for the root vertex of the junction in-tree T.

Below are two junction in-trees T_1 and T_2 that have been constructed from the two junction tree examples from the previous step upon selecting C_1 and C_3, respectively, as the root.

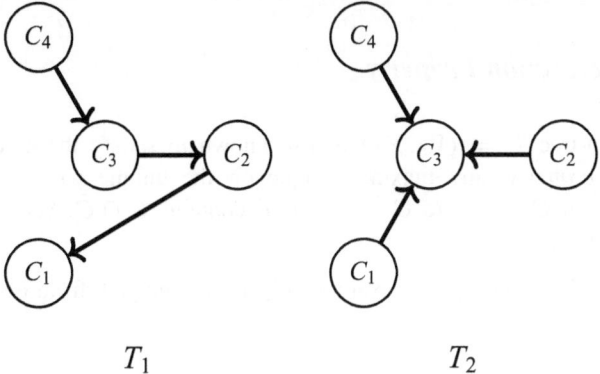

Running Intersection Property

We will use the so-called running intersection property in several arguments to follow. Fix some junction in-tree $T \in \mathcal{T}(G)$. Order the maximal cliques in G, so that the maximal clique $\tilde{C}$ comes before the maximal clique C in the ordering if the edge $(C, \tilde{C})$ is present in T. The run time to calculate such a so-called *reverse topological ordering* of T is linear in the number of vertices plus the number of edges of T, for instance, with a depth-first search algorithm. Without loss of generality, we assume that the maximal cliques are labeled, so that $C_1, C_2, \ldots, C_m$ is such an ordering, so that $j < i$ if (C_i, C_j) is an edge in T.

A reverse topological ordering $C_1, \ldots, C_m$ of maximal cliques constructed from a junction in-tree T is known to satisfy the fundamental *running intersection property*: for all edges $(C_i, C_j) \in E_T$,

$$C_i \cap (C_1 \cup \cdots \cup C_{i-1}) \subset C_j. \tag{11.5}$$

See, for example, Sections 3.3 and 3.5 in [4].

Example 11.9 The maximal cliques for our running example are $C_1 = \{X_5, X_6, X_7\}$, $C_2 = \{X_3, X_4, X_6, X_7\}$, $C_3 = \{X_2, X_3, X_6, X_7\}$, and $C_4 = \{X_1, X_2\}$. For the junction in-tree T_1, the unique reverse topological ordering of the maximal cliques is given by C_1, C_2, C_3, C_4. We have that

$$C_2 \cap C_1 = \{X_6, X_7\} \subset \{X_5, X_6, X_7\} = C_1$$

$$C_3 \cap (C_1 \cup C_2) = \{X_3, X_6, X_7\} \subset \{X_3, X_4, X_6, X_7\} = C_2$$

$$C_4 \cap (C_1 \cup C_2 \cup C_3) = \{X_2\} \subset \{X_2, X_3, X_6, X_7\} = C_3,$$

thus confirming the running intersection property.   △

Clique Intersection Property

A junction in-tree $T = (V_T, E_T)$ is also known to satisfy the so-called *clique intersection property*: any maximal clique on the unique path, ignoring edge directions, from $C \in V_T$ to $\tilde{C} \in V_T$ in T contains $C \cap \tilde{C}$. See, for example, Section 3.1 of [4].

Example 11.10 In our running example, C_2 lies on the path from C_1 to C_3 in T_1, so

$$\{X_6, X_7\} = C_1 \cap C_3 \subseteq C_2 = \{X_3, X_4, X_6, X_7\}$$

by the clique intersection property. Similarly, C_3 lies on the path from C_1 to C_2 in T_2, so

$$\{X_6, X_7\} = C_1 \cap C_2 \subseteq C_3 = \{X_2, X_3, X_6, X_7\}$$

by the clique intersection property.   △

11.4   Global Markov Property Representations

We let $\mathcal{F}_G$ denote the set of all collections $\{\psi_C\}_{C \in \mathcal{C}(G)}$, where each ψ_C is a nonnegative function with scope in C. An element of $\mathcal{F}_G$ is denoted either by $\{\psi_C\}_{C \in \mathcal{C}(G)}$ or simply by ψ. The following proposition describes a very useful sufficient condition for the global Markov property. The proof is given in Sect. 11.6.

Proposition 11.7 *Fix $\psi \in \mathcal{F}_G$, and define $p = \prod_{C \in \mathcal{C}(G)} \psi_C$. If $p \in \mathcal{P}_V$, then we have that $p \in \mathcal{M}_G$.*

Here are two applications of Proposition 11.7.

Example 11.11 Consider the graph G below. It has two maximal cliques, $C_1 = \{X_1, X_2\}$ and $C_2 = \{X_2, X_3, X_4\}$.

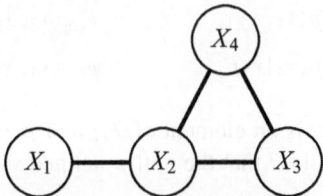

Given $\psi = \{\psi_{C_1}, \psi_{C_2}\} \in \mathcal{F}_G$, it can be directly verified that if $p = \psi_{C_1} \times \psi_{C_2}$ is a pmf, then it satisfies the global Markov property with respect to G. This amounts to saying that X_1 and (X_3, X_4) are conditionally independent given X_2 under p. Indeed, writing $\psi_{C_1}(x_2) = \sum_{x_1} \psi_{C_1}(x_1, x_2)$ and $\psi_{C_2}(x_2) = \sum_{x_3, x_4} \psi_{C_2}(x_2, x_3, x_4)$, we have that

$$p(x_1, x_3, x_4 | x_2) = \frac{\psi_{C_1}(x_1, x_2) \times \psi_{C_2}(x_2, x_3, x_4)}{\psi_{C_1}(x_2) \times \psi_{C_2}(x_2)}$$

$$= \frac{\psi_{C_1}(x_1, x_2)}{\psi_{C_1}(x_2)} \times \frac{\psi_{C_2}(x_2, x_3, x_4)}{\psi_{C_2}(x_2)}$$

$$= p(x_1 | x_2) \times p(x_3, x_4 | x_2).$$

As an example, suppose that ψ_{C_1} and ψ_{C_2} are defined via

$$\psi_{C_1}(x_1, x_2) = e^{-(x_1 + \frac{1}{2}x_2)}$$

$$\psi_{C_2}(x_2, x_3, x_4) = (1 - e^{-1})^4 \times e^{-(\frac{1}{2}x_2 + x_3 + x_4)}$$

and define the function $p(x_1, x_2, x_3, x_4)$ via

$$p(x_1, x_2, x_3, x_4) = \psi_{C_1}(x_1, x_2) \times \psi_{C_2}(x_2, x_3, x_4).$$

Proposition 11.7 states that $p \in \mathcal{M}_G$ for the graph G shown above. In fact, the random variables are independent with success probabilities $1 - e^{-1}$ under p, and so p satisfies the global Markov property for *any* graph with vertex set $\{X_1, X_2, X_3, X_4\}$ by Lemma 11.4.   △

Example 11.12 Let $V = \{X_1, \ldots, X_5\}$, and consider an arbitrary pmf $p \in \mathcal{P}_V$. Consider the graph G associated with the Markov chain of Example 11.3. It has four maximal cliques given by

$$C_1 = \{X_1, X_2\}, \ \ C_2 = \{X_2, X_3\}, \ \ C_3 = \{X_3, X_4\}, \ \ C_4 = \{X_4, X_5\}.$$

Define functions

$$\psi_{C_1}(x_1, x_2) = p(x_1, x_2), \qquad\qquad \psi_{C_2}(x_2, x_3) = p(x_3 | x_2),$$

$$\psi_{C_3}(x_3, x_4) = p(x_4 | x_3), \qquad\qquad \psi_{C_4}(x_4, x_5) = p(x_5 | x_4).$$

The collection $\psi = \{\psi_{C_i}\}_i$ is an element of $\mathcal{F}_G$ and $p = \prod_i \psi_{C_i}$, and so we may conclude from Proposition 11.7 that the pmf $\tilde{p}$ defined via

$$\tilde{p}(x_1, \ldots, x_5) = p(x_1, x_2) \times p(x_3 | x_2) \times p(x_4 | x_3) \times p(x_5 | x_4)$$

satisfies the global Markov property with respect to G. That is, $X_1, \ldots, X_5$ is a Markov chain under $\tilde{p}$, even if it is not under p.   △

Fix some junction in-tree $T \in \mathcal{T}(G)$. Let $\text{ch}_T(C)$ denote the unique child of any vertex $C \neq R_T$ in T, where we adopt the convention that $\text{ch}_T(R_T) = \emptyset$. We will use the fact that $C \backslash \text{ch}_T(C)$ is always nonempty. (Obviously, the root $C = R_T$ is nonempty.) To see why this is true, pick a maximal clique $C \in \mathcal{C}(G)$ such that $C \neq R_T$. If, to the contrary, $C \backslash \text{ch}_T(C)$ were empty, then $C \subseteq \text{ch}_T(C)$, which would then contradict the fact that C and $\text{ch}_T(C)$ are different maximal cliques.

Recall that $\mathcal{P}_C$ is the set of all joint pmfs of the random variables in C. For each maximal clique C and each $p \in \mathcal{P}_C$, define the nonnegative function $\phi_C^T(p)$ with scope C via

$$\phi_C^T(p) = p_{C \backslash \text{ch}_T(C) | C \cap \text{ch}_T(C)}, \tag{11.6}$$

where $p_{C|\emptyset}$ should be interpreted as p_C and $\phi_C^T(p)$ is understood to be zero whenever the conditional pmf is undefined. Note that $\phi_C^T(p)$ is a (possibly conditional) pmf, so it is a nonnegative function with scope C.

Example 11.13 Consider the graph G associated with the "Markov chain" of Example 11.3. The maximal cliques are $C_i = \{X_i, X_{i+1}\}$ for $1 \leq i \leq N - 1$. Below is a junction in-tree $T \in \mathcal{T}(G)$ for $N = 5$.

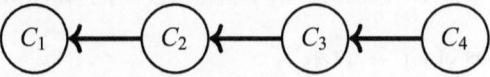

For this junction tree, we have that

$$[\phi_{C_1}^T(p_{C_1})](x_1, x_2) = p_{C_1}(x_1, x_2) = p(x_1, x_2)$$

and that

$$[\phi_{C_i}^T(p_{C_i})](x_i, x_{i+1}) = p_{C_i \setminus C_{i-1} | C_i \cap C_{i-1}}(x_{i+1} | x_i) = p(x_{i+1} | x_i)$$

for $i = 2, 3, 4$. △

For the graph G associated with the Markov chain of Example 11.3, Examples 11.12 and 11.13 show that each $p \in \mathcal{M}_G$ can be represented as $\prod_{C \in \mathcal{C}(G)} \psi_C$ where $\psi_C = \phi_C^T(p_C)$. The following proposition shows this particular representation is not by accident. The proof is given in Sect. 11.6. Keep in mind that the subscript C is used as an index in ϕ_C^T but that p_C is the marginal pmf of the variables in C under p.

Proposition 11.8 *If $p \in \mathcal{M}_G$, then we have, for all $T \in \mathcal{T}(G)$, that*

$$p = \prod_{C \in \mathcal{C}(G)} \phi_C^T(p_C).$$

Here is an application of Proposition 11.8.

Example 11.14 Assume that $p \in \mathcal{M}_G$ for the graph G from Example 11.7. The maximal cliques are $C_1 = \{X_5, X_6, X_7\}$, $C_2 = \{X_3, X_4, X_6, X_7\}$, $C_3 = \{X_2, X_3, X_6, X_7\}$ and $C_4 = \{X_1, X_2\}$. For the junction in-tree T_1 from Sect. 11.3, Proposition 11.8 states that p can be written as

$$p = p_{C_1} \times p_{C_2 \setminus C_1 | C_2 \cap C_1} \times p_{C_3 \setminus C_2 | C_3 \cap C_2} \times p_{C_4 \setminus C_3 | C_4 \cap C_3}$$

or equivalently

$$p(x_1, \ldots, x_7) = p(x_5, x_6, x_7) \times p(x_3, x_4 | x_6, x_7) \times p(x_2 | x_3, x_6, x_7) \times p(x_1 | x_2)$$

using the definitions of $C_1, \ldots, C_4$.

For the junction in-tree T_2 from Sect. 11.3, Proposition 11.8 states that p can be written as

$$p = p_{C_3} \times p_{C_1 \setminus C_3 | C_1 \cap C_3} \times p_{C_2 \setminus C_3 | C_2 \cap C_3} \times p_{C_4 \setminus C_3 | C_4 \cap C_3}$$

or equivalently

$$p(x_1, \ldots, x_7) = p(x_2, x_3, x_6, x_7) \times p(x_5 | x_6, x_7) \times p(x_4 | x_3, x_6, x_7) \times p(x_1 | x_2)$$

using the definitions of $C_1, \ldots, C_4$. $\triangle$

11.5 Marginalization, Lift, and Projection Maps

Fix some triangulated graph $G = (V, E)$. We let $\mathcal{I}_G$ denote the subset of $\mathcal{F}_G$ consisting of all collections of pmfs $\iota = \{\iota_C\}_{C \in \mathcal{C}(G)}$ for which the ι_C are consistent. Note that the subscript is used as an index here, so we are not using subscript to extract a marginal pmf. Recall that an element of $\mathcal{I}_G$ is called a G-information collection. For a junction in-tree $T \in \mathcal{T}(G)$ and $\iota \in \mathcal{I}_G$, let the nonnegative function $L_T(\iota)$ on V be defined via

$$L_T(\iota) = \prod_{C \in \mathcal{C}(G)} \phi_C^T(\iota_C), \tag{11.7}$$

where ϕ_C^T is defined in (11.6). Each $L_T(\iota)$ is thus defined as a product of elements of $\{\phi_C^T(\iota_C)\}_{C \in \mathcal{C}(G)} \in \mathcal{F}_G$.

The following proposition implies that we can take $\mathcal{P}_V$ as the codomain of L_T, i.e., $L_T : \mathcal{I}_G \to \mathcal{P}_V$, and that each element ι_C of the G-information collection $\iota = \{\iota_C\}_{C \in \mathcal{C}(G)}$ is a marginal pmf of $L_T(\iota)$. The proof is given in Sect. 11.6.

Proposition 11.9 *Let G be a triangulated graph, and consider an arbitrary G-information collection $\iota = \{\iota_C\}_{C \in \mathcal{C}(G)} \in \mathcal{I}_G$. We have that $L_T(\iota) \in \mathcal{P}_V$ and that the marginal pmfs of $L_T(\iota)$ associated with each maximal clique in G is an element of ι, i.e., for all $C \in \mathcal{C}(G)$,*

$$\sum_{V \setminus C} L_T(\iota) = \iota_C.$$

We think of the following map I_G as taking a pmf p and "marginalizing" it into its collection of marginals associated with the maximal cliques in G.

Definition 11.10 The G-*marginalization map* $I_G : \mathcal{P}_V \to \mathcal{I}_G$ is defined via

$$I_G(p) = \{p_C\}_{C \in \mathcal{C}(G)}.$$

By Proposition 11.9 and the definitions of L_T and I_G, we have, for all $\iota \in \mathcal{I}_G$, that

$$(I_G \circ L_T)(\iota) = I_G(L_T(\iota)) = \{\iota_C\}_{C \in \mathcal{C}(G)} = \iota,$$

which shows that the composition map $I_G \circ L_T$ is the *identity map on* $\mathcal{I}_G$. Now consider the (reverse) composition map $L_T \circ I_G$. We know by Propositions 11.7 and 11.9 that $L_T(\iota)$ must satisfy the global Markov property with respect to G for all $T \in \mathcal{T}(G)$, i.e.,

$$L_T(\iota) \in \mathcal{M}_G. \tag{11.8}$$

This implies that $(L_T \circ I_G)(\tilde{p}) \in \mathcal{M}_G$ for all $\tilde{p} \in \mathcal{P}_V$. With the notation introduced in this section, Proposition 11.8 states that, for all $p \in \mathcal{M}_G$,

$$p = (L_T \circ I_G)(p). \tag{11.9}$$

As a result, for any two junction in-trees $T, \tilde{T} \in \mathcal{T}(G)$, we have that

$$L_T = L_T \circ (I_G \circ L_{\tilde{T}}) = (L_T \circ I_G) \circ L_{\tilde{T}} = L_{\tilde{T}},$$

where the first equality uses the fact that $I_G \circ L_{\tilde{T}}$ is the identity map and the last equality uses (11.8) and (11.9). This establishes that all L_T maps are *identical*, which leads to the following definition.

Definition 11.11 The *G-lift map* $L_G : \mathcal{I}_G \to \mathcal{P}_V$ is defined by $L_G(\iota) = L_T(\iota)$ for some fixed choice of $T \in \mathcal{T}(G)$.

The next corollary follows from this definition and (11.8).

Corollary 11.12 *The range of L_G is a subset of $\mathcal{M}_G$.*

The maps $L_T \circ I_G$ encountered earlier are equal to $L_G \circ I_G$, and this map plays a key role in the remainder.

Definition 11.13 The *G-projection map* $\Pi_G : \mathcal{P}_V \to \mathcal{P}_V$ is defined via $\Pi_G = L_G \circ I_G$.

The next proposition justifies its name.

Proposition 11.14 Π_G *is a projection map from* $\mathcal{P}_V$ *onto* $\mathcal{M}_G$.

To see why this proposition holds, we verify the two conditions in Definition 1.2. It follows from Corollary 11.12 that

$$\Pi_G(\mathcal{P}_V) \subseteq \mathcal{M}_G \tag{11.10}$$

and from (11.9) that, for all $p \in \mathcal{M}_G$,

$$\Pi_G(p) = p. \tag{11.11}$$

Having established (11.10) and (11.11), we are justified in calling Π_G a "projection map."

The following corollary directly follows from the definitions of Π_G, I_G, and Proposition 11.9.

Corollary 11.15 *For any $p \in \mathcal{P}_V$, the marginal pmfs of p and $\Pi_G(p)$ associated with any clique in G are identical.*

Fix triangulated graphs $G_1 = (V, E_1)$ and $G_2 = (V, E_2)$ such that $E_1 \subseteq E_2$. The following corollary immediately follows from Proposition 11.14 and the fact that $\mathcal{M}_{G_1} \subseteq \mathcal{M}_{G_2}$ (Lemma 11.4).

Corollary 11.16 *For triangulated graphs $G_1 = (V, E_1)$ and $G_2 = (V, E_2)$ such that $E_1 \subseteq E_2$ we have that $\Pi_{G_2} \circ \Pi_{G_1} = \Pi_{G_1}$.*

The next proposition formulates a sufficient condition for the sum of $\Pi_G(p)$ over one variable to be an appropriate projection of the corresponding sum of p. The proof is given in Sect. 11.6.

Proposition 11.17 *Suppose that $p \in \mathcal{P}_V$. If v is a simplicial vertex of G and $\tilde{p} = \Pi_G(p)$, then we have that $\tilde{p}_{V \setminus \{v\}} = \Pi_{G - \{v\}}(p_{V \setminus \{v\}})$.*

The next proposition is an optimality result for the map Π_G and states that $\Pi_G(\hat{p})$ minimizes the Kullback-Leibler divergence between $\hat{p} \in \mathcal{P}_V$ and the set $\mathcal{M}_G$. Thus, errors can be measured with the Kullback-Leibler divergence. The proof is given in Sect. 11.6.

Proposition 11.18 *Let $G = (V, E)$ be a triangulated graph, and consider a pmf $\hat{p} \in \mathcal{P}_V$ with finite support. The solution to the minimization problem $\inf_{p \in \mathcal{M}_G} D(\hat{p} \parallel p)$ is unique and corresponds to $p^* = \Pi_G(\hat{p})$.*

11.6   Proofs

Proof of Proposition 11.7

Lemma 11.19 *Let $T \in \mathcal{T}(G)$. Suppose that the ordering $C_1, \ldots, C_m$ of the maximal cliques in G satisfies the running intersection property. For each edge $(C_i, C_j) \in E_T$, we have that*

$$C_i \setminus (C_1 \cup \cdots \cup C_{i-1}) = C_i \setminus C_j. \tag{11.12}$$

Example 11.15 Continuing Example 11.9, we find that

$$C_3 \backslash (C_1 \cup C_2) = \{X_2\} = C_3 \backslash C_2$$
$$C_4 \backslash (C_1 \cup C_2 \cup C_3) = \{X_1\} = C_4 \backslash C_3$$

as predicted by Lemma 11.19. △

Proof Fix some edge (C_i, C_j) in T. It follows from (11.5) that

$$C_i \cap (C_1 \cup \cdots \cup C_{i-1}) \subseteq C_i \cap C_j.$$

Since $j \in \{1, \ldots, i-1\}$, the reverse inclusion holds, so we have that

$$C_i \cap (C_1 \cup \cdots \cup C_{i-1}) = C_i \cap C_j. \tag{11.13}$$

Identity (11.12) now follows from (11.13) and the fact that C_i can either be represented as the disjoint union of $C_i \backslash (C_1 \cup \cdots \cup C_{i-1})$ and $C_i \cap (C_1 \cup \cdots \cup C_{i-1})$ or the disjoint union of $C_i \backslash C_j$ and $C_i \cap C_j$. □

We now turn to the proof of Proposition 11.7. It uses a standard argument; see, for example, [16, Prop. 3.8]. Fix some $\psi \in \mathcal{F}_G$, and suppose that $p = \prod_{C \in \mathcal{C}(G)} \psi_C$ is a pmf on V. Consider pairwise disjoint sets $B_1, B_2, S \subseteq V$ such that S separates B_1, and B_2 in G. Let $\hat{B}_1$ denote the union of the vertices in the components of $G - S$ that intersect B_1. Set $\hat{B}_2 = V \backslash (\hat{B}_1 \cup S)$. By construction, for $i = 1, 2$, the sets $B_i \subseteq \hat{B}_i$ and the sets $\hat{B}_1, \hat{B}_2, S$ partition the set of vertices V. Since the elements of B_1 and B_2 are vertices in different components of $G - S$, S also separates $\hat{B}_1$ and $\hat{B}_2$. Each $C \in \mathcal{C}(G)$ must therefore be a maximal clique in $G[\hat{B}_1 \cup S]$ or in $G[\hat{B}_2 \cup S]$. Consequently, we obtain that

$$p = \left[\prod_{C \in \mathcal{C}(G[\hat{B}_1 \cup S])} \psi_C \right] \times \left[\prod_{C \in \mathcal{C}(G[\hat{B}_2 \cup S])} \psi_C \right].$$

In particular, we may express the pmf p as

$$p(x_{\hat{B}_1}, x_{\hat{B}_2}, x_S) = \phi_1(x_{\hat{B}_1}, x_S) \times \phi_2(x_{\hat{B}_2}, x_S)$$

for appropriate functions ϕ_1 and ϕ_2. Writing $\tilde{\phi}_i(x_S) = \sum_{x_{\hat{B}_i}} \phi_i(x_{\hat{B}_i}, x_S)$ yields

$$p(x_{\hat{B}_1}, x_{\hat{B}_2} | x_S) = \frac{\phi_1(x_{\hat{B}_1}, x_S) \times \phi_2(x_{\hat{B}_2}, x_S)}{\tilde{\phi}_1(x_S) \times \tilde{\phi}_2(x_S)}, \tag{11.14}$$

from which we further deduce that

$$p(x_{\hat{B}_i} | x_S) = \frac{\phi_i(x_{\hat{B}_i}, x_S)}{\tilde{\phi}_i(x_S)}. \tag{11.15}$$

It now follows from (11.14) and (11.15) that $p_{\hat{B}_1 \cup \hat{B}_2 | S} = p_{\hat{B}_1 | S} \times p_{\hat{B}_2 | S}$, i.e., $\hat{B}_1$ and $\hat{B}_2$ are conditionally independent given S. By appropriate marginalization, this implies that B_1 and B_2 are conditionally independent given S, as claimed.

Proof of Proposition 11.8

Assume that there are m maximal cliques and that they have been ordered in reverse topological ordering with respect to T, so that $C_1, C_2, \ldots, C_m$ has the running intersection property. The chain rule applied to this ordering yields

$$p = p_{C_1} \times p_{C_2 \setminus C_1 | C_1} \times \cdots \times p_{C_m \setminus (C_1 \cup \cdots \cup C_{m-1}) | C_1 \cup \cdots \cup C_{m-1}}.$$

We show below that the identity

$$p_{C_i \setminus (C_1 \cup \cdots \cup C_{i-1}) | C_1 \cup \cdots \cup C_{i-1}} = p_{C_i \setminus C_j | C_i \cap C_j} = \phi_{C_i}^T (p_{C_i}) \tag{11.16}$$

holds for each $i = 1, 2, \ldots, m$, from which the result directly follows.

Identity (11.16) is trivial if $i = 1$, and so we fix $i > 1$ and let $j < i$ be such that $C_j = \mathrm{ch}_T(C_i)$. It follows directly from (11.12) that

$$p_{C_i \setminus (C_1 \cup \cdots \cup C_{i-1}) | C_1 \cup \cdots \cup C_{i-1}} = p_{C_i \setminus C_j | C_1 \cup \cdots \cup C_{i-1}}. \tag{11.17}$$

It remains to show that the right-hand side of (11.17) equals $p_{C_i \setminus C_j | C_i \cap C_j}$. It follows from [4, Lem. 4.2] that $S = C_i \cap C_j$ separates

$$B_1 = C_i \setminus (C_1 \cup \cdots \cup C_{i-1}) = C_i \setminus C_j$$

and

$$B_2 = (C_1 \cup \cdots \cup C_{i-1}) \setminus S.$$

By the assumed global Markov property $p_{B_1 | B_2 \cup S} = p_{B_1 | S}$, and so

$$p_{C_i \setminus C_j | C_1 \cup \cdots \cup C_{i-1}} = p_{C_i \setminus C_j | C_i \cap C_j},$$

as required.

Proof of Proposition 11.9

Pick some $T \in \mathcal{T}(G)$, $\iota \in \mathcal{I}_G$, and $C \in \mathcal{C}(G)$. We first argue that, without loss of generality, C can be assumed to be the root of T. Indeed, let $\tilde{T}$ be obtained from T by directing all edges toward C instead of R_T. Rewrite (11.7) as

$$L_T(\iota) = \frac{\prod_{C \in V_T} \iota_C}{\prod_{(C,\tilde{C}) \in E_T} [\iota_C]_{C \cap \tilde{C}}}. \tag{11.18}$$

Since T and $\tilde{T}$ have the same vertices, the expression in the numerator of (11.18) does not change if we replace T with $\tilde{T}$. Now consider the expression in the denominator. For each edge $(C, \tilde{C}) \in E_T$, there is either a directed edge from C to $\tilde{C}$ or a directed edge from $\tilde{C}$ to C in $\tilde{T}$. Regardless of the direction, the term corresponding to this edge in the denominator is the same, because ι is a G-information collection and therefore $[\iota_C]_{C \cap \tilde{C}} = [\iota_{\tilde{C}}]_{\tilde{C} \cap C}$. Thus, the expression in the denominator of (11.18) does not change if we replace T with $\tilde{T}$. Thus, $L_{\tilde{T}}(\iota) = L_T(\iota)$.

Next, assume that there are m maximal cliques and that $C_1, C_2, \ldots, C_m$ is a reverse topological ordering with respect to T. We know that $C_1 = C$ and that $C_1, C_2, \ldots, C_m$ have the running intersection property. Let $(C_m, \tilde{C})$ denote the unique edge in T associated with clique C_m. It follows from Lemma 11.19 for $i = m$ that the variables in $C_m \backslash \tilde{C}$ do not belong to any of the other terms on the right-hand side of (11.7). Consequently, they may be summed out, which removes the conditional pmf associated with this term. This argument can then be repeated for the term involving clique C_{m-1} and so on until what remains is the pmf $\iota_{C_1} = \iota_C$. This establishes that $L_T(\iota) \in \mathcal{P}_V$ and that the marginals of $L_T(\iota)$ indeed coincide with the marginals of ι, as claimed.

Proof of Proposition 11.17

Assume that p, $\tilde{p}$ and υ satisfy the conditions of the proposition. We know from Proposition 11.9 that $\Pi_G(p)$ and p agree on each maximal clique in G, and so $I_G(\Pi_G(p)) = I_G(p)$. Since $\tilde{p} = \Pi_G(p)$, we have that $I_G(\tilde{p}) = I_G(p)$, which implies that

$$I_{G-\{v\}}(\tilde{p}_{V \backslash \{v\}}) = I_{G-\{v\}}(p_{V \backslash \{v\}})$$

since v is a simplicial vertex. Since $\tilde{p} \in \mathcal{M}_G$, it follows from Lemma 11.5 that $\tilde{p}_{V \backslash \{v\}} \in \mathcal{M}_{G-\{v\}}$. By Proposition 11.14, we have that

$$\tilde{p}_{V \backslash \{v\}} = \Pi_{G-\{v\}}(\tilde{p}_{V \backslash \{v\}}),$$

and so

$$\begin{aligned}
\tilde{p}_{V \backslash \{v\}} &= L_{G-\{v\}}(I_{G-\{v\}}(\tilde{p}_{V \backslash \{v\}})) \\
&= L_{G-\{v\}}(I_{G-\{v\}}(p_{V \backslash \{v\}})) \\
&= \Pi_{G-\{v\}}(p_{V \backslash \{v\}}),
\end{aligned}$$

as claimed.

Proof of Proposition 11.18

If π is a pmf, then we interpret sums of the form $\sum_y \pi(y)f(y)$ for some function f as a sum over the support of π.

Lemma 11.20 *Suppose $p \in \mathcal{P}_V$ has finite support. Let $\mathcal{C}$ be a collection of subsets of a generic finite set of variables V, and suppose that $\{\psi_C^{(1)} : C \in \mathcal{C}\}$ and $\{\psi_C^{(2)} : C \in \mathcal{C}\}$ are two collections of functions, with the scope of each $\psi_C^{(i)}$ equal to C. Define $p^{(1)} = \prod_{C \in \mathcal{C}} \psi_C^{(1)}$ and $p^{(2)} = \prod_{C \in \mathcal{C}} \psi_C^{(2)}$, and suppose that $p^{(1)}, p^{(2)} \in \mathcal{P}_V$. Furthermore, suppose that the support of $p^{(1)}$ includes the support of p. We then have the identity*

$$D(p \parallel p^{(1)}) - D(p \parallel p^{(2)}) = \sum_{C \in \mathcal{C}} \sum_{x_C} p(x_C) \log \left(\frac{\psi_C^{(2)}(x_C)}{\psi_C^{(1)}(x_C)} \right),$$

where the right-hand side should be interpreted as $-\infty$ if the numerator is zero for at least one of the summands.

Proof If the numerator on the right-hand side is zero for at least one of the summands, then we have that $D(p \parallel p^{(2)}) = \infty$, which establishes the claim since $D(p \parallel p^{(1)}) < \infty$. We can thus assume, without loss of generality, that the support of $p^{(2)}$ includes the support of p. Using the definition of Kullback-Leibler divergence, we find that

$$D(p \parallel p^{(1)}) - D(p \parallel p^{(2)}) = \sum_x p(x) \log \left(\frac{p^{(2)}(x)}{p^{(1)}(x)} \right)$$

$$= \sum_x p(x) \log \left(\prod_{C \in \mathcal{C}} \frac{\psi_C^{(2)}(x_C)}{\psi_C^{(1)}(x_C)} \right)$$

$$= \sum_{C \in \mathcal{C}} \sum_x p(x) \log \left(\frac{\psi_C^{(2)}(x_C)}{\psi_C^{(1)}(x_C)} \right)$$

$$= \sum_{C \in \mathcal{C}} \sum_{x_C} p(x_C) \log \left(\frac{\psi_C^{(2)}(x_C)}{\psi_C^{(1)}(x_C)} \right),$$

as claimed. $\square$

We now turn to the proof of Proposition 11.18. Fix some pmf $\hat{p} \in \mathcal{P}_V$ with finite support and set $p^* = \Pi_G(\hat{p})$. Let $p \in \mathcal{M}_G$. We can assume, without loss of generality, that the support of p includes the support of $\hat{p}$ since otherwise $D(\hat{p} \parallel p) = \infty$. We know from the characterization of Π_G (Proposition 11.14) that

$p^* \in \mathcal{M}_G$. Using the junction in-tree representation (Proposition 11.8), both p and p^* can be written in the form

$$p = \prod_{C \in \mathcal{C}(G)} \phi_C^T(p_C), \quad p^* = \prod_{C \in \mathcal{C}(G)} \phi_C^T(p_C^*)$$

for an arbitrary fixed junction in-tree $T \in \mathcal{T}(G)$. Using Lemma 11.20 and the fact that $p_C^* = \hat{p}_C$ for all $C \in \mathcal{C}(G)$ (Proposition 11.9), we obtain that

$$D(\hat{p} \parallel p) - D(\hat{p} \parallel p^*) = \sum_{C \in \mathcal{C}(G)} \sum_{x_C} \hat{p}(x_C) \log \left(\frac{[\phi_C^T(p_C^*)](x_C)}{[\phi_C^T(p_C)](x_C)} \right)$$

$$= \sum_{C \in \mathcal{C}(G)} \sum_{x_C} p^*(x_C) \log \left(\frac{[\phi_C^T(p_C^*)](x_C)}{[\phi_C^T(p_C)](x_C)} \right). \quad (11.19)$$

Using Lemma 11.20 once again (now with $p = p^{(2)}$), we see that the right-hand side of (11.19) equals $D(p^* \parallel p)$, which is nonnegative and equal to zero if and only if $p = p^*$.

Chapter 12
Information Structure

In Chap. 10, we showed how to overcome the curse of dimensionality for counters by imposing independence of the counter and label variables corresponding to different subsystems. This chapter lays the foundation for incorporating more flexibility with respect to the independence assumptions imposed between the counter and label variables. Specifically, we will use the global Markov property to define the conditional independence relationships between these variables, specified by a modeler via *information structure graphs*. The choice of information structure affects the resulting approximation as well as the computational requirements. The information structure perspective generalizes the subsystem setting of Chap. 10. In fact, the latter corresponds to a special class of information structure graphs. We will exploit the conditional independence relationships imposed via information structure graphs to develop the graphical QPLEX calculus in Chaps. 13 and 14.

In this chapter, we consider arbitrary counter and label variables. We fix time t throughout and assume that we are given a size function for each label variable. Therefore, we suppress the superscript of the size functions to simply write x_m instead of $x_m^{(t)}$ for $m = 1, \ldots, M$. The counter and label variables as well as the size functions vary from example to example.

12.1 Information Structure Graphs

In this section, we define an *information structure graph*, an undirected graph $G = (V, E)$ for which each vertex represents either a counter or label variable. When stressing that these variables correspond to vertices in G, we also refer to them as *counter vertices* and *label vertices*, respectively.

© The Author(s) 2025

A. B. Dieker, S. T. Hackman, *QPLEX: A Computational Modeling and Analysis Methodology for Stochastic Systems*, Springer Series in Operations Research and Financial Engineering, https://doi.org/10.1007/978-3-031-74870-7_12

Let N_m denote the set of counter vertices that are adjacent to label vertex L_m, a set we call the *neighbor set* of L_m. Also let S_m be the set of counter vertices corresponding to the scope of the size function x_m. We slightly abuse terminology and refer to S_m as the *scope* of x_m. We also abuse notation to write $x_m(z_{S_m})$ instead of $x_m(z)$ throughout, where z_{S_m} denotes the subvector of z corresponding to the counter vertices in S_m. We let G_Z and G_L denote, respectively, the subgraphs of G associated with the counter and label vertices. The requirements imposed for G to be an information graph are provided in the following definition.

Definition 12.1 A graph G on counter and label variables is an *information structure* graph compatible with the size function vector x if it satisfies the following requirements:

(R1) Subgraph G_Z is recursively simplicial, and each label vertex L_m is simplicial.
(R2) The scope S_m of each size function x_m is a subset of N_m.

When the size function vector is clear from the context, we simply say that G is an information structure graph.

Requirements (R1) and (R2) imply several graph-theoretic properties that we collect in the next lemma.

Lemma 12.2 *If G is an information structure graph, then we have that:*

(a) *Each component K of subgraph G_L is a complete graph, and all neighbor sets N_m such that label vertex L_m lies in K are identical.*
(b) *For each m, N_m separates $\{L_m\}$ and the set of counter vertices not in N_m.*
(c) *For each m, $N_m \cup \{L_m\}$ is a clique in G, and N_m is a clique in G_Z.*
(d) *G is triangulated.*

To see why (a) holds, fix some component K of G_L. We first establish that K is complete, which is trivially true if K contains only one or two label vertices, and so we assume that K contains at least three label vertices. Pick two distinct vertices L_m and $L_{\tilde{m}}$ in K. We show that there is an edge connecting these two vertices. There must be a shortest path between L_m and $L_{\tilde{m}}$ with all vertices in this path contained in K. Suppose there were an internal vertex on this path. Since this is a label vertex, it must be simplicial, so there must be an edge between its neighbors on the path. This, however, contradicts that this path is the shortest path. Thus, there cannot be internal vertices on this path, and consequently L_m and $L_{\tilde{m}}$ must be adjacent. The fact that the neighbor sets N_m for those m with label vertex L_m in K are identical is an immediate consequence of the requirement that each label vertex is simplicial.

To see why (b) holds, fix L_m, and let K denote the component of G_L of which L_m is a vertex. We establish the claim by contradiction. If it is false, then there must exist a counter vertex $Z \notin N_m$ such that some path between Z and L_m does not contain counter vertices in N_m. First, suppose that this path contains only label vertices (except Z itself), and let $L_{\tilde{m}}$ denote the vertex on this path adjacent to Z. Since K is complete by part (a), each of the label vertices of this path must belong to K, in particular, the vertex $L_{\tilde{m}}$. Since $Z \in N_{\tilde{m}}$ and $N_{\tilde{m}} = N_m$, this shows that

$Z \in N_m$, contradicting the definition of Z. Now, suppose this path contains at least one counter vertex. Let $\tilde{Z}$ denote the "last" counter vertex in the path such that the subpath connecting it to L_m only contains label vertices. We may repeat the argument above to deduce that $\tilde{Z} \in N_m$, which contradicts the construction of the path.

Next, consider (c). The first claim is a direct consequence of the definition of N_m and Requirement (R1), and the second claim is an immediate consequence. For future reference, we note that, as a consequence of this property, for any $\mu \in \mathcal{P}_{Z,L}$, the pmf $\mu(z_{N_m}, \ell_m)$ can be extracted from the collection $I_G(\mu)$ of marginal pmfs corresponding to the maximal cliques in G.

To see why (d) holds, recall that G is triangulated if and only if it admits a simplicial ordering. *Any* ordering of the label vertices appended with the simplicial ordering of the counter vertices from (R1) is a simplicial ordering. Note that since G is triangulated, we may invoke all of the results of Chap. 11.

As a result of Lemma 12.2, an information structure graph can equivalently be specified as follows. Specify a partition $\{U_k\}_k$ of the set of label vertices and a collection $\{R_k\}_k$ of sets of counter vertices such that the scope S_m of the size function x_m satisfies $S_m \subseteq R_k$ for all m with L_m in U_k. We emphasize that the R_k can overlap and even be identical, but the U_k cannot overlap. Starting with an empty graph on the counter and label variables, we plant the cliques U_k, which correspond to the complete components of G_L, and the cliques R_k, which are the neighbor sets of each label vertex in U_k. Additional edges between the counter vertices can then be inserted as long as the resulting graph G_Z is triangulated.

We now describe several examples of information structure graphs.

Example 12.1 A complete graph on the set of vertices representing all counter and label variables in the model is an information structure graph regardless of the scopes of the size functions, since N_m is the set of all counter vertices for every m. This information structure graph imposes no structure on the counter and label variables in the model, as in Part I.   △

Example 12.2 Consider the graph below.

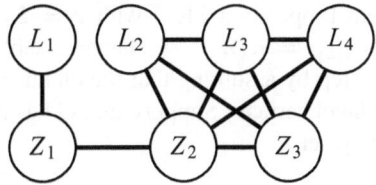

This graph satisfies Requirement (R1) of Definition 12.1. We have that $N_1 = \{Z_1\}$ and $N_2 = N_3 = N_4 = \{Z_2, Z_3\}$. Note that N_1 is not a maximal clique in G_Z, but it is contained in the maximal clique $\{Z_1, Z_2\}$. Requirement (R2) means that the scopes S_m of the size functions x_m must satisfy $S_1 \subseteq \{Z_1\}$ and $S_2, S_3, S_4 \subseteq \{Z_2, Z_3\}$.   △

Example 12.3 Consider the graph below.

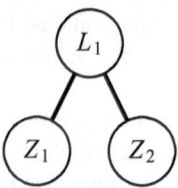

This graph cannot be an information structure graph, since the vertex L_1 is not simplicial, and therefore Requirement (R1) cannot hold.   △

Example 12.4 Consider the graph below.

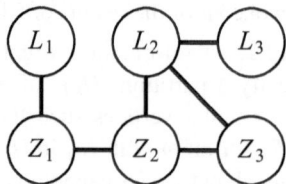

This graph cannot be an information structure graph since the vertex L_2 is not simplicial, and therefore Requirement (R1) cannot hold.   △

The next lemma establishes a key property of the G-projection map Π_G if G is an information structure graph.

Lemma 12.3 *If* $\mu \in \mathcal{P}_{Z,L}$, *then we have that*

$$[\Pi_G(\mu)](\mathbf{Z} = \cdot) = \Pi_{G_Z}(\mu(\mathbf{Z} = \cdot)).$$

To see why this lemma holds, fix some $\mu \in \mathcal{P}_{Z,L}$. We will establish the claim by induction on the number of label vertices. For the base case, suppose that G contains exactly one label vertex L_1. Since the vertex L_1 is simplicial, the claim follows immediately from Proposition 11.17 with $v = L_1$ because in this setting $G - \{L_1\} = G_Z$ and $\mu_{V \setminus \{L_1\}} = \mu(\mathbf{Z} = \cdot)$. Having established the base case, we proceed to the inductive step by assuming that the claim holds for all information structure graphs with M label vertices. Suppose that G has $M + 1$ label vertices and let $\tilde{G} = G - \{L_{M+1}\}$. Also set

$$\tilde{\mu} = \mu(\mathbf{Z} = \cdot, L_1 = \cdot, \ldots, L_M = \cdot).$$

Note that L_{M+1} is a simplicial vertex of G. The following identities hold:

$$[\Pi_G(\mu)](z) = \sum_{\ell_1, \ldots, \ell_M} [\Pi_G(\mu)](z, \ell_1, \ldots, \ell_M)$$

$$= \sum_{\ell_1,\dots,\ell_M} [\Pi_{\tilde{G}}(\tilde{\mu})](z, \ell_1, \dots, \ell_M)$$

$$= [\Pi_{\tilde{G}}(\tilde{\mu})](z)$$

$$= [\Pi_{\tilde{G}_z}(\tilde{\mu}(Z = \cdot))](z)$$

$$= [\Pi_{G_z}(\mu(Z = \cdot))](z),$$

where the first line follows by definition of marginal pmfs, the second line applies Proposition 11.17 with $v = L_{M+1}$, the third line follows by definition of marginal pmfs, the fourth line follows by the inductive hypothesis, and the last line follows because $G_z = \tilde{G}_z$ and $\tilde{\mu}(Z = \cdot) = \mu(Z = \cdot)$. The inductive step has been established.

For our next result, we suppose that $\mu \in \mathcal{M}_G$. The following proposition states several facts we use repeatedly in the remainder of this book.

Proposition 12.4 *If G is an information structure graph, then we have, for every $\mu \in \mathcal{M}_G$, that*

1. $\mu(Z = \cdot) \in \mathcal{M}_{G_z}$.
2. $\mu(\ell_m | z) = \mu(\ell_m | z_{N_m})$ *if $\mu(z) > 0$.*
3. $[\mathcal{L}(\mu)](z, h)$ *can be written as*

$$[\mathcal{L}(\mu)](z, h) = \mu(z) \times \prod_m \Pr[\text{mult}(x_m(z_{S_m}), \mu(L_m = \cdot | z_{N_m})) = h_m].$$

$$(12.1)$$

To see why (a) holds, we first use Lemma 12.3 to conclude that

$$\mu(Z = \cdot) = \Pi_{G_z}(\mu(Z = \cdot))$$

and then use the fact that Π_{G_z} is a projection map onto $\mathcal{M}_{G_z}$ to deduce that $\mu(Z = \cdot) \in \mathcal{M}_{G_z}$.

We use Lemma 12.2(b) to establish (b). The set of counter vertices N_m separates $\{L_m\}$ and the set of counter vertices that are not in N_m, so (11.2) immediately yields (b). Note that $\mu(z) > 0$ implies $\mu(z_{N_m}) > 0$ for all m.

Now consider (c), which substitutes the identity in (b) into the expression of $[\mathcal{L}(\mu)](z, h)$ from Definition 16.2. Recall that it is understood that the multiplicands in (12.1) are only relevant if $\mu(z) > 0$. It also makes the scopes of the size functions explicit.

12.2   Label Conditional Independence Graphs

We will see in Sect. 17.6 that choosing a label conditional independence graph can provide computational savings without compromising approximation quality. Here is a formal definition.

Definition 12.5 A *label conditional independence* graph is an information structure graph that has no edges between label vertices.

Since there are no edges between any two label vertices in a label conditional independence graph G, the sets N_m consist of all vertices that are adjacent to L_m. Coupled with Lemma 12.2(c), this means that each $N_m \cup \{L_m\}$ is necessarily a *maximal* clique in G and these are the only maximal cliques that contain a label vertex. All other maximal cliques in G are necessarily maximal cliques in G_Z, so the set $\mathcal{C}(G)$ of maximal cliques in G can be expressed as

$$\mathcal{C}(G) = \mathcal{C}(G_Z) \cup \bigcup_m \{N_m \cup \{L_m\}\}.$$

A label conditional independence graph can equivalently be specified via a collection $\{N_m\}_m$ of neighbor sets of the label vertices and a triangulated graph on counter vertices in which each N_m is a clique.

Example 12.5 A complete graph G is a label conditional independence graph if and only if there is at most one label vertex. $\triangle$

Example 12.6 Consider the graph below.

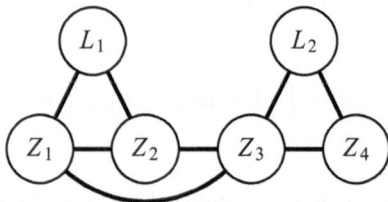

This graph is a label conditional independence graph with two label vertices and four counter vertices in which $N_1 = \{Z_1, Z_2\}$ and $N_2 = \{Z_3, Z_4\}$ provided $S_1 \subseteq N_1$ and $S_2 \subseteq N_2$. Note that N_1 is a clique in the subgraph G_Z induced by the counter vertices, but it is not a maximal clique. $\triangle$

The following proposition records a structural property of pmfs satisfying the Markov property with respect to a label conditional independence graph. It implies that the label variables are conditionally independent given the counter variables for any pmf with the global Markov property with respect to G, which justifies the name. Recall that the product of $\mu(z)$ and an undefined quantity should be interpreted as zero when $\mu(z) = 0$.

Proposition 12.6 *If G is a label conditional independence graph, then we have, for every $\mu \in \mathcal{M}_G$, that*

$$\mu(z, \ell) = \mu(z) \times \prod_m \mu(\ell_m | z_{N_m}).$$

To see why this proposition holds, fix some label conditional independence graph G and some $\mu \in \mathcal{M}_G$. Since there are no edges between any two label vertices, the set of all counter vertices separates each pair of the singletons $\{L_m\}$. Fix some z, and suppose that $\mu(z) > 0$. By the global Markov property (11.1), we have that $\mu(\ell|z) = \prod_m \mu(\ell_m|z)$. We furthermore have that $\mu(\ell_m|z) = \mu(\ell_m|z_{N_m})$ by Proposition 12.4(b), yielding the claim.

Example 12.7 Consider the graph G below.

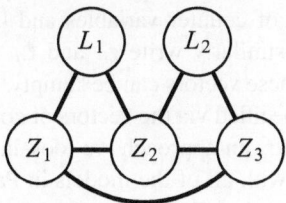

This graph is a label conditional independence graph if $S_1 \subseteq \{Z_1, Z_2\}$ and $S_2 \subseteq \{Z_2, Z_3\}$. Any $\mu \in \mathcal{M}_G$ satisfies

$$\mu(z_1, z_2, z_3, \ell_1, \ell_2) = \mu(z_1, z_2, z_3) \times \mu(\ell_1|z_1, z_2) \times \mu(\ell_2|z_2, z_3)$$

by Proposition 12.6. $\triangle$

We next show that it is possible to construct a label conditional independence graph from an arbitrary information structure graph. We need the following definition.

Definition 12.7 Given some information structure graph $G = (V, E)$, the *label conditional independence graph derived from* G is the graph $\mathrm{lci}(G) = (V, \tilde{E})$, where $\tilde{E}$ is the subset of E defined by removing all edges in E between label vertices.

The following lemma confirms that $\mathrm{lci}(G)$ is indeed an information structure graph if G is.

Lemma 12.8 *If G is an information structure graph, then so is $\mathrm{lci}(G)$.*

To see why this lemma holds, we show that $\mathrm{lci}(G)$ satisfies the two requirements in Definition 12.1 by virtue of the fact that G satisfies these requirements. Indeed, it satisfies Requirement (R1), because the subgraphs G_Z and $\mathrm{lci}(G)_Z$ are identical. Since the set N_m denotes the set of counter vertices that are adjacent to label vertex L_m in an information structure graph, the sets N_m for G and $\mathrm{lci}(G)$ are identical. As a direct consequence, Requirement (R2) is satisfied for $\mathrm{lci}(G)$ if it is satisfied for G.

12.3 Independent Subsystem Graphs

Often, systems can be divided into parts that can be thought of as subsystems. This motivates the following class of information structure graphs, which is consistent with the discussion in Chap. 10.

Definition 12.9 An *independent subsystem graph* is an information structure graph for which each component is a complete graph.

This definition implies that the counter and label vertices of an independent subsystem graph can be partitioned according to the component of the graph in which they lie. Each component corresponds (loosely speaking) to a "subsystem."

We index the components of an independent subsystem graph by σ and write Z_σ and L_σ for the vectors of counter variables and label variables, respectively, lying in component σ. We similarly write z_σ and ℓ_σ for possible values of these vectors. Note that some of these vectors can be "empty." An independent subsystem graph can equivalently be specified via the vectors of counter and label variables Z_σ and L_σ in each component σ, an approach we took in Chap. 10. The information structure graphs associated with all of the models in Part A can each be viewed as an independent subsystem graph with exactly one subsystem.

Example 12.8 Consider the graph below.

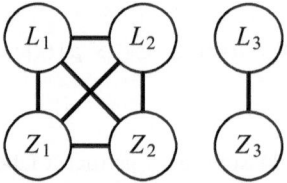

This graph is an independent subsystem graph with two subsystems as long as the scopes of the size functions satisfy $S_1, S_2 \subseteq \{Z_1, Z_2\}$ and $S_3 \subseteq \{Z_3\}$.   △

Example 12.9 Consider the graph below.

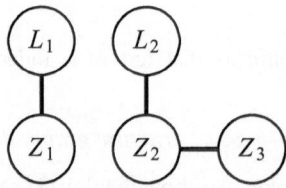

This graph can be an information structure graph as long as the scopes of the size functions satisfy $S_1 \subseteq \{Z_1\}$ and $S_2 \subseteq \{Z_2\}$, but it *cannot* be an independent subsystem graph.   △

Let G be an independent subsystem graph. Since the empty set separates the vertex sets of each pair of components in this graph, we have the following representation by the global Markov property (11.1).

Proposition 12.10 *Let G be an independent subsystem graph. For each $\mu \in \mathcal{M}_G$, we have that*

$$\mu(z, \ell) = \prod_\sigma \mu(z_\sigma, \ell_\sigma).$$

12.4   Model Examples

In the examples to follow, for convenience, all label variables represent "ages," although this is by no means necessary.

Multiserver Queueing Models

We first consider the multiserver queueing model with renewal arrivals from Sect. 5.1. The model has two counter variables Z and Z_A and one label variable L. Since the scope of the size function is $\{Z\}$, there are only two connected information structure graphs, as depicted below.

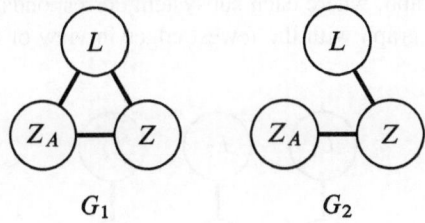

$$G_1 \qquad\qquad\qquad G_2$$

Since G_1 is complete, $\mathcal{M}_{G_1} = \mathcal{P}_V$, i.e., $\mu \in \mathcal{M}_{G_1}$ does not imply any (conditional) independence relationships between any of the model variables under μ. Now, consider the conditional independence assumptions imposed by graph G_2. Under any $\mu \in \mathcal{M}_{G_2}$, the number of periods since the last customer arrived to the system (the variable Z_A) and the time spent in service so far by a uniformly chosen at random customer in service (the variable L) are conditionally independent given the number of customers in the system (the variable Z). This is the only (conditional) independence imposed.

Next we turn to the multiserver queueing model with abandonments from Sect. 7.1. The model has two label variables L_S and L_B and one counter variable Z. The scopes of the size functions x_S and x_B are $\{Z\}$. Here is the only label conditional independence graph G for this model.

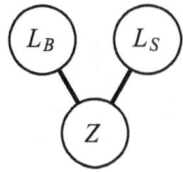

Under any $\mu \in \mathcal{M}_G$, the time spent in service so far by a uniformly chosen at random customer in service (the variable L_S) and the time spent waiting so far by a uniformly chosen at random customer in the buffer (the variable L_B) are conditionally independent given the number of customers in the system (the variable Z). This is the only (conditional) independence imposed.

Flow Line

Consider a flow line model with three stations, as described in Sect. 5.4. The model has three label variables L_1, L_2, L_3 and three counter variables Z_1, Z_2, Z_3. The scopes of the size functions x_1, x_2, and x_3 are $\{Z_1\}$, $\{Z_2\}$, and $\{Z_3\}$, respectively. We consider different label conditional independence graphs that vary with respect to the degree of conditional independence imposed on the variables. This example is readily extended to an arbitrary number of stations.

The graph below is both a label conditional independence graph and an independent subsystem graph, where each subsystem corresponds to a station. It is the information structure graph with the fewest edges in view of Requirement (R2) in Definition 12.1.

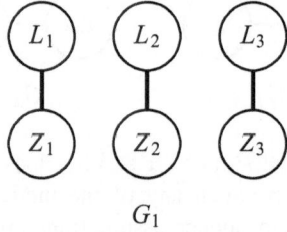

G_1

Under any $\mu \in \mathcal{M}_{G_1}$, the model variables for each station, corresponding to the time spent in service so far by a uniformly chosen at random customer in service (the variables L_1, L_2, and L_3) and the number of customers in the station (the variables Z_1, Z_2, and Z_3), are independent across the stations. This example will be revisited in subsequent chapters.

The next information structure graph is similar to the previous example, except that here, the subgraph of counter variables is a path graph.

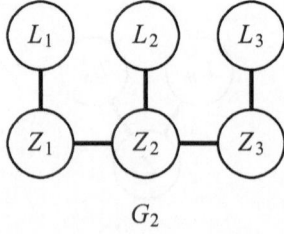

G_2

As we know from Example 11.3, this means that the counter variables form a Markov chain under any $\mu \in \mathcal{M}_{G_2}$. In other words, given the model variables for any station, the model variables corresponding to the upstream and downstream stations are independent. Moreover, given the number of customers at any station, the age of a uniformly chosen customer in service at that station is independent of all other model variables. We will revisit this information structure graph in subsequent chapters and refer to this graph as the *fishbone* graph.

The information structure graph G_3 depicted below combines the two previous examples.

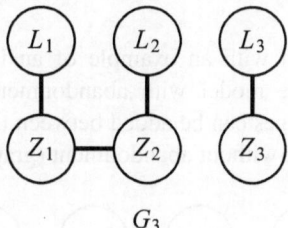

G_3

It has two components, each corresponding to a subsystem. The subgraph associated with the first subsystem is a fishbone graph. In addition to the (conditional) independence imposed in the fishbone graph, the model variables corresponding to each subsystem are independent. Note that G_3 is not an independent subsystem graph, but edges can be added to make it so; see Example 12.8.

The label conditional independence graph G_4 depicted below imposes no (conditional) independence assumptions on the number of customers at each station in the model, but it still imposes the assumption that the age of a uniformly chosen at random customer in service at a station given the number of customers at that station is independent of all the other variables.

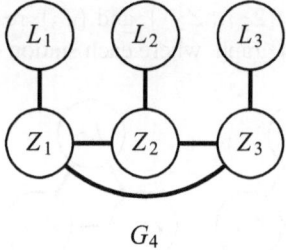

G_4

The label conditional independence graph G_5 depicted below imposes no assumptions on the counter variables. It does impose the assumption that the age of a randomly chosen customer in service at each station is independent of the rest of the variables given the number of customers at that station and all upstream stations, e.g., the variable L_2 is conditionally independent of the variables Z_3, L_1, and L_3 given Z_1 and Z_2.

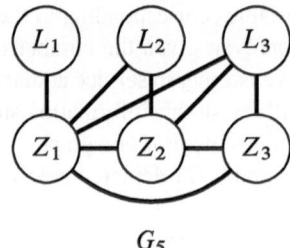

$$G_5$$

We close this discussion with an example of an information structure graph associated with a flow line model with abandonments, where each subsystem corresponds to a station. Edges can be added between the counter vertices much in the same way as for the case without abandonment (graphs G_2, G_3, and G_4 above).

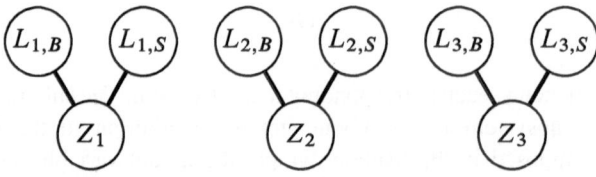

Queueing Networks

We first consider the queueing network model with blocking from Sect. 5.7 with three stations. The label variables are L_1, L_2, and L_3, while the counter variables are $Z_{1,C}$, $Z_{1,S}$, $Z_{2,C}$, $Z_{2,S}$, and Z_3, respectively. The scopes of the size functions x_1, x_2, and x_3 are $\{Z_{1,C}, Z_{1,S}\}$, $\{Z_{2,C}, Z_{2,S}\}$, and $\{Z_3\}$, respectively. The graph below is an independent subsystem graph, where each station corresponds to a subsystem.

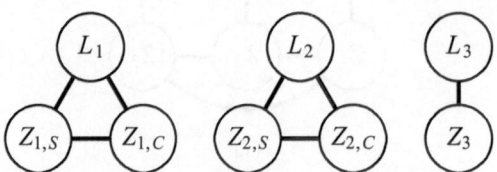

Next, we consider the model consisting of six multiserver single stations depicted below. This is a queueing network model with probabilistic routing; see Sect. 5.5. The server pools are shown vertically, and the buffers are shown horizontally. Circles represent capacity. Arrows in this picture stand for internal flow. Not shown are flows into and out of the network because these are irrelevant for purposes of the discussion here. Similarly, we do not show the routing probabilities.

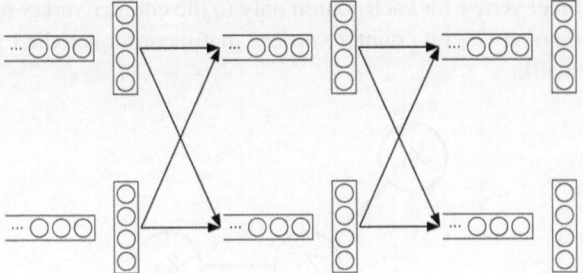

Consider the label conditional independence graph in which each label vertex for each of the stations is only connected to the counter vertex for that station. This is a minimal choice in view of the scope of each size function. A natural choice for the subgraph G_Z is to mimic the topology of the system flows:

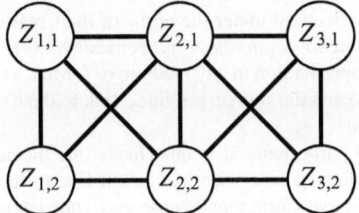

This subgraph is isomorphic to the "strong product" Markov chain of Example 11.4.

The model depicted below consists of four multiserver stations with buffers that have infinite capacity such that two stations feed a two-station flow line. Again, we only show the internal flow. This is again a queueing network model covered in Sect. 5.5, but each routing probability is either 0 or 1.

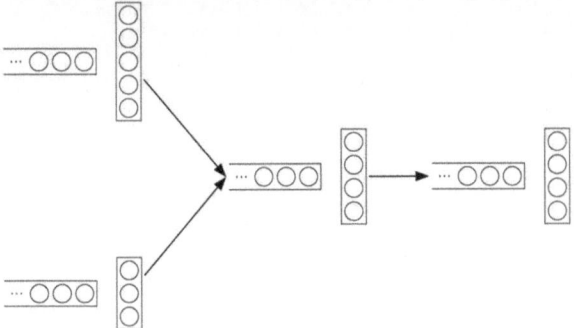

Consider a label conditional independence graph where we again let each label vertex be minimally connected in view of the scopes of the size functions, i.e., we connect each label vertex for each station only to the counter vertex for that station. Here is the subgraph G_Z on counter vertices that incorporates the topology of the underlying network.

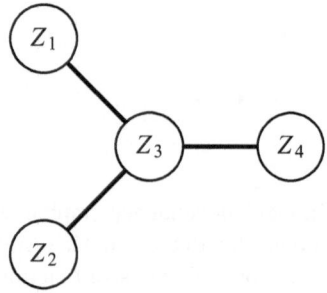

Chapter 13
Graphical QPLEX Calculus with Distributional Programs

This chapter shows how to represent graphical QPLEX maps via a collection of "efficient" distributional programs. Here is an overview.

Consider a QPLEX map $Q^{(t)}$ and two information structure graphs G and G' at times t and $t+1$, respectively. We reproduce the bottom part of the diagram from Sect. 9.2, but we now insert the G'-marginalization and G-lift maps $I_{G'}$ and L_G from Definitions 11.10 and 11.11, respectively. Starting from the lower left corner with a G-information collection $\iota = \{\iota_C\}_{C \in \mathcal{C}(G)}$, the *graphical QPLEX map* $Q^{(t)}_{G \to G'}$ outputs a G'-information collection $\iota' = \{\iota'_{C'}\}_{C' \in \mathcal{C}(G)}$ by going "up, over, and down" in this diagram. (A formal definition of $Q^{(t)}_{G \to G'}$ requires careful consideration of the domains and codomains of the three maps used in the composition; see Definition 17.4.)

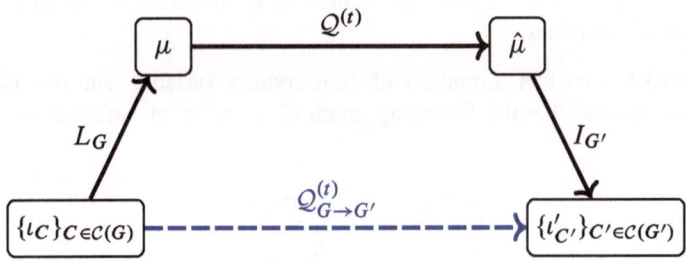

The G'-information collection $\iota' = Q^{(t)}_{G \to G'}(\iota)$ in the above diagram consists of pmfs $\iota'_{C'}$ associated with maximal cliques C' in G', and, for each C', we seek an efficient distributional program that outputs the pmf $\iota'_{C'}$. We refer to the map that outputs element $C' \in \mathcal{C}(G')$ of $Q^{(t)}_{G \to G'}(\iota)$ given input ι as *element C' of the graphical QPLEX map* $Q^{(t)}_{G \to G'}$. If we use the definition of the graphical QPLEX map for evaluating element C' of $Q^{(t)}_{G \to G'}$, then we must first calculate the *full* pmf

© The Author(s) 2025

A. B. Dieker, S. T. Hackman, *QPLEX: A Computational Modeling and Analysis Methodology for Stochastic Systems*, Springer Series in Operations Research and Financial Engineering, https://doi.org/10.1007/978-3-031-74870-7_13

μ from ι, thus encountering the curse of dimensionality for counters. In this chapter, we instead represent element C' of $\mathcal{Q}^{(t)}_{G \to G'}$ via a distributional program that no longer requires μ to be calculated (but instead uses ι directly), does not introduce variables that are subsequently unused, and suppresses unused arguments. We call this program the *efficient refactored distributional program*, which can be viewed as "going directly across" the bottom level of our diagram.

Allowing the information structure graphs G and G' to be different can be practically useful for modeling different dependencies. For example, if G' is constructed from G by adding (removing) edges, then there are fewer (more) dependencies modeled via G' than G. For convenience, however, we will assume that $G' = G$ in the examples of this chapter. Since the graph G will be understood from the context, we simply call $\iota \in \mathcal{I}_G$ an *information collection*.

13.1 Distributional Program Representation

This section develops a distributional program for each element of the graphical QPLEX map $\mathcal{Q}^{(t)}_{G \to G'}$, which serves as the starting point for this chapter. It is this program we seek to refactor to obtain a more efficient representation under certain assumptions to be articulated momentarily. We also introduce some notation and terminology.

Queries Given an information collection ι, element C' of $\mathcal{Q}^{(t)}_{G \to G'}(\iota)$ is a marginal pmf of $\hat{\mu}$ in the above diagram. Therefore, we refer to such a set C' as a *query*; this is consistent with terminology used in the literature on probabilistic graphical models [14]. Thus, any query C' is a maximal clique in G' but with its elements replaced with "primed" variables.

Example 13.1 Consider a model with four counter variables and two label variables, and suppose that the following graph $G = G'$ is an information structure graph:

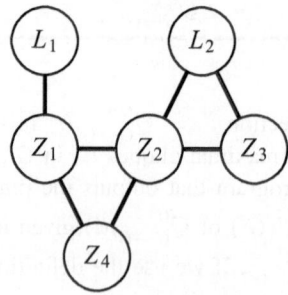

The set of maximal cliques is given by

$$\mathcal{C}(G) = \{\{Z_1, L_1\}, \{Z_1, Z_2, Z_4\}, \{Z_2, Z_3, L_2\}\},$$

so the queries are $\{Z_1', L_1'\}, \{Z_1', Z_2', Z_4'\}, \{Z_2', Z_3', L_2'\}$. Thus, the graphical QPLEX map is represented by three distributional programs, one for each of the queries. $\triangle$

Distributional Program for Element C' of $\mathcal{Q}_{G \to G'}^{(t)}$ We focus on QPLEX chains with simple transition dynamics and use the following generic distributional program for element C' of $\mathcal{Q}_{G \to G'}^{(t)}$ as the starting point for this chapter. It uses the definition of $\mathcal{Q}_{G \to G'}^{(t)}$, with the generic program for $\mathcal{Q}^{(t)}$ taken from Sect. 4.3. In the return statement (Line 10), $\mathrm{proj}_{C'}$ stands for the projection on the coordinates in C', i.e., only the values corresponding to the variables in C' are returned; this is where the G'-marginalization map $I_{G'}$ appears. We use the query as a subscript in the distributional program and omit the subscript "$G \to G'$," since it will be clear from the context in our examples which graph $G = G'$ we use.

1: **function** GRAPHICALQPLEXMAPELEMENT$_{C'}^{(t)}(\iota)$
2: $\quad \mu \leftarrow L_G(\iota)$
3: $\quad z \sim \mu(\mathbf{Z} = \cdot)$
4: $\quad$ **for** $m = 1, \ldots, M$ **do**
5: $\qquad d_m \sim$ NUMBEROFCOMPLETIONS$_{\mu,m}^{(t)}(z)$
6: $\quad z' \sim \pi^{(t)}(\mathbf{Z}' = \cdot | z, d)$
7: $\quad$ **for** $m = 1, \ldots, M$ **do**
8: $\qquad k_m' \sim$ ENTITYTYPE$_m^{(t)}(z, d_m, z')$
9: $\qquad \ell_m' \sim$ NEXTLABEL$_{\mu,m}^{(t)}(z, z', k_m')$
10: $\quad$ **return** $\mathrm{proj}_{C'}(z', \ell')$

Assumptions We have chosen to restrict our attention to a narrower class of QPLEX maps with simple transition dynamics for ease of exposition, although the procedure we describe in this chapter applies more generally. We make three simplifying assumptions, each of which can be relaxed relatively easily. These assumptions have a bearing on the distributional programs NUMBEROF-COMPLETIONS$_{\mu,m}^{(t)}$, NEXTLABEL$_{\mu,m}^{(t)}$, and ENTITYTYPE$_m^{(t)}$ introduced in Sect. 4.3. Henceforth, we call these programs *utility programs*.

Our first assumption is that the scopes of each utility program NUMBEROFCOM-PLETIONS$_{\mu,m}^{(t)}$ and NEXTLABEL$_{\mu,m}^{(t)}$ are as small as possible given the information structure graph. Refer to Sect. 1.1 for the definition of scope. Specifically, throughout, we assume that the completion probability functions $\gamma_m^{(t)}(z, \ell_m)$, the relabeling pmfs $\pi_m^{(t),\mathrm{rel}}(\ell_m' | z, \ell_m, z')$, and the join pmfs $\pi_m^{(t),\mathrm{join}}(\ell_m' | z, z')$ do not include any of the current or next counters (if applicable) in their scope, except for those current counters in the neighbor set N_m of L_m in G. In view of this assumption, we write these model primitives as

$$\gamma_m^{(t)}(z_{N_m}, \ell_m),$$

$$\pi_m^{(t),\text{rel}}(\ell_m' | z_{N_m}, \ell_m), \tag{13.1}$$

$$\pi_m^{(t),\text{join}}(\ell_m' | z_{N_m}).$$

The purpose of this assumption is to avoid having to introduce notation for the scopes of these utility programs, as their scopes can be expressed in terms of the N_m.

Our second assumption is that the size functions $x_m^{(t)}$ and $x_m^{(t+1)}$ have the same scope, which we denote by S_m. The purpose of this assumption is to similarly be able to express the scope of the utility program ENTITYTYPE$_m^{(t)}$ in terms of existing notation, in this case S_m.

Our third assumption is that the routing pmf primitive is represented via a distributional program and that this program has a single exit point (i.e., the only return statement is at the end of the program). The purpose of this assumption is that the instructions of this program can then be inserted into the above generic distributional program for the graphical QPLEX map. All distributional programs for the routing pmfs in Chaps. 5 and 7 have already been written in this form.

Notation To make it easier to work with marginal and conditional pmfs encoded in an information collection $\iota \in \mathcal{I}_G$, we will use the symbol ι with appropriate arguments to select elements of this information collection (as well as marginal or conditional pmfs thereof). These conventions are best understood with examples.

In Example 13.1, for instance, we write $\iota(z_1, z_2, z_4)$ for the pmf $\iota_{\{Z_1, Z_2, Z_4\}}$ (or its value evaluated at (z_1, z_2, z_4)), and we also write $\iota(z_2, z_4)$ and $\iota(z_2)$ for the corresponding marginal pmfs of $\iota_{\{Z_1, Z_2, Z_4\}}$. Here, $\iota(z_2)$ can also be viewed as a marginal pmf of $\iota_{\{Z_2, Z_3, L_2\}}$, but this is the same pmf by the consistency requirement on the information collection ι. Thus, this notation can only be used if the arguments of ι correspond to a clique in G. For instance, $\iota(z_3, z_4)$ is *not* defined, as $\{Z_3, Z_4\}$ is not a clique in G. We use similar notation for conditional pmfs. For instance, $\iota(z_1|z_2)$ and $\iota(\ell_2|z_2, z_3)$ are defined via this convention, but $\iota(z_3|z_4)$ is not. In the same vein, we let ι_C be well-defined for all cliques C in G, not necessarily *maximal* cliques. Moreover, we let $\iota_{A|B}$ be well-defined as long as A and B are disjoint and $A \cup B$ is a clique in G.

Procedure We convert the distributional program for element C' of $\mathcal{Q}_{G \to G'}^{(t)}$ to an efficient refactored distributional program in three steps. The first step replaces the instructions of the utility programs (NUMBEROFCOMPLETIONS$_{\mu,m}^{(t)}$, ENTITY-TYPE$_m^{(t)}$, NEXTLABEL$_{\mu,m}^{(t)}$) in such a way that unused arguments are suppressed. The second step eliminates *dead variables*, namely, variables that are irrelevant to the program, because they are assigned values or sampled but are subsequently unused. The third step rewrites the sampling instruction of the counter variables in a way that the new instruction(s) use pmfs from the information collection ι instead of the full pmf μ. These steps are worked out in Sects. 13.2–13.4, after which we illustrate this three-step procedure for a number of examples.

13.2 Utility Program Modifications

The generic distributional program from the previous section uses utility programs in three places, specifically in Lines 5, 8, and 9. Two of these programs use μ. Since we seek to convert this program to an equivalent program free of μ, these instructions need to be replaced. We also need to omit arguments in these utility programs that are unused in the body of these programs. We discuss the three utility programs one by one, exploiting the specifics of the scopes of the label dynamics primitives. Although we will use the same program names, the presence of ι instead of μ in the subscript (and possibly the argument) indicates that these programs are, in fact, different. Keep in mind that ι here stands for a *full* information collection, which contains one pmf for each maximal clique in G.

Number of Completions Utility Program Reproduced below is the definition of the number of completions utility program.

1: **function** NUMBEROFCOMPLETIONS$_{\mu,m}^{(t)}(z)$

2: $\quad d_m \sim \text{bin}\left(x_m^{(t)}(z), \sum_{\ell_m} \mu(\ell_m|z) \times \gamma_m^{(t)}(z, \ell_m)\right)$

3: $\quad$ **return** d_m

The term $x_m^{(t)}(z)$ can be replaced with $x_m^{(t)}(z_{S_m})$ by definition of the scope S_m of $x_m^{(t)}$. We next use several properties stemming from the fact that G is an information structure graph. In view of Proposition 12.4(b), we have that

$$\mu(\ell_m|z) = \mu(\ell_m|z_{N_m}).$$

The right-hand side can be expressed in terms of ι as $\iota(\ell_m|z_{N_m})$, since $N_m \cup \{L_m\}$ is a clique in G (Lemma 12.2(c)), so $\mu(\ell_m|z)$ in the above program can be replaced with $\iota(\ell_m|z_{N_m})$. Moreover, $\gamma_m^{(t)}(z, \ell_m)$ can be replaced with $\gamma_m^{(t)}(z, \ell_m)$ in view of our assumption on the scope of the completion probability function.

Since $S_m \subseteq N_m$ is a requirement of an information structure graph (see Definition 12.1), we may replace the sampling instruction

$$d_m \sim \text{NUMBEROFCOMPLETIONS}_{\mu,m}^{(t)}(z)$$

with the sampling instruction

$$d_m \sim \text{NUMBEROFCOMPLETIONS}_{\iota,m}^{(t)}(z_{N_m})$$

where the program for NUMBEROFCOMPLETIONS$_{\iota,m}^{(t)}$ is given below. Note the use of the subscript ι instead of μ as well as the (possibly) fewer arguments.

1: **function** NUMBEROFCOMPLETIONS$_{\iota,m}^{(t)}(z_{N_m})$

2: $\quad d_m \sim \text{bin}\left(x_m^{(t)}(z_{S_m}), \sum_{\ell_m} \iota(\ell_m|z_{N_m}) \times \gamma_m^{(t)}(z_{N_m}, \ell_m)\right)$

3: $\quad$ **return** d_m

Example 13.2 In Example 13.1, suppose that the size functions are given by

$$x_1^{(t)}(z_1, z_2, z_3, z_4) = z_1$$

$$x_2^{(t)}(z_1, z_2, z_3, z_4) = \max(z_2, z_3).$$

The scopes of these respective functions are given by $S_1 = \{Z_1\}$ and $S_2 = \{Z_2, Z_3\}$. From the graph G, we furthermore obtain that $N_1 = \{Z_1\}$ and $N_2 = \{Z_2, Z_3\}$.

The sampling instruction

$$d_m \sim \text{NUMBEROFCOMPLETIONS}_{\mu,m}^{(t)}(z_1, z_2, z_3, z_4)$$

in the distributional program for element C' of $\mathcal{Q}_{G \to G'}^{(t)}$ can be replaced with

$$d_1 \sim \text{NUMBEROFCOMPLETIONS}_{t,1}^{(t)}(z_1)$$
$$d_2 \sim \text{NUMBEROFCOMPLETIONS}_{t,2}^{(t)}(z_2, z_3)$$

for $m = 1$ and $m = 2$, respectively. $\triangle$

Entity Type Utility Program The entity type utility program is the easiest to modify, because it does not depend on μ, and the only model primitives that appear in the body are the size functions $x_m^{(t)}$ and $x_m^{(t+1)}$. The arguments z and z' can be replaced with z_{S_m} and z'_{S_m}, respectively. As a result, the sampling instruction

$$k_m' \sim \text{ENTITYTYPE}_m^{(t)}(z, d_m, z')$$

can be replaced with the sampling instruction

$$k_m' \sim \text{ENTITYTYPE}_m^{(t)}(z_{S_m}, d_m, z'_{S_m})$$

The modified distributional program is shown below. Note that we again use the same name for this program, but we change the arguments in the signature of this program.

```
1: function ENTITYTYPE_m^(t)(z_{S_m}, d_m, z'_{S_m})
2:     if x_m^(t+1)(z'_{S_m}) = 0 then
3:         return n/a
4:     probabilityOld ← (x_m^(t)(z_{S_m}) − d_m)/x_m^(t+1)(z'_{S_m})
5:     entityIsOld ~ ber(probabilityOld)
6:     if entityIsOld then
7:         return old
8:     else
9:         return new
```

Example 13.3 Continuing Example 13.2, the sampling instruction

$$k_m' \sim \text{ENTITYTYPE}_m^{(t)}(z_1, z_2, z_3, z_4, d_m, z_1', z_2', z_3', z_4')$$

in the distributional program for element C' of $\mathcal{Q}_{G \to G'}^{(t)}$ can be replaced with

$$k_1' \sim \text{ENTITYTYPE}_1^{(t)}(z_1, d_1, z_1')$$
$$k_2' \sim \text{ENTITYTYPE}_2^{(t)}(z_2, z_3, d_2, z_2', z_3')$$

for $m = 1$ and $m = 2$, respectively.   △

Next Label Utility Program Replacing $\gamma_m^{(t)}(z, \ell_m)$, $\pi_m^{(t),\text{rel}}(\ell_m'|z, \ell_m)$, and $\pi_m^{(t),\text{join}}(\ell_m'|z)$ with (13.1) and $\mu(\ell_m|z)$ with $\iota(\ell_m|z_{N_m})$ yields the following modified distributional program for this utility program. Note that we have also updated the signature of this program from $\text{NEXTLABEL}_{\mu,m}^{(t)}(z, z', k_m')$.

1: **function** $\text{NEXTLABEL}_{\iota,m}^{(t)}(z_{N_m}, k_m')$
2:     **switch** k_m' **do**
3:         **case** new
4:             $\ell_m' \sim \pi_m^{(t),\text{join}}(L_m' = \cdot|z_{N_m})$
5:         **case** old
6:             $\ell_m \propto \iota(L_m = \cdot|z_{N_m}) \times (1 - \gamma_m^{(t)}(z_{N_m}, L_m = \cdot))$
7:             $\ell_m' \sim \pi_m^{(t),\text{rel}}(L_m' = \cdot|z_{N_m}, \ell_m)$
8:         **case** n/a
9:             $\ell_m' \sim \theta_m^0(L_m = \cdot)$
10:    **return** ℓ_m'

Example 13.4 Continuing Example 13.2, the sampling instruction

$$\ell_m' \sim \text{NEXTLABEL}_{\mu,m}^{(t)}(z_1, z_2, z_3, z_4, z_1', z_2', z_3', z_4', k_m')$$

in the distributional program for element C' of $\mathcal{Q}_{G \to G'}^{(t)}$ can be replaced with

$$\ell_1' \sim \text{NEXTLABEL}_{\iota,1}^{(t)}(z_1, k_1')$$
$$\ell_2' \sim \text{NEXTLABEL}_{\iota,2}^{(t)}(z_2, z_3, k_2')$$

for $m = 1$ and $m = 2$, respectively.   △

In the equivalent distributional program for element C' of $\mathcal{Q}_{G \to G'}^{(t)}$ shown below, we have replaced each of the three utility programs with their modified counterparts. Note that μ only appears in Lines 2 and 3 as a result of the changes made in this section.

1: **function** $\text{GRAPHICALQPLEXMAPELEMENT}_{C'}^{(t)}(\iota)$
2:    $\mu \leftarrow L_G(\iota)$
3:    $z \sim \mu(Z = \cdot)$
4:    **for** $m = 1, \ldots, M$ **do**
5:         $d_m \sim \text{NUMBEROFCOMPLETIONS}_{\iota,m}^{(t)}(z_{N_m})$
6:    $z' \sim \pi^{(t)}(Z' = \cdot|z, d)$
7:    **for** $m = 1, \ldots, M$ **do**
8:         $k_m' \sim \text{ENTITYTYPE}_m^{(t)}(z_{S_m}, d_m, z_{S_m}')$
9:         $\ell_m' \sim \text{NEXTLABEL}_{\iota,m}^{(t)}(z_{N_m}, k_m')$
10:   **return** $\text{proj}_{C'}(z', \ell')$

13.3   Dead Variable Elimination

Dead variable elimination is a standard technique used in computing by compilers in order to optimize code. The distributional program for the QPLEX map $\mathcal{Q}_{G \rightarrow G'}^{(t)}$ derived in the previous section returns $\text{proj}_{C'}(z', \ell')$, corresponding to a subset of the vectors ℓ' and z' generated in the body of the program. Returning only this subset can create unused variables called *dead variables*, and the instructions where such variables are assigned or sampled can be removed from the program. Each time a dead variable is eliminated, it may create additional dead variables and thus further opportunities to eliminate variables.

When we eliminate one or more dead variables from a sampling instruction where the right-hand side is a pmf (not a common random variable or a call to another distributional program), we replace the right-hand side with the appropriate marginal pmf, which requires calculating an appropriate sum on the right-hand side. When the right-hand side is a "general" pmf with no known structure, then the requisite summation becomes implicit. When the right-hand side is a common random variable, however, it may be possible to exploit known properties of its distribution to avoid such a sum, which could be expensive to calculate. For example, suppose that a sampling instruction involved a multinomial pmf of many variables. If all but one of the variables are dead, then the marginal pmf of the remaining variable is a binomial pmf. Finally, if the right-hand side is a call to another distributional program, then remove the dead variable(s) from its return statement(s) and (if possible) subsequently eliminate any dead variables from this program itself.

Different queries can result in different dead variables, so the development below (and subsequent steps to be discussed) needs to be carried out for each query separately. (How to identify, organize, and perhaps automate dead variable elimination we leave as an implementation detail.)

Example 13.5 We extend Example 13.1 in two ways. First, suppose that z_2 and z_4 are dead variables for the query $C' = \{Z_1', L_1'\}$, while z_1 and z_3 are not. The second instruction in the program for element C' of $\mathcal{Q}_{G \rightarrow G'}^{(t)}$ is the following sampling instruction:

$$(z_1, z_2, z_3, z_4) \sim \mu(Z_1 = \cdot, Z_2 = \cdot, Z_3 = \cdot, Z_4 = \cdot)$$

After eliminating the dead variables z_2 and z_4, this sampling instruction becomes the following sampling instruction:

$$(z_1, z_3) \sim \mu(Z_1 = \cdot, Z_3 = \cdot)$$

Now suppose that the program for element C' of $\mathcal{Q}_{G \rightarrow G'}^{(t)}$ contains the sampling instruction

$$(\phi_1, \phi_2, \phi_3) \sim \text{mult}(z_1, (\beta_1, \beta_2, \beta_3))$$

and that ϕ_1 and ϕ_2 are dead variables for the given query but ϕ_3 is not. After eliminating the dead variables ϕ_1 and ϕ_2, this instruction becomes the following sampling instruction:

$$\phi_3 \sim \text{bin}(z_1, \beta_3)$$

These changes lead to an equivalent distributional program with fewer variables. $\triangle$

13.4 Sampling Counters Using Information Collections

Line 3 of the distributional program for element C' of $\mathcal{Q}_{G \to G'}^{(t)}$ is a sampling instruction from a marginal pmf of μ of (a subset of) counters. It is our next objective to replace this instruction with one or more instructions that (1) only use pmfs in $\iota = \{\iota_C\}_{C \in \mathcal{C}(G)}$ and (2) reduce the number of other counters in the program. We rely on some machinery from Chap. 11 to do so, and the reader may wish to review this chapter first.

We fix some query C'. We let V_{alive} denote the set of counter vertices that have not been eliminated, so V_{alive} is a subset of the vertex set of G_Z. Let $\tilde{G}_Z = (\tilde{V}_Z, \tilde{E}_Z)$ be the subgraph of G_Z that remains after recursively eliminating as many simplicial vertices from G_Z as possible without eliminating vertices in V_{alive}.

Since $\tilde{G}_Z$ is constructed from G_Z by repeatedly removing simplicial vertices, the marginal pmf $\mu_{\tilde{V}_Z}$ of μ must satisfy the global Markov property with respect to $\tilde{G}_Z$ by Proposition 12.4(a) and Lemma 11.5. First, suppose that $\tilde{G}_Z$ is connected. Proposition 11.8 then states that, for any $\tilde{T} \in \mathcal{T}(\tilde{G}_Z)$, $\mu_{\tilde{V}_Z}$ can be represented as

$$\mu_{\tilde{V}_Z} = \prod_{C \in \mathcal{C}(\tilde{G}_Z)} \mu_{C \setminus \text{ch}_{\tilde{T}}(C) | C \cap \text{ch}_{\tilde{T}}(C)}. \tag{13.2}$$

The instruction for sampling the counters in V_{alive} can now be replaced with one or more sampling instructions, namely, one instruction per term in this product. Each maximal clique in $\tilde{G}_Z$ is a clique in G, so $\mu_C = \iota_C$ for each $C \in \mathcal{C}(\tilde{G}_Z)$ in view of $\mu = L_G(\iota)$ and Proposition 11.9. The instruction for $C \in \mathcal{C}(\tilde{G}_Z)$ is

$$z_{C \setminus \text{ch}_{\tilde{T}}(C)} \sim \iota_{C \setminus \text{ch}_{\tilde{T}}(C) | C \cap \text{ch}_{\tilde{T}}(C)}. \tag{13.3}$$

A reverse topological ordering of the maximal cliques in $\mathcal{C}(\tilde{G}_Z)$ results in an ordering of the associated instructions. As a consequence of Lemma 11.19, the sequence of sampling instructions so constructed has the property that no variable is used before it is sampled.

If $\tilde{G}_Z$ is not connected, then the procedure described above is applied to each of its components. More specifically, each component gives rise to a junction in-tree, a representation (13.2), and an ordered sequence of instructions (13.3). An arbitrary order can be used to order the resulting per-component sequence of instructions,

and the instructions from different components can even be interleaved as long as
the ordering of the instructions from each junction in-tree is maintained.

Since the representation in (13.2) and therefore the instructions (13.3) depend on
the choice of $\tilde{T} \in \mathcal{T}(\tilde{G})$, the distributional program that results may not be unique,
even in the single component case and with the same distributional program for
element C' of $\mathcal{Q}^{(t)}_{G \to G'}$ as a starting point. However, each such program is equivalent
in the sense that it outputs the same pmf.

Example 13.6 In Example 13.5, sampling (z_1, z_3) requires knowledge of the
marginal pmf $\mu(z_1, z_3)$. The set V_{alive} is given by $\{Z_1, Z_3\}$. To find the graph $\tilde{G}_{\mathbf{Z}}$,
we first eliminate the simplicial vertex Z_4 from $G_{\mathbf{Z}}$. The only remaining simplicial
vertices are Z_1 and Z_3, but these cannot be eliminated, because they belong to
V_{alive}. This yields the graph $\tilde{G}_{\mathbf{Z}}$ shown below. Also shown is a junction in-tree
$\tilde{T} \in \mathcal{T}(\tilde{G}_{\mathbf{Z}})$.

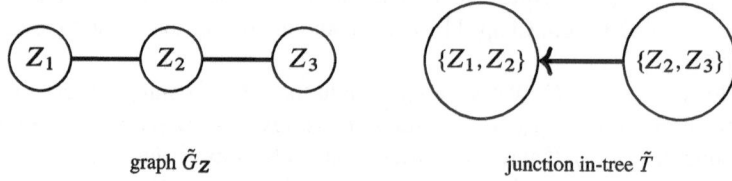

graph $\tilde{G}_{\mathbf{Z}}$ junction in-tree $\tilde{T}$

With this choice of $\tilde{T}$, (13.2) yields the representation

$$\mu(z_1, z_2, z_3) = \mu(z_1, z_2) \times \mu(z_3|z_2),$$

with the terms on the right-hand side ordered according to the unique reverse topo-
logical ordering of the vertices of $\tilde{T}$. Since $\mu(z_1, z_2) = \iota(z_1, z_2)$ and $\mu(z_3|z_2) =
\iota(z_3|z_2)$, the sampling instruction

$(z_1, z_3) \sim \mu(Z_1 = \cdot, Z_3 = \cdot)$

is thus replaced with the following pseudocode fragment:

$(z_1, z_2) \sim \iota(Z_1 = \cdot, Z_2 = \cdot)$
$z_3 \sim \iota(Z_3 = \cdot|z_2)$

Note that while z_2 was previously classified as a dead variable, it is "resurrected" in
these instructions due to the form of the information structure graph. The variable
z_2 is used to sample z_3 but unused in the rest of the distributional program. $\triangle$

13.5 Examples

Multiserver Queueing Model with Renewal Arrivals

We consider the multiserver queueing model with renewal arrivals from Sect. 5.1
with the information structure graph G depicted below. Recall that there are two
counter variables, Z and Z_A, and one label variable L. Also recall that Z_A represents

the number of time periods since the last customer arrival. The size function $x(z, z_A)$ is given by $x(z, z_A) = \min(z, n)$, so its scope is $\{Z\}$. The set of counter vertices that are adjacent to the label vertex L is $\{Z\}$. The two queries associated with the information structure graph $G' = G$ are $C' = \{Z', L'\}$ and $C' = \{Z', Z'_A\}$.

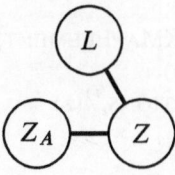

The distributional program for element C' of $\mathcal{Q}^{(t)}_{G \to G'}$ is shown below, and this program serves as our starting point. It incorporates the routing pmf primitive from Sect. 5.1 (Lines 5–7 below).

```
1: function GRAPHICALQPLEXMAPELEMENT_{C'}^{(t)}(\iota)
2:     μ ← L_G(\iota)
3:     (z, z_A) ~ μ(Z = ·, Z_A = ·)
4:     d ~ NUMBEROFCOMPLETIONS_μ^{(t)}(z, z_A)
5:     a ~ ber( η(z_A+1) / Σ_{z̄_A ≥ z_A+1} η(z̄_A) )
6:     z' ← z − d + a
7:     z'_A ← 1(a = 0) × (z_A + 1)
8:     k' ~ ENTITYTYPE^{(t)}(z, z_A, d, z'_A, z')
9:     ℓ' ~ NEXTLABEL_μ^{(t)}(z, z_A, z'_A, z', k')
10:    return proj_{C'}(z', z'_A, ℓ')
```

Query $C' = \{Z', L'\}$ We first derive the efficient refactored distributional program for the query $C' = \{Z', L'\}$.

Utility Program Modification The neighbor set of the label vertex L is $\{Z\}$, and the scope of the size function is also $\{Z\}$. The following pseudocode fragment replaces the corresponding lines in the above program:

```
4:     d ~ NUMBEROFCOMPLETIONS_\iota^{(t)}(z)
8:     k' ~ ENTITYTYPE^{(t)}(z, d, z')
9:     ℓ' ~ NEXTLABEL_\iota^{(t)}(z, k')
```

Dead Variable Elimination For this query, z'_A is the only dead variable in the resulting program.

Sampling Counters from ι In Line 3, we sample (z, z_A) using μ. The collection ι contains the pmf $\iota(z, z_A)$ since $\{Z, Z_A\} \in \mathcal{C}(G)$, so (z, z_A) can be sampled directly from this pmf instead. We reach the same conclusion with the procedure from Sect. 13.4. We find that $\tilde{G}_Z = G_Z$ because $V_{alive} = \{Z, Z_A\}$ and no simplicial vertex can be eliminated. Since $\{Z, Z_A\} \in \mathcal{C}(G)$, the junction in-tree has one vertex representing the set of vertices in G_Z. Therefore, we can replace the sampling instruction in Line 3 with a single sampling instruction of the form (13.3), and its right-hand side equals ι_C with $C = \{Z, Z_A\}$.

Here is the efficient refactored distributional program corresponding to element $C' = \{Z', L'\}$ of $\mathcal{Q}_{G \to G'}^{(t)}$.

1: **function** GRAPHICALQPLEXMAPELEMENT$_{\{Z', L'\}}^{(t)}(\iota)$
2:     $(z, z_A) \sim \iota(Z = \cdot, Z_A = \cdot)$
3:     $d \sim$ NUMBEROFCOMPLETIONS$_{\iota}^{(t)}(z)$
4:     $a \sim \text{ber}\left(\frac{\eta(z_A+1)}{\sum_{\tilde{z}_A \geq z_A+1} \eta(\tilde{z}_A)} \right)$
5:     $z' \leftarrow z - d + a$
6:     $k' \sim$ ENTITYTYPE$^{(t)}(z, d, z')$
7:     $\ell' \sim$ NEXTLABEL$_{\iota}^{(t)}(z, k')$
8:     **return** (z', ℓ')

Query $C' = \{Z', Z_A'\}$ We next derive the efficient refactored distributional program for the query $C' = \{Z', Z_A'\}$. The utility program modification step is exactly as before, so we do not repeat it.

Dead Variable Elimination For this query, k' and ℓ' are the dead variables in the resulting program for element C' of $\mathcal{Q}_{G \to G'}^{(t)}$.

Sampling Counters from ι Once again, z and z_A are not dead variables, so the instruction in Line 3 is modified as before. This observation results in the efficient refactored distributional program corresponding to element $C' = \{Z', Z_A'\}$ of $\mathcal{Q}_{G \to G'}^{(t)}$.

1: **function** GRAPHICALQPLEXMAPELEMENT$_{\{Z', Z_A'\}}^{(t)}(\iota)$
2:     $(z, z_A) \sim \iota(Z = \cdot, Z_A = \cdot)$
3:     $d \sim$ NUMBEROFCOMPLETIONS$_{\iota}^{(t)}(z)$
4:     $a \sim \text{ber}\left(\frac{\eta(z_A+1)}{\sum_{\tilde{z}_A \geq z_A+1} \eta(\tilde{z}_A)} \right)$
5:     $z' \leftarrow z - d + a$
6:     $z_A' \leftarrow 1(a = 0) \times (z_A + 1)$
7:     **return** (z', z_A')

Queueing Network with Probabilistic Routing

We consider the queueing network model with probabilistic routing from Sect. 5.5 in which the information structure graph is an independent subsystem graph where each component corresponds to a station. It is depicted below for the case when $M = 5$. Recall that $\Phi_{m \to \tilde{m}}$ denotes the flow of customers from station m to station

$\tilde{m}$. Station $m = 0$ denotes both the source of external customers and the sink for those customers who leave the system. The size functions are given by $x_m^{(t)}(z) = \min(z_m, n_m)$. The M queries associated with the information structure graph $G' = G$ are $C' = \{Z_m', L_m'\}$, $m = 1, \ldots, M$.

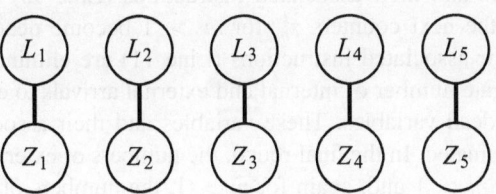

The distributional program for element C' of the graphical QPLEX map $\mathcal{Q}_{G \to G'}^{(t)}$ is shown below. It incorporates the routing pmf primitive from Sect. 5.5 (Lines 6–11 below). The first two for-loops can obviously be merged to make the presentation of this program more compact. We have chosen instead to preserve the routing pmf instructions verbatim in this program to emphasize that they have been copied. This choice does not affect the computational effort of the algorithms we discuss later.

1: **function** GRAPHICALQPLEXMAPELEMENT$_{C'}^{(t)}(\iota)$
2: $\quad \mu \leftarrow L_G(\iota)$
3: $\quad (z_1, \ldots, z_M) \sim \mu(Z_1 = \cdot, \ldots, Z_M = \cdot)$
4: $\quad$ **for** $m = 1, \ldots, M$ **do**
5: $\quad\quad d_m \sim$ NUMBEROFCOMPLETIONS$_{\mu,m}^{(t)}(z_1, \ldots, z_M)$
6: $\quad$ **for** $m = 1, \ldots, M$ **do**
7: $\quad\quad \phi_{0 \to m} \sim \alpha_m^{(t)}(A = \cdot)$
8: $\quad\quad (\phi_{m \to 0}, \ldots, \phi_{m \to M}) \sim \text{mult}\left(d_m, \beta_{m \to}^{(t)}\right)$
9: $\quad$ **for** $m = 1, \ldots, M$ **do**
10: $\quad\quad a_m \leftarrow \phi_{0 \to m} + \sum_{\tilde{m} \notin \{0, m\}} \phi_{\tilde{m} \to m}$
11: $\quad\quad z_m' \leftarrow z_m - d_m + \phi_{m \to m} + a_m$
12: $\quad$ **for** $m = 1, \ldots, M$ **do**
13: $\quad\quad k_m' \sim$ ENTITYTYPE$_m^{(t)}(z_1, \ldots, z_M, d_m, z_1', \ldots, z_M')$
14: $\quad\quad \ell_m' \sim$ NEXTLABEL$_{\mu,m}^{(t)}(z_1, \ldots, z_M, z_1', \ldots, z_M', k_m')$
15: $\quad$ **return** proj$_{C'}(z_1', \ldots, z_M', \ell_1', \ldots, \ell_M')$

We show how to construct the efficient refactored distributional program for the query $C' = \{Z_1', L_1'\}$, as the other queries are the same by symmetry.

Utility Program Modification The neighbor set of each L_m is $\{Z_m\}$, and the scope of each size function $x_m^{(t)}$ is also $\{Z_m\}$. The following pseudocode fragment replaces the corresponding lines in the above program:

5: $\quad\quad d_m \sim$ NUMBEROFCOMPLETIONS$_{\iota,m}^{(t)}(z_m)$
13: $\quad\quad k_m' \sim$ ENTITYTYPE$_m^{(t)}(z_m, d_m, z_m')$
14: $\quad\quad \ell_m' \sim$ NEXTLABEL$_{\iota,m}^{(t)}(z_m, k_m')$

Dead Variable Elimination The dead variable elimination step consists of five rounds. In the first round, the next labels ℓ'_m for $m > 1$ are dead variables. These variables and their associated instructions (Line 14 of the distributional program) are eliminated. In the second round, the corresponding variables k'_m for $m > 1$ are dead variables and their associated instructions (Line 13) are eliminated. In the third round, the next counters z'_m for $m > 1$ become dead variables. These variables and their associated instructions (Line 11) are eliminated. In the fourth round, the aggregate number of internal and external arrivals to each station a_m for $m > 1$ become dead variables. These variables and their associated instructions (Line 10) are eliminated. In the final round, the numbers of external arrivals to each station $\phi_{0 \to m}$ for $m > 1$ and, again for $\tilde{m} > 1$, the numbers of customers $\phi_{m \to \tilde{m}}$ from station m routed to station $\tilde{m}$ all become dead variables, too. As a result, the instruction in Line 8 is replaced with the following sampling instruction:

$$\phi_{m \to 1} \sim \text{bin}(d_m, \beta^{(t)}_{m \to 1})$$

At this point, with the z_m, d_m, $\phi_{m \to 1}$ as well as $\phi_{0 \to 1}$, a_1, z'_1, k'_1, and ℓ'_1 remaining, there are no more dead variables.

Sampling Counters from ι The sampling instruction in Line 3 generates the vector $(z_1, \ldots, z_M)$ using μ. To translate this instruction into one or more instructions using ι, we first note that all counters remain used in the program, so $V_{\text{alive}} = \{Z_1, \ldots, Z_M\}$. In particular, no simplicial vertices can be eliminated and $\tilde{G}_Z = G_Z$. The marginal pmf $\mu_{\tilde{V}_Z}$ of μ satisfies the global Markov property with respect to $\tilde{G}_Z$, and this graph has M components. Each of these components has an associated junction in-tree with a single vertex representing a component. We thus find that $\mu(z_1, \ldots, z_M) = \prod_{m=1}^{M} \mu(z_m)$. Each of these junction in-trees gives rise to an instruction of the form $z_m \sim \iota(Z_m = \cdot)$ using (13.3). In particular, the sampling instruction in Line 3 is replaced with the M sampling instructions $z_m \sim \iota(Z_m = \cdot)$ for $m = 1, \ldots, M$.

Here is the efficient refactored distributional program corresponding to element $C' = \{Z'_1, L'_1\}$ of $Q^{(t)}_{G \to G'}$ after the first three steps. In this program, we recognize the QPLEX calculus instructions for a single station in Lines 1, 2, 12, and 13. In fact, this program is identical to the subsystem QPLEX map from Chap. 10, in which each subsystem corresponds to a station. Lines 4–11 are the instructions that define the subsystem routing pmf there.

1: **function** GRAPHICALQPLEXMAPELEMENT$^{(t)}_{\{Z'_1, L'_1\}}(\iota)$

2:     $z_1 \sim \iota(Z_1 = \cdot)$

3:     $d_1 \sim$ NUMBEROFCOMPLETIONS$^{(t)}_{\iota, 1}(z_1)$

4:     **for** $m = 2, \ldots, M$ **do**

5:         $z_m \sim \iota(Z_m = \cdot)$

6:         $d_m \sim$ NUMBEROFCOMPLETIONS$^{(t)}_{\iota, m}(z_m)$

7:         $\phi_{m \to 1} \sim \text{bin}\left(d_m, \beta^{(t)}_{m \to 1}\right)$

8:     $\phi_{1 \to 1} \leftarrow \text{bin}\left(d_1, \beta^{(t)}_{1 \to 1}\right)$

9:     $\phi_{0\to1} \sim \alpha_1^{(t)}(A = \cdot)$
10:    $a_1 \leftarrow \phi_{0\to1} + \sum_{\tilde{m}\notin\{0,1\}} \phi_{m\to1}$
11:    $z_1' \leftarrow z_1 - d_1 + \phi_{1\to1} + a_1$
12:    $k_1' \sim \text{ENTITYTYPE}_1^{(t)}(z_1, d_1, z_1')$
13:    $\ell_1' \sim \text{NEXTLABEL}_{\ell,1}^{(t)}(z_1, k_1')$
14:    **return** (z_1', ℓ_1')

Flow Line with Fishbone Information Structure

We consider the flow line of M multiserver stations of Sect. 5.4 in which the information structure graph is a fishbone information structure graph depicted below for the case when $M = 5$. The flow line model is a special case of a queueing network model with probabilistic routing, so the size functions are defined as $x_m^{(t)}(z) = \min(z_m, n_m)$ and their scopes are $S_m = \{Z_m\}$.

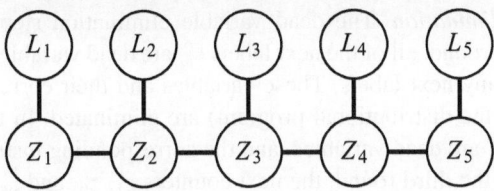

There are two differences with the previous example specialized to the flow line. The most important one is that we use a different information structure graph even though the other model primitives have not changed. The $2M - 1$ queries associated with the information structure graph $G' = G$ are $C' = \{Z_m', L_m'\}$, $m = 1, \ldots, M$, and $C' = \{Z_m', Z_{m+1}'\}$, $m = 1, \ldots, M - 1$. The second is that we use the routing pmf program for the flow line system from Sect. 5.4, which yields slightly more compact distributional programs.

The distributional program for the graphical QPLEX map element C' of $\mathcal{Q}_{G\to G'}^{(t)}$ is shown below. It incorporates the routing pmf primitive from Sect. 5.4 (Lines 6–8).

1: **function** GRAPHICALQPLEXMAPELEMENT$_{C'}^{(t)}(\iota)$
2:    $\mu \leftarrow L_G(\iota)$
3:    $(z_1, \ldots, z_M) \sim \mu(Z_1 = \cdot, \ldots, Z_M = \cdot)$
4:    **for** $m = 1, \ldots, M$ **do**
5:       $d_m \sim \text{NUMBEROFCOMPLETIONS}_{\mu,m}^{(t)}(z_1, \ldots, z_M)$
6:    $d_0 \sim \alpha_1^{(t)}(A = \cdot)$
7:    **for** $m = 1, \ldots, M$ **do**
8:       $z_m' \leftarrow z_m - d_m + d_{m-1}$
9:    **for** $m = 1, \ldots, M$ **do**

10:     $k'_m \sim \text{ENTITYTYPE}_m^{(t)}(z_1, \ldots, z_M, d_m, z'_1, \ldots, z'_M)$

11:     $\ell'_m \sim \text{NEXTLABEL}_{\mu,m}^{(t)}(z_1, \ldots, z_M, z'_1, \ldots, z'_M, k'_m)$

12:     **return** $\text{proj}_{C'}(z'_1, \ldots, z'_M, \ell'_1, \ldots, \ell'_M)$

Query $C' = \{Z'_3, Z'_4\}$ Assuming $M \geq 4$, we show how to construct the efficient refactored distributional programs for the two queries $C' = \{Z'_3, Z'_4\}$ and $C' = \{Z'_3, L'_3\}$, as all other queries for this information structure graph follow from identical arguments. We start with the query $C' = \{Z'_3, Z'_4\}$.

Utility Program Modification This step is identical to the one from the previous section, because the neighbor set of each label variable L_m is again $\{Z_m\}$, and the scope of each size function $x_m^{(t)}$ is again $\{Z_m\}$. The following pseudocode fragment thus replaces the corresponding lines in the above program:

5:      $d_m \sim \text{NUMBEROFCOMPLETIONS}_{\iota,m}^{(t)}(z_m)$

10:     $k'_m \sim \text{ENTITYTYPE}_m^{(t)}(z_m, d_m, z'_m)$

11:     $\ell'_m \sim \text{NEXTLABEL}_{\iota,m}^{(t)}(z_m, k'_m)$

Dead Variable Elimination The dead variable elimination step proceeds in five rounds. In the first round, all of the next labels ℓ'_m are dead variables, since this query does not involve any next labels. These variables and their corresponding instructions (Line 11 of the distributional program) are eliminated. In the second round, the entity types k'_m are dead variables, and the corresponding instructions (Line 10) are eliminated. In the third round, the next counters z'_1, z'_2, and $z'_5, \ldots, z'_M$ become dead variables. These variables and their corresponding instructions (Line 8) are eliminated. In the fourth round, the numbers of activity completions $d_1, d_5, \ldots, d_M$ and the external number of arrivals d_0 become dead variables. These variables and their corresponding instructions (Lines 5 and 6) are eliminated. In the fifth round, the counters z_1 and $z_5, \ldots, z_M$ become dead variables and are eliminated (Line 3). At this point, there are no more dead variables.

Sampling Counters from ι We have that $V_{\text{alive}} = \{Z_2, Z_3, Z_4\}$. To find the graph $\tilde{G}_Z$, we eliminate the counter vertices Z_1 and $Z_M, \ldots, Z_5$, after which we can no longer eliminate simplicial vertices. Therefore, $\tilde{G}_Z$ is the subgraph of G_Z induced by V_{alive}. It is shown below, together with a junction in-tree $\tilde{T} \in \mathcal{T}(\tilde{G}_Z)$.

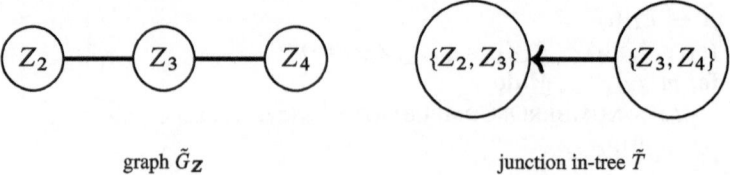

graph $\tilde{G}_Z$ junction in-tree $\tilde{T}$

With this choice of $\tilde{T}$, Proposition 11.8 then states that, since μ satisfies the global Markov property with respect to $\tilde{G}_Z$,

$$\mu(z_2, z_3, z_4) = \mu(z_2, z_3) \times \mu(z_4 | z_3).$$

The terms on the right-hand side are ordered according to the unique reverse topological ordering of the vertices of $\tilde{T}$. We thus obtain with (13.3) that the sampling instruction in Line 3 needs to be replaced with the following pseudocode fragment:

$(z_2, z_3) \sim \iota(Z_2 = \cdot, Z_3 = \cdot)$
$z_4 \sim \iota(Z_4 = \cdot | z_3)$

Here is the efficient refactored distributional program corresponding to element $C' = \{Z'_3, Z'_4\}$ of $\mathcal{Q}^{(t)}_{G \to G'}$.

1: **function** GRAPHICALQPLEXMAPELEMENT$^{(t)}_{\{Z'_3, Z'_4\}}(\iota)$
2:     $(z_2, z_3) \sim \iota(Z_2 = \cdot, Z_3 = \cdot)$
3:     $z_4 \sim \iota(Z_4 = \cdot | z_3)$
4:     **for** $m = 2, 3, 4$ **do**
5:         $d_m \sim$ NUMBEROFCOMPLETIONS$^{(t)}_{\iota, m}(z_m)$
6:     **for** $m = 3, 4$ **do**
7:         $z'_m \leftarrow z_m - d_m + d_{m-1}$
8:     **return** (z'_3, z'_4)

Query $C' = \{Z'_3, L'_3\}$ We now turn to the derivation of the efficient refactored distributional program for the query $C' = \{Z'_3, L'_3\}$. The utility program modification step is exactly as before, so we do not repeat it.

Dead Variable Elimination   The dead variable elimination step consists of five rounds. In the first round, the next labels ℓ'_1, ℓ'_2, and $\ell'_4, \ldots, \ell'_M$ are dead variables. These variables and their corresponding instructions (Line 11 of the program) are eliminated. The second round eliminates the corresponding variables k'_1, k'_2, and $k'_4, \ldots, k'_M$ and instructions (Line 10). In the third round, the next counters z'_1, z'_2, and $z'_4, \ldots, z'_M$ become dead variables. These variables and their corresponding instructions (Lines 8) are eliminated. In the fourth round, the numbers of activity completions $d_1, d_4, \ldots, d_M$ and the number of external arrivals d_0 all become dead variables. These variables and their corresponding instructions (Lines 5 and 6) are eliminated. In the fifth round, the counters z_1, z_4, and z_5 become dead variables (Line 3). At this point, no further variables can be eliminated.

Sampling Counters from ι   We have that $V_{\text{alive}} = \{Z_2, Z_3\}$. Applying the same logic as for the query $\{Z'_3, Z'_4\}$ shows that $\tilde{G}_Z$ is the subgraph of G_Z induced by V_{alive}. Since $V_{\text{alive}} \in \mathcal{C}(G)$, we can directly sample from the pmf represented by $\tilde{G}_Z$.

Here is the efficient refactored distributional program corresponding to element $C' = \{Z_3', L_3'\}$ of $\mathcal{Q}_{G \to G'}^{(t)}$.

```
1: function GRAPHICALQPLEXMAPELEMENT(t)_{Z_3', L_3'}(ι)
2:     (z_2, z_3) ~ ι(Z_2 = ·, Z_3 = ·)
3:     for m = 2, 3 do
4:         d_m ~ NUMBEROFCOMPLETIONS(t)_{ι,m}(z_m)
5:     z_3' ← z_3 − d_3 + d_2
6:     k_3' ~ ENTITYTYPE_3^(t)(z_3, d_3, z_3')
7:     ℓ_3' ~ NEXTLABEL_{ι,3}^(t)(z_3, k_3')
8:     return (z_3', ℓ_3')
```

1: **function** GRAPHICALQPLEXMAPELEMENT$^{(t)}_{\{Z_3', L_3'\}}(\iota)$
2: $\quad (z_2, z_3) \sim \iota(Z_2 = \cdot, Z_3 = \cdot)$
3: $\quad$ **for** $m = 2, 3$ **do**
4: $\quad\quad d_m \sim$ NUMBEROFCOMPLETIONS$^{(t)}_{\iota,m}(z_m)$
5: $\quad z_3' \leftarrow z_3 - d_3 + d_2$
6: $\quad k_3' \sim$ ENTITYTYPE$_3^{(t)}(z_3, d_3, z_3')$
7: $\quad \ell_3' \sim$ NEXTLABEL$^{(t)}_{\iota,3}(z_3, k_3')$
8: $\quad$ **return** (z_3', ℓ_3')

Load Balancing

We consider the load balancing model from Sect. 5.6 in which the information structure graph is an independent subsystem graph where each component corresponds to a station. We consider the load balancing policy where all external arrivals go to the same station based on the shortest cycle time percentile; see Example 5.2.

The distributional program for element C' of $\mathcal{Q}_{G \to G'}^{(t)}$ is shown below. It incorporates the instructions of the routing pmf from Example 5.2 in Lines 6–9. Recall from Sect. 4.5 that the dependence of the routing pmf on μ puts us outside the setting of simple (and advanced) transition dynamics.

1: **function** GRAPHICALQPLEXMAPELEMENT$^{(t)}_{C'}(\iota)$
2: $\quad \mu \leftarrow L_G(\iota)$
3: $\quad (z_1, \ldots, z_M) \sim \mu(Z_1 = \cdot, \ldots, Z_M = \cdot)$
4: $\quad$ **for** $m = 1, \ldots, M$ **do**
5: $\quad\quad d_m \sim$ NUMBEROFCOMPLETIONS$^{(t)}_{\mu,m}(z_1, \ldots, z_M)$
6: $\quad a \sim \alpha^{(t)}(A = \cdot)$
7: $\quad m^* \sim$ unif$(\arg\min(t_{\mu(L_1 = \cdot | z_1), 1}(z_1), \ldots, t_{\mu(L_M = \cdot | z_M), M}(z_M)))$
8: $\quad$ **for** $m = 1, \ldots, M$ **do**
9: $\quad\quad z_m' \leftarrow z_m - d_m + a \times 1(m^* = m)$
10: $\quad$ **for** $m = 1, \ldots, M$ **do**
11: $\quad\quad k_m' \sim$ ENTITYTYPE$_m^{(t)}(z_1, \ldots, z_M, d_m, z_1', \ldots, z_M')$
12: $\quad\quad \ell_m' \sim$ NEXTLABEL$^{(t)}_{\mu,m}(z_1, \ldots, z_M, z_1', \ldots, z_M', k_m')$
13: $\quad$ **return** proj$_{C'}(z_1', \ldots, z_M', \ell_1', \ldots, \ell_M')$

We show how to construct the efficient refactored distributional programs for the query $C' = \{Z_1', L_1'\}$, as all other queries for this information structure graph follow from identical arguments by symmetry.

Utility Program Modification This step is identical to the one for the queueing network model with probabilistic routing, because the neighbor set of each label vertex L_m is $\{Z_m\}$, and the scope of each size function $x_m^{(t)}$ is also $\{Z_m\}$. The following pseudocode fragment replaces the corresponding lines in the above program:

5:    $d_m \sim \text{NUMBEROFCOMPLETIONS}_{\iota,m}^{(t)}(z_m)$

10:    $k_m' \sim \text{ENTITYTYPE}_m^{(t)}(z_m, d_m, z_m')$

11:    $\ell_m' \sim \text{NEXTLABEL}_{\iota,m}^{(t)}(z_m, k_m')$

Dead Variable Elimination The dead variable elimination step consists of three rounds. First, the next labels ℓ_m' for $m > 1$ are dead variables. After eliminating these variables and their corresponding instructions, the next counters z_m' for $m > 1$ then become dead variables. After eliminating these variables and their corresponding instructions, the number of completions d_m for $m > 1$ then become dead variables. These variables and their corresponding instructions are eliminated. There are no other dead variables. In particular, unlike other examples, *none of the counters z_m for $m > 1$ can be eliminated* as they are needed for the determination of m^*.

Sampling Counters from ι This step is also identical to the one used for the queueing network model with probabilistic routing. We have that $V_{\text{alive}} = \{z_1, \ldots, z_M\}$, so no simplicial vertices can be eliminated and $\tilde{G}_Z = G_Z$. The sampling instruction in Line 3 of the distributional program is replaced with the M sampling instructions $z_m \sim \iota(Z_m = \cdot)$ for $m = 1, \ldots, M$.

Here is the efficient refactored distributional program corresponding to element $C' = \{Z_1', L_1'\}$ of $\mathcal{Q}_{G \to G'}^{(t)}$. Note that we have also replaced the $\mu(L_m = \cdot | z_m)$ in Line 7 with $\iota(L_m = \cdot | z_m)$.

1: **function** $\text{GRAPHICALQPLEXMAPELEMENT}_{\{Z_1', L_1'\}}^{(t)}(\iota)$

2:    **for** $m = 1, \ldots, M$ **do**

3:        $z_m \sim \iota(Z_m = \cdot)$

4:    $d_1 \sim \text{NUMBEROFCOMPLETIONS}_{\iota,1}^{(t)}(z_1)$

5:    $a \sim \alpha^{(t)}(A = \cdot)$

6:    $m^* \sim \text{unif}(\arg\min(t_{\iota(L_1 = \cdot | z_1),1}(z_1), \ldots, t_{\iota(L_M = \cdot | z_M),M}(z_M)))$

7:    $z_1' \leftarrow z_1 - d_1 + a \times 1(m^* = 1)$

8:    $k_1' \sim \text{ENTITYTYPE}_1^{(t)}(z_1, d_1, z_1')$

9:    $\ell_1' \sim \text{NEXTLABEL}_{\iota,1}^{(t)}(z_1, k_1')$

10:    **return** (z_1', ℓ_1')

Queueing Network with Blocking

We consider the queueing network model with blocking from Sect. 5.7 in which the information structure graph is an independent subsystem graph with each station corresponding to a subsystem. We reproduce this information structure graph below for $M = 2$; see Sect. 12.4.

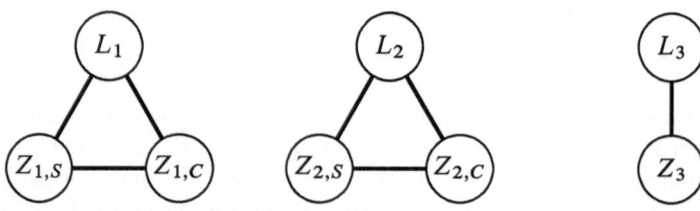

We construct the efficient refactored distributional programs for the two elements of the graphical QPLEX map with both the priority and randomized allocation policies from Sect. 5.7.

Priority Allocation Policy

The distributional program for element C' of the graphical QPLEX map $\mathcal{Q}_{G \to G'}^{(t)}$ is shown below for the priority allocation policy. It incorporates the instructions of the routing pmf in Lines 6–13, albeit in a different form than in Sect. 5.7. We have substituted the distributional program PRIORITYDEMANDSPLIT into the routing pmf given in Sect. 5.7 and made the following additional modifications. First, we substitute s_0 for s and eliminate s. Then, we eliminate the demand $\hat{d}_m$ for slots at station $M + 1$ from station m by substituting $\hat{d}_m = z_{m,C} + d_m$. We also eliminate the unmet demand u_m from station m by substituting $u_m = z'_{m,C}$.

1: **function** GRAPHICALQPLEXMAPELEMENT$_{C'}^{(t)}(\iota)$
2:     $\mu \leftarrow L_G(\iota)$
3:     $(z_{1,S}, z_{1,C}, \ldots, z_{M+1}) \sim \mu(Z_{1,S} = \cdot, Z_{1,C} = \cdot, \ldots, Z_{M+1} = \cdot)$
4:     **for** $m = 1, \ldots, M + 1$ **do**
5:         $d_m \sim$ NUMBEROFCOMPLETIONS$_{\mu,m}^{(t)}(z_{1,S}, z_{1,C}, \ldots, z_{M+1})$
6:     $s_0 \leftarrow n_{M+1} + b_{M+1} - (z_{M+1} - d_{M+1})$
7:     **for** $m = 1, \ldots, M$ **do**
8:         $a_m \sim \alpha_m^{(t)}(A = \cdot)$
9:         $z'_{m,S} \leftarrow z_{m,S} - d_m + a_m$
10:        $\phi_m \leftarrow \min(z_{m,C} + d_m, s_{m-1})$
11:        $z'_{m,C} \leftarrow z_{m,C} + d_m - \phi_m$
12:        $s_m \leftarrow s_{m-1} - \phi_m$
13:    $z'_{M+1} \leftarrow z_{M+1} - d_{M+1} + \sum_m \phi_m$
14:    **for** $m = 1, \ldots, M + 1$ **do**
15:        $k'_m \sim$ ENTITYTYPE$_m^{(t)}(z_{1,S}, z_{1,C}, \ldots, z_{M+1}, d_m, z'_{1,S}, z'_{1,C}, \ldots, z'_{M+1})$
16:        $\ell'_m \sim$ NEXTLABEL$_{\mu,m}^{(t)}(z_{1,S}, z_{1,C}, \ldots, z_{M+1}, z'_{1,S}, z'_{1,C}, \ldots, z'_{M+1}, k'_m)$
17:    **return** proj$_{C'}(z'_{1,S}, z'_{1,C}, \ldots, z'_{M+1}, \ell'_1, \ldots, \ell'_{M+1})$

Query $C' = \{Z'_{m,S}, Z'_{m,C}, L'_m\}$ The steps to modify the utility programs, eliminate dead variables, and sample counters using ι follow the usual logic. We only work out the utility modification step here. For $1 \leq m \leq M$, the neighbor set N_m of each L_m is $\{Z_{m,S}, Z_{m,C}\}$, and the scope S_m of each size function $x_m^{(t)}$ is also $\{Z_{m,S}, Z_{m,C}\}$. The following pseudocode fragment replaces the corresponding lines in the above program for such m:

5:     $d_m \sim \text{NUMBEROFCOMPLETIONS}_{\iota,m}^{(t)}(z_{m,S}, z_{m,C})$

15:    $k'_m \sim \text{ENTITYTYPE}_m^{(t)}(z_{m,S}, z_{m,C}, d_m, z'_{m,S}, z'_{m,C})$

16:    $\ell'_m \sim \text{NEXTLABEL}_{\iota,m}^{(t)}(z_{m,S}, z_{m,C}, k'_m)$

The neighbor set of L_{M+1} is $\{Z_{M+1}\}$ and so is the scope of $x_{M+1}^{(t)}$. The following pseudocode fragment replaces the corresponding lines in the above program for $m = M + 1$:

5:     $d_{M+1} \sim \text{NUMBEROFCOMPLETIONS}_{\iota,M+1}^{(t)}(z_{M+1})$

15:    $k'_{M+1} \sim \text{ENTITYTYPE}_{M+1}^{(t)}(z_{M+1}, d_{M+1}, z'_{M+1})$

16:    $\ell'_{M+1} \sim \text{NEXTLABEL}_{\iota,M+1}^{(t)}(z_{M+1}, k'_{M+1})$

Here is the efficient refactored distributional program corresponding to element $C' = \{Z'_{m,S}, Z'_{m,C}, L'_m\}$ of $\mathcal{Q}_{G \to G'}^{(t)}$.

1: **function** $\text{GRAPHICALQPLEXMAPELEMENT}_{\{Z'_{m,S}, Z'_{m,C}, L'_m\}}^{(t)}(\iota)$

2:     **for** $\tilde{m} = 1, \ldots, m$ **do**

3:         $(z_{\tilde{m},S}, z_{\tilde{m},C}) \sim \iota(Z_{\tilde{m},S} = \cdot, Z_{\tilde{m},C} = \cdot)$

4:         $d_{\tilde{m}} \sim \text{NUMBEROFCOMPLETIONS}_{\iota,\tilde{m}}^{(t)}(z_{\tilde{m},S}, z_{\tilde{m},C})$

5:     $z_{M+1} \sim \iota(Z_{M+1} = \cdot)$

6:     $d_{M+1} \sim \text{NUMBEROFCOMPLETIONS}_{\iota,M+1}^{(t)}(z_{M+1})$

7:     $s_0 \leftarrow n_{M+1} + b_{M+1} - (z_{M+1} - d_{M+1})$

8:     **for** $\tilde{m} = 1, \ldots, m$ **do**

9:         $\phi_{\tilde{m}} \leftarrow \min(z_{\tilde{m},C} + d_{\tilde{m}}, s_{\tilde{m}-1})$

10:        $s_{\tilde{m}} \leftarrow s_{\tilde{m}-1} - \phi_{\tilde{m}}$

11:    $a_m \sim \alpha_m^{(t)}(A = \cdot)$

12:    $z'_{m,S} \leftarrow z_{m,S} - d_m + a_m$

13:    $z'_{m,C} \leftarrow z_{m,C} + d_m - \phi_m$

14:    $k'_m \sim \text{ENTITYTYPE}_m^{(t)}(z_{m,S}, z_{m,C}, d_m, z'_{m,S}, z'_{m,C})$

15:    $\ell'_m \sim \text{NEXTLABEL}_{\iota,m}^{(t)}(z_{m,S}, z_{m,C}, k'_m)$

16:    **return** $(z'_{m,S}, z'_{m,C}, \ell'_m)$

Although this is the efficient refactored distributional program obtained using the steps from this chapter, it is possible to eliminate more variables. Indeed, Lines 8–10 can be removed, and Line 13 can be replaced with

$$z'_{m,C} \leftarrow \max\left(z_{m,C} + d_m - \left(s_0 - \sum_{\tilde{m} < m}(z_{\tilde{m},C} + d_{\tilde{m}})\right), 0\right)$$

as the right-hand side represents the excess demand for slots at station $M + 1$ from station m given the leftover supply after allocating slots to stations $\tilde{m} < m$. After this change, this is the only instruction where the stations with higher priorities than m appear.

Query $C' = \{Z'_{M+1}, L'_{M+1}\}$ The efficient refactored distributional program corresponding to element $C' = \{Z'_{M+1}, L'_{M+1}\}$ of $\mathcal{Q}^{(t)}_{G \to G'}$ can be found similarly. It is given below, but we display only the lines that need to be changed relative to the above distributional program. Lines 8–14 below replace Lines 8–16 above. Note that station $M + 1$ needs the demand for its slots by all stations in order to determine its next counter.

1: **function** GRAPHICALQPLEXMAPELEMENT$^{(t)}_{\{Z'_{M+1}, L'_{M+1}\}}(\iota)$

2:     **for** $\tilde{m} = 1, \ldots, M$ **do**

8:     **for** $m = 1, \ldots, M$ **do**

9:         $\phi_m \leftarrow \min(z_{m,C} + d_m, s_{m-1})$

10:         $s_m \leftarrow s_{m-1} - \phi_m$

11:     $z'_{M+1} \leftarrow z_{M+1} - d_{M+1} + \sum_m \phi_m$

12:     $k'_{M+1} \sim$ ENTITYTYPE$^{(t)}_{M+1}(z_{M+1}, d_{M+1}, z'_{M+1})$

13:     $\ell'_{M+1} \sim$ NEXTLABEL$^{(t)}_{\iota, M+1}(z_{M+1}, k'_{M+1})$

14:     **return** (z'_{M+1}, ℓ'_{M+1})

As for the query $C' = \{Z'_{m,S}, Z'_{m,C}, L'_m\}$, it is possible to eliminate variables beyond the steps presented in this chapter. Indeed, Lines 8–11 can be replaced with

$$z'_{M+1} \leftarrow z_{M+1} - d_{M+1} + \min\left(\sum_{m=1}^{M}(z_{m,C} + d_m), s_0\right)$$

as the minimum on the right-hand side represents the met demand from the first M stations for slots at station $M + 1$.

Randomized Allocation Policy

The distributional program for element C' of the graphical QPLEX map $\mathcal{Q}^{(t)}_{G \to G'}$, assuming the randomized allocation policy, is shown below. It includes instructions for the randomized routing policy (Lines 6–14), which have been obtained by substituting the RANDOMDEMANDSPLIT program into the routing pmf program developed in Sect. 5.7. As we did for the priority allocation rule, we eliminate the unmet demand u_m from station m by substituting $u_m = z'_{m,C}$. Unlike before, we do not substitute the definition of $\hat{d}_m$, since it appears in multiple places.

1: **function** GRAPHICALQPLEXMAPELEMENT$^{(t)}_{C'}(\iota)$

2:     $\mu \leftarrow L_G(\iota)$

3:     $(z_{1,S}, z_{1,C}, \ldots, z_{M+1}) \sim \mu(Z_{1,S} = \cdot, Z_{1,C} = \cdot, \ldots, Z_{M+1} = \cdot)$

4:     **for** $m = 1, \ldots, M + 1$ **do**

5:         $d_m \sim$ NUMBEROFCOMPLETIONS$^{(t)}_{\mu,m}(z_{1,S}, z_{1,C}, \ldots, z_{M+1})$

6:     $s \leftarrow n_{M+1} + b_{M+1} - (z_{M+1} - d_{M+1})$

7:     **for** $m = 1, \ldots, M$ **do**

8:         $a_m \sim \alpha^{(t)}_m(A = \cdot)$

9:   $z'_{m,S} \leftarrow z_{m,S} - d_m + a_m$

10:   $\hat{d}_m \leftarrow d_m + z_{m,C}$

11:   $(\phi_1, \ldots, \phi_M) \sim \text{multivariateHypergeometric}(\hat{d}_1, \ldots, \hat{d}_M; s)$

12:   **for** $m = 1, \ldots, M$ **do**

13:    $z'_{m,C} \leftarrow \hat{d}_m - \phi_m$

14:   $z'_{M+1} \leftarrow z_{M+1} - d_{M+1} + \sum_{m=1}^{M} \phi_m$

15:   **for** $m = 1, \ldots, M + 1$ **do**

16:    $k'_m \sim \text{ENTITYTYPE}_m^{(t)}(z_{1,S}, z_{1,C}, \ldots, z_{M+1}, d_m, z'_{1,S}, z'_{1,C}, \ldots, z'_{M+1})$

17:    $\ell'_m \sim \text{NEXTLABEL}_{\mu,m}^{(t)}(z_{1,S}, z_{1,C}, \ldots, z_{M+1}, z'_{1,S}, z'_{1,C}, \ldots, z'_{M+1}, k'_m)$

18:   **return** $\text{proj}_{C'}(z'_{1,S}, z'_{1,C}, \ldots, z'_{M+1}, \ell'_1, \ldots, \ell'_{M+1})$

Query $C' = \{Z'_{m,S}, Z'_{m,C}, L'_m\}$ Consider $m \leq M$. The utility program modification step is the same as for the priority routing policy, and we omit the details of the other steps. We would like to highlight one feature of the dead variable elimination step, which is reminiscent of the second part of Example 13.5. From the perspective of station m, it is not necessary to know the specific demands for slots at station $M + 1$ from all of the other stations. What is necessary (and sufficient) is to know the total demand for slots at station $M + 1$ from all of the other stations. We can exploit this fact by replacing the sampling instruction that uses the multivariate hypergeometric random variable in Line 11 of the distributional program above with one that only uses a (univariate) hypergeometric random variable. (Note, however, that the hypergeometric and multivariate hypergeometric random variables are parameterized differently.)

1:   **function** $\text{GRAPHICALQPLEXMAPELEMENT}_{\{Z'_{m,S}, Z'_{m,C}, L'_m\}}^{(t)}(\iota)$

2:   **for** $\tilde{m} = 1, \ldots, M$ **do**

3:    $(z_{\tilde{m},S}, z_{\tilde{m},C}) \sim \iota(Z_{\tilde{m},S} = \cdot, Z_{\tilde{m},C} = \cdot)$

4:    $d_{\tilde{m}} \sim \text{NUMBEROFCOMPLETIONS}_{\iota,\tilde{m}}^{(t)}(z_{\tilde{m},S}, z_{\tilde{m},C})$

5:   $z_{M+1} \sim \iota(Z_{M+1} = \cdot)$

6:   $d_{M+1} \sim \text{NUMBEROFCOMPLETIONS}_{\iota,M+1}^{(t)}(z_{M+1})$

7:   $s \leftarrow n_{M+1} + b_{M+1} - (z_{M+1} - d_{M+1})$

8:   **for** $\tilde{m} = 1, \ldots, M$ **do**

9:    $\hat{d}_{\tilde{m}} = z_{\tilde{m},C} + d_{\tilde{m}}$

10:   $a_m \sim \alpha_m^{(t)}(A = \cdot)$

11:   $z'_{m,S} \leftarrow z_{m,S} - d_m + a_m$

12:   $\phi_m \sim \text{hypergeometric}(\sum_{\tilde{m}} \hat{d}_{\tilde{m}}, \hat{d}_m, s)$

13:   $z'_{m,C} \leftarrow \hat{d}_m - \phi_m$

14:   $k'_m \sim \text{ENTITYTYPE}_m^{(t)}(z_{m,S}, z_{m,C}, d_m, z'_{m,S}, z'_{m,C})$

15:   $\ell'_m \sim \text{NEXTLABEL}_{\iota,m}^{(t)}(z_{m,S}, z_{m,C}, k'_m)$

16:   **return** $(z'_{m,S}, z'_{m,C}, \ell'_m)$

Query $C' = \{Z'_{M+1}, L'_{M+1}\}$ The efficient refactored distributional program for the query $C' = \{z'_{M+1}, \ell'_{M+1}\}$ is shown below. Lines 2–9 are the same as for the first query and are not displayed. Lines 10–13 below replace Lines 10–16 above.

1: **function** GRAPHICALQPLEXMAPELEMENT$^{(t)}_{\{Z'_{M+1}, L'_{M+1}\}}(t)$

10:    $z'_{M+1} \leftarrow (z_{M+1} - d_{M+1}) + \min(\sum_m \hat{d}_m, s)$

11:    $k'_{M+1} \sim$ ENTITYTYPE$^{(t)}_{M+1}(z_{M+1}, d_{M+1}, z'_{M+1})$

12:    $\ell'_{M+1} \sim$ NEXTLABEL$^{(t)}_{t,M+1}(z_{M+1}, k'_{M+1})$

13:    **return** (z'_{M+1}, ℓ'_{M+1})

Chapter 14
Efficient Calculation for Distributional Programs

This chapter shows how to calculate the output of a distributional program efficiently. This computational task is a so-called exact computational inference task at its core, and we carry out this task with the so-called *sum-product message passing algorithm* [14, 19]. The resulting procedure can be automated for the most part, but we do not discuss implementation details.

The procedure consists of four main steps. In the first step, a query-specific graph dubbed the *query graph* is created from the variables and instructions in the program. The second step defines functions we call *factors*, which are constructed from the query graph and the instructions in the distributional program. In the third step, a junction in-tree is created using the maximal cliques in the query graph and the factors obtained from the second step. In the fourth step, the sum-product message-passing algorithm is applied to the junction in-tree, which calculates a marginal pmf of a set of variables from which the desired query pmf is extracted.

At the end of this chapter, we illustrate the entire procedure for three of the examples from the previous chapter. Each of these examples gives rise to one or more distributional programs to which we apply the techniques from the current chapter. This chapter is foundational and self-contained (apart from mathematical preliminaries), but the examples at the end of this chapter arise from QPLEX maps using simple transition dynamics.

14.1 Structured Distributional Programs

In this chapter, we assume that the distributional program satisfies the following two requirements, in which case we call the distributional program *structured*. The first requirement is that each instruction is a sampling statement except for the last instruction, which must be a return statement. The right-hand side of each sampling instruction can be viewed as a pmf that abstracts out the details of the

© The Author(s) 2025
A. B. Dieker, S. T. Hackman, *QPLEX: A Computational Modeling and Analysis Methodology for Stochastic Systems*, Springer Series in Operations Research and Financial Engineering, https://doi.org/10.1007/978-3-031-74870-7_14

instruction. For instance, an assignment can be viewed as a sampling statement from a degenerate pmf. It is only relevant for this chapter which variable(s) appear in the pmf expression on the right-hand side of the sampling statement and which variable(s) appear on the left-hand side, although the details could be relevant for implementation purposes. The second requirement is that each variable is only sampled once in the program. As a result, variables cannot be overwritten.

Such a structured program outputs a pmf $\prod_k \pi_k$ on the variables in the distributional program, where π_k is the (conditional) pmf representing the k-th sampling statement in the program. The variable(s) returned by the distributional program corresponds to a set of random variables denoted by C', and we refer to this set as a *query*. The output of the program is the sum of $\prod_k \pi_k$ over the variables not represented in C', which is a pmf. Of course, it is our objective to calculate this pmf efficiently. In view of our intended application to graphical QPLEX maps, we denote the output of the program by ι'.

A given distributional program may need to undergo some preprocessing to convert it to a structured distributional program. For instance, an "if" statement and a subsequent block of code may need to be rewritten as a single sampling statement. It is always possible to refactor a program into a structured distributional program, since the body of the program can trivially be converted into a single sampling statement for all the variables appearing in the program. In general, however, the more structure is exposed, the more potential computational benefit the tools from this chapter offer.

We consider the following distributional program as a running example to illustrate the general methodology. As we will do throughout this chapter, the set of variables that appears in the instruction is shown on each line of the instruction. For instance, in Line 3, we see that x_2 is generated using x_1, and the set of variables is shown as $\{x_1, x_2\}$. The query is $C' = \{X_4, X_6\}$, so this program outputs a pmf $\iota'(x_4, x_6)$.

```
1: function PROGRAM( )
2:     x₁ ~ π₁(X₁ = ·)                          ▷ {x₁}
3:     x₂ ~ π₂(X₂ = ·|x₁)                       ▷ {x₁, x₂}
4:     x₃ ~ π₃(X₃ = ·|x₂)                       ▷ {x₂, x₃}
5:     x₄ ~ π₄(X₄ = ·|x₂, x₃)                   ▷ {x₃, x₄}
6:     x₅ ~ π₅(X₅ = ·|x₃)                       ▷ {x₃, x₅}
7:     x₆ ~ π₆(X₆ = ·|x₅)                       ▷ {x₅, x₆}
8:     return (x₄, x₆)                          ▷ {x₄, x₆}
```

14.2   Query Graphs

Given a distributional program, we construct a graph with vertex set equal to the set of random variables corresponding to each variable appearing in the program as follows. For each instruction of the distributional program, including the return statement, we plant a clique representing all of the variables that are used in that instruction. If the resulting graph is not triangulated, which can be determined as discussed in Sect. 12.1, we add edges to triangulate it. We call the resulting triangulated graph the *query graph* and denote it by G_{query}. The query variables in C' correspond to elements of its vertex set.

A word on notation. We use the same symbol C for a generic maximal clique in an information structure graph G as we do for a generic maximal clique in a query graph G_{query}. We also use the symbols $\tilde{C}$, $\widehat{C}$, C_1, C_2, etc., for maximal cliques in G_{query}. The set of query variables C', however, is a clique but not necessarily a maximal clique in G_{query}.

For our running example, the graph produced by planting cliques associated with each instruction is shown below.

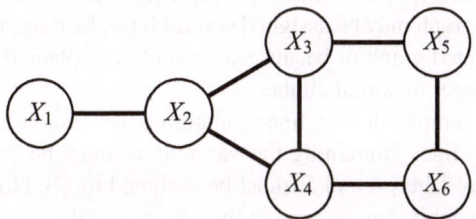

This graph is not triangulated, since the cycle $X_3 - X_5 - X_6 - X_4 - X_3$ does not have a chord. One way to add one or more edges so that the graph becomes triangulated is to add the edge $X_3 - X_6$, as shown in blue below.

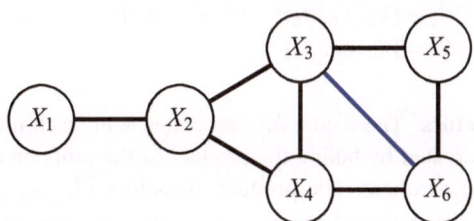

This graph is triangulated, since it is recursively simplicial with respect to the ordering X_1, X_2, X_4, X_3, X_5, X_6. Therefore, it is a query graph corresponding to the above distributional program.

14.3 Factors

The next step is to identify a collection of *factors*

$$\psi = \{\psi_C\}_{C \in \mathcal{C}(G_{\text{query}})} \in \mathcal{F}_{G_{\text{query}}}$$

so that $\prod_{C \in \mathcal{C}(G_{\text{query}})} \psi_C = \prod_k \pi_k$ is the pmf represented by the distributional program *without* the return statement. Note that this form reveals that this pmf has the global Markov property with respect to G_{query} by Proposition 11.7. The procedure to find the factors ψ_C consists of four steps.

Step 1: Find Maximal Cliques in Query Graph In our running example, the maximal cliques are

$$C_1 = \{X_1, X_2\}, \qquad\qquad C_2 = \{X_2, X_3, X_4\},$$
$$C_3 = \{X_3, X_5, X_6\}, \qquad\qquad C_4 = \{X_3, X_4, X_6\}.$$

Step 2: Assign Line Numbers to Maximal Cliques Each line in the distributional program except the line with the return statement is assigned to exactly one of the maximal cliques in the query graph. The only requirement is that the variables used in that line are represented in the maximal clique to which it is assigned. A maximal clique in the query graph may be assigned several lines, be it because of edges added to create a triangulated graph or because several cliques planted in the query graph together create a larger maximal clique.

In our running example, the two lines containing the variable x_1 must be assigned to C_1, and the two lines containing the variable x_5 must be assigned to C_3. The remaining two lines, Lines 4 and 5, must be assigned to C_2. No lines are assigned to C_4. This assignment is summarized in the following table.

Maximal Clique	Line Number(s) Assigned
$C_1 = \{X_1, X_2\}$	2, 3
$C_2 = \{X_2, X_3, X_4\}$	4, 5
$C_3 = \{X_3, X_5, X_6\}$	6, 7
$C_4 = \{X_3, X_4, X_6\}$	–

Step 3: Create Factors The factor ψ_C associated with each maximal clique C of the query graph is created by taking the product of the pmfs on the right-hand side of the assigned line numbers. The product of factors $\prod_{C \in \mathcal{C}(G_{query})} \psi_C$ is the pmf represented by the distributional program when the return statement is modified to return all variables in the program.

In our running example, the definition of the following factors is based on the unique feasible assignment of line numbers to maximal cliques.

Factor	Definition
$\psi_{C_1}(x_1, x_2)$	$\pi_1(x_1) \times \pi_2(x_2|x_1)$
$\psi_{C_2}(x_2, x_3, x_4)$	$\pi_3(x_3|x_2) \times \pi_4(x_4|x_2, x_3)$
$\psi_{C_3}(x_3, x_5, x_6)$	$\pi_5(x_5|x_3) \times \pi_6(x_6|x_5)$
$\psi_{C_4}(x_3, x_4, x_6)$	1

Note how the factor for C_4 is the constant function 1 because it corresponds to an empty product. The product of factors

$$\psi_{C_1}(x_1, x_2) \times \psi_{C_2}(x_2, x_3, x_4) \times \psi_{C_3}(x_3, x_5, x_6) \times \psi_{C_4}(x_3, x_4, x_6)$$

equals

$$\pi_1(x_1) \times \pi_2(x_2|x_1) \times \pi_3(x_3|x_2) \times \pi_4(x_4|x_2, x_3) \times \pi_5(x_5|x_3) \times \pi_6(x_6|x_5),$$

which is indeed the pmf of $(x_1, \ldots, x_6)$ represented by the body of the distributional program.

14.4   Query Junction In-Trees

Section 11.3 shows how to construct a junction in-tree from an arbitrary triangulated graph G. Here, we apply this methodology to the query graph G_{query} but impose the requirement that the root of the resulting in-tree is chosen, so that it contains all of the query variables in C'. We call a junction in-tree so constructed from the query graph G_{query} a *query junction in-tree* and denote it by $T = (V_T, E_T)$. We write R_T for the root of T.

In our running example, we construct the weighted clique intersection graph from the cardinalities of the intersections of the maximal cliques:

$$C_1 \cap C_2 = \{X_2\}, \qquad C_1 \cap C_3 = \emptyset, \qquad C_1 \cap C_4 = \emptyset,$$
$$C_2 \cap C_3 = \{X_3\}, \qquad C_2 \cap C_4 = \{X_3, X_4\}, \qquad C_3 \cap C_4 = \{X_3, X_6\}.$$

To construct a query junction in-tree, we need to find a maximum weight spanning tree in this graph and direct all edges toward $R_T = C_4$, since it is the unique maximal clique that contains C'. The weighted clique intersection graph and the (unique) query junction in-tree are shown below.

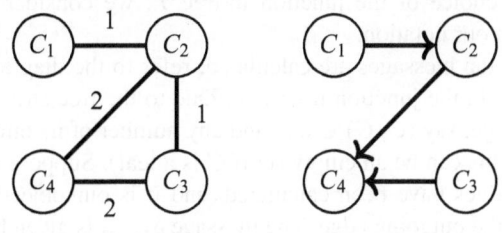

weighted clique intersection graph   query junction in-tree

14.5   Message Passing Algorithm

Given a distributional program, we have shown how to construct:

(1) A query graph G_{query}, with each variable in the program corresponding to a vertex of G_{query}.
(2) A collection of factors $\{\psi_C\}_{C \in \mathcal{C}(G_{\text{query}})}$, the scopes of which correspond to the maximal cliques in G_{query} and the product of which is the same as $\prod_k \pi_k$.
(3) A query junction in-tree $T \in \mathcal{T}(G_{\text{query}})$ for which the root R_T is a superset of the query C'.

We now show how to calculate the marginal pmf of the variables in R_T under the pmf $\prod_{C \in \mathcal{C}(G_{\text{query}})} \psi_C$. The desired marginal pmf ι' is then calculated by summing out the variables in $R_T \backslash C'$.

The task at hand in this section differs fundamentally from the one studied in Chap. 11. Section 11.5 shows how a collection of consistent marginal pmfs can be "lifted" to define a joint pmf with the global Markov property. This section goes in the *opposite* direction: We start with a product of factors, i.e., a joint pmf with the global Markov property, and seek to calculate a marginal pmf. Junction in-trees for the associated graphical model play a key role in each case.

Suppose we are given a junction in-tree $T = (V_T, E_T)$ for which the vertex set V_T is the set of maximal cliques in some triangulated graph and a collection of nonnegative functions $\{\psi_C\}_{C \in V_T}$, where ψ_C has scope C. The *sum-product message-passing algorithm* efficiently calculates the function

$$\sum_{V \backslash R_T} \prod_{C \in V_T} \psi_C, \tag{14.1}$$

where $V = \bigcup_{C \in V_T} C$ is the set of all variables that appear in the distributional program. If $\prod_{C \in V_T} \psi_C$ is a pmf, then (14.1) is the marginal pmf of the variables in R_T.

The algorithm works by passing *messages* along the directed edges toward the root R_T of the in-tree T, starting from the leaves. Each message $\delta_{\tilde{C} \to C}$ along the directed edge $\tilde{C} \to C$ is a function of the variables in $\tilde{C} \cap C$. Although the messages depend on the choice of the junction in-tree T, we consider T to be fixed and suppress it from our notation.

Here is how the messages are calculated; refer to the diagram below. Fix some vertex $\tilde{C} \neq R_T$ in the junction in-tree T. Due to the tree structure, $\tilde{C}$ has exactly one outgoing edge, say $(\tilde{C}, C) \in E_T$, and any number of incoming edges $\{(\widehat{C}, \tilde{C}) : \widehat{C} \in \text{pa}_T(\tilde{C})\}$ (this can be an empty set if $\tilde{C}$ is a leaf). Suppose the messages along the incoming edges have been calculated, and it is our objective to calculate the message along the outgoing edge. The message $\delta_{\tilde{C} \to C}$ is given by

$$\delta_{\tilde{C} \to C} = \sum_{\tilde{C} \backslash C} \left(\psi_{\tilde{C}} \times \prod_{\widehat{C} \in \text{pa}_T(\tilde{C})} \delta_{\widehat{C} \to \tilde{C}} \right), \tag{14.2}$$

which is obtained by taking the product of the factor associated with the vertex $\tilde{C}$, multiplying it by the incoming messages, and then summing out all variables *not* in the intersection $\tilde{C} \cap C$. Note that the sum is taken over all the variables in the scope $\tilde{C}$ of the summand that are *not* in the scope $\tilde{C} \cap C$ of the outgoing message $\delta_{\tilde{C} \to C}$.

If $\tilde{C}$ is a leaf, i.e., $\mathrm{pa}_T(\tilde{C}) = \emptyset$, the product is empty, and the message $\delta_{\tilde{C} \to C}$ is simply $\sum_{\tilde{C} \backslash C} \psi_{\tilde{C}}$. The sum and product in this expression are what gives the sum-product message passing algorithm its name, although we will henceforth suppress "sum-product" for brevity.

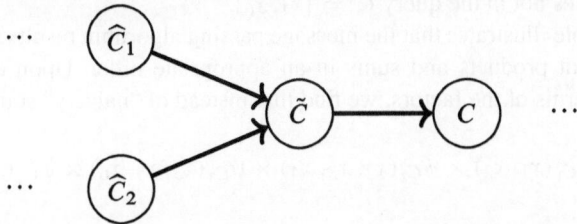

The following proposition shows that the sum (14.1) can be calculated by calculating the messages along each of the incoming edges of the root, which in turn requires recursively calculating the messages along the other edges using the message passing algorithm. The proof of this proposition is provided in Sect. 14.8. Its main ingredient is a property of a junction in-trees, which relies on the fact that the query graph is triangulated.

Proposition 14.1 *We have the following identity:*

$$\sum_{V \backslash R_T} \prod_{C \in V_T} \psi_C = \psi_{R_T} \times \prod_{\tilde{C} \in \mathrm{pa}_T(R_T)} \delta_{\tilde{C} \to R_T}.$$

In our running example, the query junction in-tree has two leaves, C_1 and C_3. As (C_1, C_2) and (C_3, C_4) are the outgoing edges, respectively, we need to sum over the variables in $C_1 \backslash C_2 = \{x_1\}$ and $C_3 \backslash C_4 = \{x_5\}$ of the corresponding factors ψ_{C_1} and ψ_{C_3}. This leads to the following messages:

$$\delta_{C_1 \to C_2}(x_2) = \sum_{x_1} \psi_{C_1}(x_1, x_2),$$

$$\delta_{C_3 \to C_4}(x_3, x_6) = \sum_{x_5} \psi_{C_3}(x_3, x_5, x_6).$$

The message along the edge (C_2, C_4) is given by

$$\delta_{C_2 \to C_4}(x_3, x_4) = \sum_{x_2} \psi_{C_2}(x_2, x_3, x_4) \times \delta_{C_1 \to C_2}(x_2),$$

which is obtained by multiplying the factor ψ_{C_2} corresponding to C_2 by the incoming message $\delta_{C_1 \to C_2}$ and then summing out the variable in $C_2 \backslash C_4 = \{x_2\}$.

The root C_4 has $\delta_{C_2 \to C_4}$ and $\delta_{C_3 \to C_4}$ for its two incoming messages. The desired pmf $\iota'(x_4, x_6)$ is given by

$$\iota'(x_4, x_6) = \sum_{x_3} \psi_{C_4}(x_3, x_4, x_6) \times \delta_{C_2 \to C_4}(x_3, x_4) \times \delta_{C_3 \to C_4}(x_3, x_6),$$

which is obtained by multiplying these messages with the factor ψ_{C_4} and summing out the variables not in the query $C' = \{x_4, x_6\}$.

This example illustrates that the message passing algorithm results in efficiencies by carrying out products and sums in an appropriate order. Upon expressing all messages in terms of the factors, we find that instead of "naively" summing

$$\sum_{x_1, x_2, x_3, x_5} \psi_{C_1}(x_1, x_2) \times \psi_{C_2}(x_2, x_3, x_4) \times \psi_{C_3}(x_3, x_5, x_6) \times \psi_{C_4}(x_3, x_4, x_6),$$

the message passing algorithm calculates this sum via

$$\sum_{x_3} \left(\sum_{x_2} \left(\sum_{x_1} \psi_{C_1}(x_1, x_2) \right) \times \psi_{C_2}(x_2, x_3, x_4) \right)$$

$$\times \left(\sum_{x_5} \psi_{C_3}(x_3, x_5, x_6) \right) \times \psi_{C_4}(x_3, x_4, x_6).$$

14.6   Message Passing Using Distributional Programs

It is often possible to refactor a distributional program in such a way that it incorporates the message passing algorithm. This amounts to writing separate distributional programs for each message and then appropriately sampling from these programs in a manner that is consistent with the message passing algorithm. This step does not yield any computational benefits per se but affords an opportunity to represent the message passing algorithm via distributional programs.

We have done so below for our running example; it is readily verified that the output of the program PROGRAMREVISITED is the same as that of PROGRAM. The distributional programs corresponding to the messages $\delta_{C_1 \to C_2}$, $\delta_{C_3 \to C_4}$, and $\delta_{C_2 \to C_4}$ are called MESSAGE$_{C_1 \to C_2}$, MESSAGE$_{C_3 \to C_4}$, and MESSAGE$_{C_2 \to C_4}$, respectively. Note that all six sampling instructions in the body of the program PROGRAM appear exactly once below, with their ordering inherited within the distributional program for each message. It is worth pointing out that x_3 is *input* to the distributional program for $\delta_{C_3 \to C_4}$, which is not immediate from its algebraic form.

```
1: function MESSAGE_{C_1 → C_2}()
2:     x_1 ~ π_1(X_1 = ·)
3:     x_2 ~ π_2(X_2 = ·|x_1)
4:     return x_2
```

```
1: function MESSAGE_{C_3 → C_4}(x_3)
2:     x_5 ~ π_5(X_5 = ·|x_3)
3:     x_6 ~ π_6(X_6 = ·|x_5)
4:     return x_6
```

```
1: function MESSAGE_{C_2 → C_4}()
2:     x_2 ~ MESSAGE_{C_1 → C_2}()
3:     x_3 ~ π_3(X_3 = ·|x_2)
4:     x_4 ~ π_4(X_4 = ·|x_2, x_3)
5:     return (x_3, x_4)
```

```
1: function PROGRAMREVISITED()
2:     (x_3, x_4) ~ MESSAGE_{C_2 → C_4}()
3:     x_6 ~ MESSAGE_{C_3 → C_4}(x_3)
4:     return (x_4, x_6)
```

It is not *always* possible to refactor a distributional program such that each message becomes a distributional program. Consider the following distributional program.

```
1: function ANOTHERPROGRAM()
2:     x_1 ~ π_1(X_1 = ·)                               ▷ {x_1}
3:     (x_2, x_3) ~ π_2(X_2 = ·, X_3 = ·)               ▷ {x_2, x_3}
4:     x_4 ~ π_3(X_4 = ·|x_1, x_3)                       ▷ {x_1, x_3, x_4}
5:     x_5 ~ π_4(X_5 = ·|x_2, x_3, x_4)                  ▷ {x_2, x_3, x_4, x_5}
6:     return (x_1, x_5)                                 ▷ {x_1, x_5}
```

The message passing algorithm takes the following form. There are two maximal cliques in the resulting query graph: $C = \{X_1, X_3, X_4, X_5\}$ and $\tilde{C} = \{X_2, X_3, X_4, X_5\}$. There is a unique assignment of line numbers to maximal cliques, namely, by assigning Lines 2, 4, and 6 to C and Lines 3 and 5 to $\tilde{C}$.

The query $C' = \{X_1, X_5\}$ is contained C, so message passing amounts to passing a message from $\tilde{C}$ to C. Since $\tilde{C} \backslash C = \{X_2\}$, this message is calculated via

$$\delta_{\tilde{C} → C}(x_3, x_4, x_5) = \sum_{x_2} π_2(x_2, x_3) \times π_4(x_5|x_2, x_3, x_4). \tag{14.3}$$

The desired pmf is then found via

$$\iota'(x_1, x_5) = \sum_{x_3, x_4} \delta_{\tilde{C} → C}(x_3, x_4, x_5) \times π_1(x_1) \times π_3(x_4|x_1, x_3). \tag{14.4}$$

Here is the distributional program representation of $\delta_{\tilde{C} → C}$.

```
1: function MESSAGE_C̃→C(x4)
2:     (x2, x3) ~ π2(X2 = ·, X3 = ·)
3:     x5 ~ π4(X5 = ·|x2, x3, x4)
4:     return (x3, x5)
```

Now suppose we try to write down a distributional program corresponding to (14.4). We need to place the instruction $x_4 \sim \pi_3(X_4 = \cdot|x_1, x_3)$ before or after sampling the message. The two possibilities are shown below. Neither program is a distributional program because each program requires an input that has not yet been sampled—x_3 in the first program and x_4 in the second program. Thus, the original distributional program cannot be refactored in a manner that is consistent with the message passing algorithm.

```
1: function ANOTHERPROGRAMOPTION1( )
2:     x1 ~ π1(X1 = ·)
3:     x4 ~ π3(X4 = ·|x1, x3)
4:     (x3, x5) ~ MESSAGE_C̃→C(x4)
5:     return (x1, x5)
```

```
1: function ANOTHERPROGRAMOPTION2( )
2:     x1 ~ π1(X1 = ·)
3:     (x3, x5) ~ MESSAGE_C̃→C(x4)
4:     x4 ~ π3(X4 = ·|x1, x3)
5:     return (x1, x5)
```

The original distributional program ANOTHERPROGRAM *can* be refactored if it is modified by splitting the sampling instruction $(x_2, x_3) \sim \pi_2(X_2 = \cdot, X_3 = \cdot)$ into two instructions: a sampling instruction for x_3 and then a sampling instruction for x_2 given x_3, as shown below.

```
1: function ANOTHERPROGRAMSPLIT( )
2:     x1 ~ π1(X1 = ·)                                      ▷ {x1}
3:     x3 ~ π2A(X3 = ·)                                     ▷ {x3}
4:     x2 ~ π2B(X2 = ·|x3)                                  ▷ {x2, x3}
5:     x4 ~ π3(X4 = ·|x1, x3)                               ▷ {x1, x3, x4}
6:     x5 ~ π4(X5 = ·|x2, x3, x4)                           ▷ {x2, x3, x4, x5}
7:     return (x1, x5)                                      ▷ {x1, x5}
```

Line 3 can now be assigned to C. The query graph is unaffected by this change, but the message passing equations (14.3) and (14.4) are replaced with

$$\hat{\delta}_{\tilde{C}\to C}(x_3, x_4, x_5) = \sum_{x_2} \pi_{2B}(x_2|x_3) \times \pi_4(x_5|x_2, x_3, x_4)$$

and

$$\iota'(x_1, x_5) = \sum_{x_3, x_4} \hat{\delta}_{\tilde{C}\to C}(x_3, x_4, x_5) \times \pi_1(x_1) \times \pi_{2A}(x_3) \times \pi_3(x_4|x_1, x_3).$$

Here is the distributional program representation of $\hat{\delta}_{\tilde{C} \to C}$.

```
1: function HATMESSAGE_{\tilde{C} \to C}(x_3, x_4)
2:     x_2 \sim \pi_{2B}(X_2 = \cdot | x_3)
3:     x_5 \sim \pi_4(X_5 = \cdot | x_2, x_3, x_4)
4:     return x_5
```

Here is the refactored distributional program:

```
1: function ANOTHERPROGRAMSPLITREVISITED()
2:     x_1 \sim \pi_1(X_1 = \cdot)
3:     x_3 \sim \pi_{2A}(X_3 = \cdot)
4:     x_4 \sim \pi_3(X_4 = \cdot | x_1, x_3)
5:     x_5 \sim HATMESSAGE_{\tilde{C} \to C}(x_3, x_4)
6:     return (x_1, x_5)
```

14.7 Examples

We apply the above techniques to some of the efficient refactored distributional programs of the elements of the graphical QPLEX map constructed in the previous chapter. We do not carry out the refactoring discussed in Sect. 14.6, even though it can be done in each of these examples. We note that the distributional programs that serve as starting points in these examples do *not* necessarily maximize the computational efficiency of the message passing algorithm as derived in this section; it may be possible to start with different equivalent distributional programs that result in more efficient message passing.

Some words on notation. We represent each of the three utility programs as their corresponding algebraic pmf, namely, we write $q_{\iota,m}^{(t)}(d_m | z_{N_m})$, $q_m^{(t)}(k_m' | z_{S_m}, d_m, z_{S_m}')$, and $q_{\iota,m}^{(t)}(\ell_m' | z_{N_m})$. We can thus view distributional programs that call these utility programs as being structured. If there is only one activity ($M = 1$), then we omit the subscripts m. Note that we use the information collection ι, not the pmf μ. Moreover, instead of numbering the pmfs used in a structured distributional program, we use the generic symbol π for such pmfs and suppress the subscript.

Multiserver Queueing Model with Renewal Arrivals

We consider the multiserver queueing model with renewal arrivals example from Sect. 13.5. We first consider the query $C' = \{Z', L'\}$ and then the query $C' = \{Z', Z_A'\}$.

Query $C' = \{Z', L'\}$ The efficient refactored distributional program for element $C' = \{Z', L'\}$ of the graphical QPLEX map is shown below. The variables in the distributional program are z, z_A, d, a, z', k', and ℓ'. The cliques associated with each instruction are shown on each line of the instruction.

1: **function** GRAPHICALQPLEXMAPELEMENT$^{(t)}_{\{Z',L'\}}(\iota)$

2:    $(z, z_A) \sim \iota(Z = \cdot, Z_A = \cdot)$ $\triangleright \{z, z_A\}$

3:    $d \sim$ NUMBEROFCOMPLETIONS$^{(t)}_{\iota}(z)$ $\triangleright \{z, d\}$

4:    $a \sim \mathrm{ber}\left(\frac{\eta(z_A+1)}{\sum_{\tilde{z}_A \geq z_A+1} \eta(\tilde{z}_A)}\right)$ $\triangleright \{z_A, a\}$

5:    $z' \leftarrow z - d + a$ $\triangleright \{z, d, a, z'\}$

6:    $k' \sim$ ENTITYTYPE$^{(t)}(z, d, z')$ $\triangleright \{z, d, z', k'\}$

7:    $\ell' \sim$ NEXTLABEL$^{(t)}_{\iota}(z, k')$ $\triangleright \{z, k', \ell'\}$

8:    **return** (z', ℓ') $\triangleright \{z', \ell'\}$

Query Graph The graph produced by planting cliques associated with each instruction is shown below.

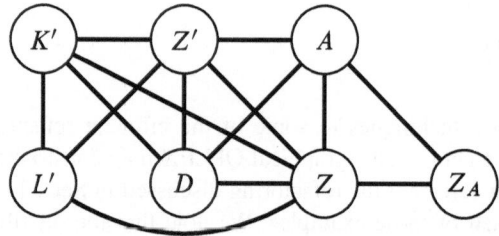

This graph is triangulated, since it is recursively simplicial with respect to the ordering Z_A, A, Z', L', K', D, Z. Therefore, it is the query graph associated with the distributional program.

Factors The maximal cliques in the query graph are

$$C_1 = \{Z, Z_A, A\}, \qquad\qquad C_2 = \{Z, D, A, Z'\},$$
$$C_3 = \{Z, D, Z', K'\}, \qquad\qquad C_4 = \{Z, Z', K', L'\}.$$

Note how a maximal clique in a query graph (e.g., C_1) can strictly include a clique planted in its construction (e.g., $\{Z, Z_A\}$).

The next step is to assign each line number in the body of the program to a maximal clique. Lines 2 and 4 must be assigned to C_1; Line 5 must be assigned to C_2; Line 6 must be assigned to C_3; Line 7 must be assigned to C_4; and Line 3 can either be assigned to C_2 or C_3. So, one feasible assignment of line numbers to maximal cliques is given in the following table.

Maximal Clique	Line Number(s) Assigned
$C_1 = \{Z, Z_A, A\}$	2, 4
$C_2 = \{Z, D, A, Z'\}$	3, 5
$C_3 = \{Z, D, Z', K'\}$	6
$C_4 = \{Z, Z', K', L'\}$	7

After defining the pmfs

$$\pi(a|z_A) = \Pr\left[\text{ber}\left(\eta(z_A) \Big/ \sum_{\tilde{z}_A > z_A} \eta(\tilde{z}_A)\right) = a \right]$$

$$\pi(z'|z, d, a) = 1(z' = z - d + a),$$

we obtain the following factors associated with the query graph.

Factor	Definition
$\psi_{C_1}(z, z_A, a)$	$\iota(z, z_A) \times \pi(a|z_A)$
$\psi_{C_2}(z, d, a, z')$	$q_t^{(t)}(d|z) \times \pi(z'|z, d, a)$
$\psi_{C_3}(z, d, z', k')$	$q_t^{(t)}(k'|z, d, z')$
$\psi_{C_4}(z, z', k', \ell')$	$q_t^{(t)}(\ell'|z, k')$

Query Junction In-Tree We construct the weighted clique intersection graph from the pairwise intersections of the maximal cliques:

$$C_1 \cap C_2 = \{Z, A\}, \qquad C_1 \cap C_3 = \{Z\}, \qquad C_1 \cap C_4 = \{Z\},$$
$$C_2 \cap C_3 = \{Z, D, Z'\}, \qquad C_2 \cap C_4 = \{Z, Z'\}, \qquad C_3 \cap C_4 = \{Z, Z', K'\}.$$

We need to find a maximum weight spanning tree in the weighted clique intersection graph and direct all edges toward $R_T = C_4$, since it is the unique maximal clique that contains $C' = \{Z', L'\}$. This weighted clique intersection graph and the (unique) query junction in-tree are shown below.

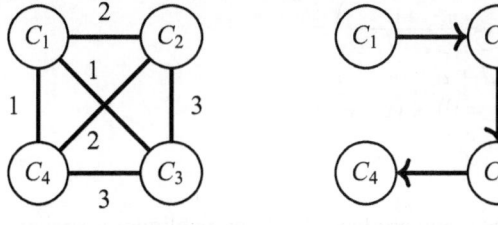

weighted clique intersection graph   query junction in-tree

Message Passing The structure of the query junction in-tree is special in the sense that none of its vertices have multiple incoming edges. The vertex C_1 is the unique leaf, and its parent is C_2, so we first need to calculate the message $\delta_{C_1 \to C_2}$. Since Z_A is the only element of C_1 that does not lie in the intersection $C_1 \cap C_2$, we must sum over z_A to obtain this message via

$$\delta_{C_1 \to C_2}(z, a) = \sum_{z_A} \psi_{C_1}(z, z_A, a).$$

The next message to be calculated is $\delta_{C_2 \to C_3}$. Since A is the only element of C_2 that does not lie in the intersection $C_2 \cap C_3$, we must sum over a to obtain this message via

$$\delta_{C_2 \to C_3}(z, d, z') = \sum_a \psi_{C_2}(z, d, a, z') \times \delta_{C_1 \to C_2}(z, a).$$

The next message to be calculated is $\delta_{C_3 \to C_4}$. Since D is the only element of C_3 that does not lie in the intersection $C_3 \cap C_4$, we must sum over d to obtain this message via

$$\delta_{C_3 \to C_4}(z, z', k') = \sum_d \psi_{C_3}(z, d, z', k') \times \delta_{C_2 \to C_3}(z, d, z').$$

Proposition 14.1 implies that the sought pmf can be found via

$$\iota'(z', \ell') = \sum_{z, k'} \psi_{C_4}(z, z', k', \ell') \times \delta_{C_3 \to C_4}(z, z', k').$$

Query $C' = \{Z', Z'_A\}$ Reproduced below is the efficient refactored distributional program for element $C' = \{Z', Z'_A\}$ of the graphical QPLEX map from Sect. 13.5. The cliques associated with each instruction are shown on each line of the instruction.

1: **function** GRAPHICALQPLEXMAPELEMENT$_{\{Z', Z'_A\}}^{(t)}(\iota)$
2:     $(z, z_A) \sim \iota(Z = \cdot, Z_A = \cdot)$ $\triangleright \{z, z_A\}$
3:     $d \sim$ NUMBEROFCOMPLETIONS$_\iota^{(t)}(z)$ $\triangleright \{z, d\}$
4:     $a \sim \text{ber}\left(\dfrac{\eta(z_A+1)}{\sum_{\tilde{z}_A \geq z_A+1} \eta(\tilde{z}_A)} \right)$ $\triangleright \{z, a\}$
5:     $z' \leftarrow z - d + a$ $\triangleright \{z, d, a, z'\}$
6:     $z'_A \leftarrow 1(a = 0) \times (z_A + 1)$ $\triangleright \{z_A, a, z'_A\}$
7:     **return** (z', z'_A) $\triangleright \{z', z'_A\}$

Query Graph The graph produced by planting cliques associated with each instruction is shown below.

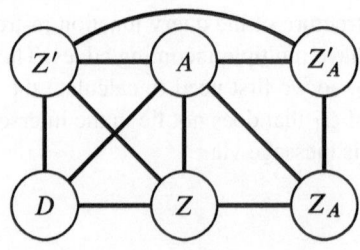

This graph is *not* triangulated, since the cycle $Z_A - Z - Z' - Z'_A - Z_A$ does not have a chord. We add the edge $Z - Z'_A$ to triangulate the graph. The query graph is shown below with the additional edge in blue.

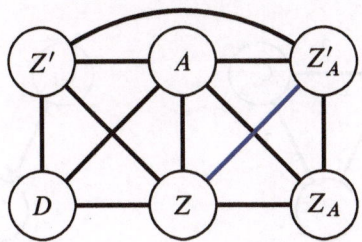

This graph is triangulated, since it is recursively simplicial with respect to the ordering Z_A, Z'_A, D, A, Z', Z.

Factors The maximal cliques in the query graph are

$$C_1 = \{Z, Z_A, A, Z'_A\}, \quad C_2 = \{Z, D, A, Z'\}, \quad C_3 = \{Z, A, Z'_A, Z'\}.$$

The next step is to assign each line number in the body of the program to a maximal clique. All lines using the variable z_A must be assigned to C_1, and all lines using the variable d must be assigned to C_2. No lines are assigned to C_3.

Maximal Clique	Line Number(s) Assigned
$C_1 = \{Z, Z_A, A, Z'_A\}$	2, 4, 6
$C_2 = \{Z, D, A, Z'\}$	3, 5
$C_3 = \{Z, A, Z'_A, Z'\}$	–

After defining the pmf

$$\pi(z'_A | z_A, a) = 1(z'_A = 1(a = 0) \times (z_A + 1)),$$

we obtain the following factors associated with the query graph.

Factor	Definition
$\psi_{C_1}(z, z_A, a, z'_A)$	$\iota(z, z_A) \times \pi(a|z_A) \times \pi(z'_A|z_A, a)$
$\psi_{C_2}(z, d, a, z')$	$q_t^{(t)}(d|z) \times \pi(z'|z, d, a)$
$\psi_{C_3}(z, a, z'_A, z')$	1

Query Junction In-Tree We construct the weighted clique intersection graph from the pairwise intersections of the maximal cliques:

$$C_1 \cap C_2 = \{Z, A\}, \quad C_1 \cap C_3 = \{Z, A, Z'_A\}, \quad C_2 \cap C_3 = \{Z, A, Z'\}.$$

We need to find a maximum weight spanning tree in the weighted clique intersection graph and direct all edges toward $R_T = C_3$, since it is the unique maximal clique that contains $C' = \{Z', Z'_A\}$. This weighted clique intersection graph and the (unique) query junction in-tree are shown below.

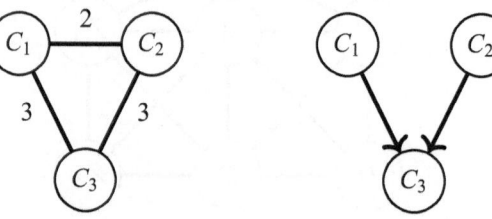

weighted clique intersection graph   query junction in-tree

Message Passing The messages are shown below:

$$\delta_{C_1 \to C_3}(z, a, z'_A) = \sum_{z_A} \psi_{C_1}(z, z_A, a, z'_A),$$

$$\delta_{C_2 \to C_3}(z, a, z') = \sum_{d} \psi_{C_2}(z, d, a, z').$$

Proposition 14.1 implies that the sought pmf can be found via

$$\iota'(z', z'_A) = \sum_{z,a} \delta_{C_1 \to C_3}(z, a, z'_A) \times \delta_{C_2 \to C_3}(z, a, z'),$$

where we have omitted the term $\psi_{C_3}(z, a, z'_A, z')$ in the summand as it equals 1.

Queueing Network with Probabilistic Routing

We consider the queueing network model with probabilistic routing from Sect. 13.5. We show the entire procedure for the query $C' = \{Z'_1, L'_1\}$ for the special case $M = 3$.

The efficient refactored distributional program for element $C' = \{Z'_1, L'_1\}$ of the graphical QPLEX map is shown below, with the for-loop expanded. The variables in the distributional program are $z_1, z_2, z_3, d_1, d_2, d_3, a_1, \phi_{0 \to 1}, \phi_{1 \to 1}, \phi_{2 \to 1}, \phi_{3 \to 1}, z'_1, k'_1,$ and ℓ'_1. The cliques associated with each instruction are shown on each line of the instruction.

1: **function** GRAPHICALQPLEXMAPELEMENT$^{(t)}_{\{Z'_1, L'_1\}}(\iota)$
2:     $z_1 \sim \iota(Z_1 = \cdot)$ $\triangleright \{z_1\}$
3:     $d_1 \sim$ NUMBEROFCOMPLETIONS$^{(t)}_{\iota, 1}(z_1)$ $\triangleright \{z_1, d_1\}$
4:     $z_2 \sim \iota(Z_2 = \cdot)$ $\triangleright \{z_2\}$

5:    $d_2 \sim \text{NUMBEROFCOMPLETIONS}_{\iota,2}^{(t)}(z_2)$ $\triangleright \{z_2, d_2\}$

6:    $\phi_{2\to1} \sim \text{bin}\left(d_2, \beta_{2\to1}^{(t)}\right)$ $\triangleright \{d_2, \phi_{2\to1}\}$

7:    $z_3 \sim \iota(Z_3 = \cdot)$ $\triangleright \{z_3\}$

8:    $d_3 \sim \text{NUMBEROFCOMPLETIONS}_{\iota,3}^{(t)}(z_3)$ $\triangleright \{z_3, d_3\}$

9:    $\phi_{3\to1} \sim \text{bin}\left(d_3, \beta_{3\to1}^{(t)}\right)$ $\triangleright \{d_3, \phi_{3\to1}\}$

10:    $\phi_{1\to1} \sim \text{bin}\left(d_1, \beta_{1\to1}^{(t)}\right)$ $\triangleright \{d_1, \phi_{1\to1}\}$

11:    $\phi_{0\to1} \sim \alpha_1^{(t)}(A = \cdot)$ $\triangleright \{\phi_{0\to1}\}$

12:    $a_1 \leftarrow \phi_{0\to1} + \phi_{2\to1} + \phi_{3\to1}$ $\triangleright \{\phi_{0\to1}, \phi_{2\to1}, \phi_{3\to1}, a_1\}$

13:    $z_1' \leftarrow z_1 - d_1 + \phi_{1\to1} + a_1$ $\triangleright \{z_1, d_1, \phi_{1\to1}, a_1, z_1'\}$

14:    $k_1' \sim \text{ENTITYTYPE}_1^{(t)}(z_1, d_1, z_1')$ $\triangleright \{z_1, d_1, z_1', k_1'\}$

15:    $\ell_1' \sim \text{NEXTLABEL}_{\iota,1}^{(t)}(z_1, k_1')$ $\triangleright \{z_1, k_1', \ell_1'\}$

16:    **return** (z_1', ℓ_1') $\triangleright \{z_1', \ell_1'\}$

Query Graph The graph produced by planting cliques associated with each instruction is shown below.

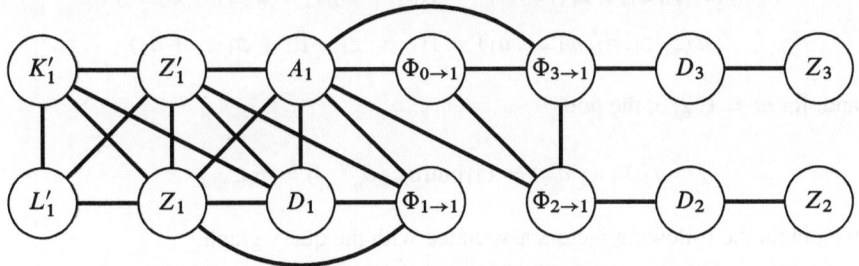

This graph is triangulated, since it is recursively simplicial with respect to the ordering Z_2, D_2, Z_3, D_3, $\Phi_{3\to1}$, $\Phi_{2\to1}$, $\Phi_{1\to1}$, $\Phi_{0\to1}$, A_1, L_1', K_1', D_1, Z_1, Z_1'. Therefore, it is the query graph associated with the distributional program.

Factors The maximal cliques in the query graph are:

$$C_1 = \{Z_2, D_2\}, \qquad\qquad\qquad C_2 = \{D_2, \Phi_{2\to1}\},$$

$$C_3 = \{Z_3, D_3\}, \qquad\qquad\qquad C_4 = \{D_3, \Phi_{3\to1}\},$$

$$C_5 = \{\Phi_{0\to1}, \Phi_{2\to1}, \Phi_{3\to1}, A_1\}, \qquad C_6 = \{Z_1, D_1, \Phi_{1\to1}, A_1, Z_1'\},$$

$$C_7 = \{Z_1, D_1, Z_1', K_1'\}, \qquad\qquad C_8 = \{Z_1, Z_1', K_1', L_1'\}.$$

The next step is to assign each line number in the body of the program to a maximal clique. Lines 4 and 5 must be assigned to C_1; Line 6 must be assigned to C_2; Lines 7 and 8 must be assigned to C_3; Line 9 must be assigned to C_4; Lines 11 and 12 must be assigned to C_5; Lines 10 and 13 must be assigned to C_6; Line 14 must be assigned to C_7; Line 15 and 16 must be assigned to C_8; Line 2 can either be assigned to C_6, C_7 or C_8; and Line 3 can either be assigned to C_6 or C_7. So, one feasible assignment of line numbers to maximal cliques is given in the following table.

Maximal Clique	Line Number(s) Assigned
$C_1 = \{Z_2, D_2\}$	4, 5
$C_2 = \{D_2, \Phi_{2\to1}\}$	6
$C_3 = \{Z_3, D_3\}$	7, 8
$C_4 = \{D_3, \Phi_{3\to1}\}$	9
$C_5 = \{\Phi_{0\to1}, \Phi_{2\to1}, \Phi_{3\to1}, A_1\}$	11, 12
$C_6 = \{Z_1, D_1, \Phi_{1\to1}, A_1, Z'_1\}$	2, 3, 10, 13
$C_7 = \{Z_1, D_1, Z'_1, K'_1\}$	14
$C_8 = \{Z_1, Z'_1, K'_1, L'_1\}$	15

After defining the pmfs

$$\pi(\phi_{0\to1}) = \alpha_1^{(t)}(A = \phi_{0\to1})$$

$$\pi(a_1|\phi_{0\to1}, \phi_{2\to1}, \phi_{3\to1}) = 1(a_1 = \phi_{0\to1} + \phi_{2\to1} + \phi_{3\to1})$$

$$\pi(z'_1|z_1, d_1, \phi_{1\to1}, a_1) = 1(z'_1 = z_1 - d_1 + \phi_{1\to1} + a_1)$$

and, for $m = 1, 2, 3$, the pmfs

$$\pi(\phi_{m\to1}|d_m) = \Pr[\text{bin}(d_m, \beta_{m\to1}^{(t)}) = \phi_{m\to1}],$$

we obtain the following factors associated with the query graph.

Factor	Definition
$\psi_{C_1}(z_2, d_2)$	$\iota(z_2) \times q_{i,2}^{(t)}(d_2|z_2)$
$\psi_{C_2}(d_2, \phi_{2\to1})$	$\pi(\phi_{2\to1}|d_2)$
$\psi_{C_3}(z_3, d_3)$	$\iota(z_3) \times q_{i,3}^{(t)}(d_3|z_3)$
$\psi_{C_4}(d_3, \phi_{3\to1})$	$\pi(\phi_{3\to1}|d_3)$
$\psi_{C_5}(\phi_{0\to1}, \phi_{2\to1}, \phi_{3\to1}, a_1)$	$\pi(\phi_{0\to1}) \times \pi(a_1|\phi_{0\to1}, \phi_{2\to1}, \phi_{3\to1})$
$\psi_{C_6}(z_1, d_1, \phi_{1\to1}, a_1, z'_1)$	$\iota(z_1) \times q_{i,1}^{(t)}(d_1|z_1) \times \pi(\phi_{1\to1}|d_1) \times \pi(z'_1|z_1, d_1, \phi_{1\to1}, a_1)$
$\psi_{C_7}(z_1, d_1, z'_1, k'_1)$	$q_1^{(t)}(k'_1|z_1, d_1, z'_1)$
$\psi_{C_8}(z_1, z'_1, k'_1, \ell'_1)$	$q_{i,1}^{(t)}(\ell'_1|z_1, k'_1)$

Query Junction In-Tree We construct the weighted clique intersection graph from the pairwise intersections of the maximal cliques (showing only the nonempty intersections):

$$C_1 \cap C_2 = \{D_2\}, \qquad\qquad C_2 \cap C_5 = \{\Phi_{2\to1}\},$$

$$C_3 \cap C_4 = \{D_3\}, \qquad\qquad C_4 \cap C_5 = \{\Phi_{3\rightarrow 1}\},$$

$$C_5 \cap C_6 = \{A_1\}, \qquad\qquad C_6 \cap C_7 = \{Z_1, D_1, Z'_1\},$$

$$C_6 \cap C_8 = \{Z_1, Z'_1\}, \qquad\qquad C_7 \cap C_8 = \{Z_1, Z'_1, K'_1\}.$$

We need to find a maximum weight spanning tree in the weighted clique intersection graph and direct all edges toward $R_T = C_8$, since it is the unique maximal clique that contains $C' = \{Z'_1, L'_1\}$. This weighted clique intersection graph and the (unique) query junction in-tree are shown below.

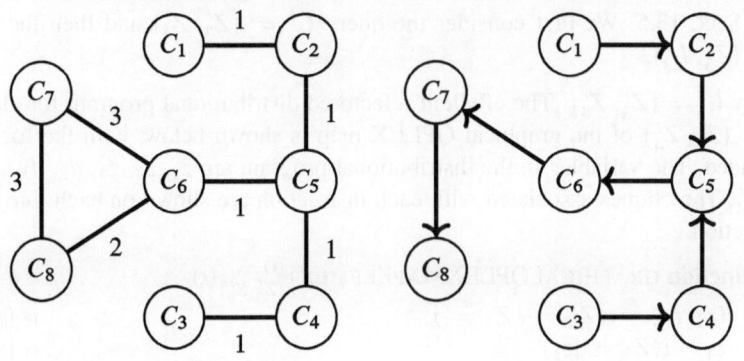

weighted clique intersection graph   query junction in-tree

Message Passing The messages are shown below:

$$\delta_{C_1 \rightarrow C_2}(d_2) = \sum_{z_2} \psi_{C_1}(z_2, d_2),$$

$$\delta_{C_2 \rightarrow C_5}(\phi_{2\rightarrow 1}) = \sum_{d_2} \psi_{C_2}(d_2, \phi_{2\rightarrow 1}) \times \delta_{C_1 \rightarrow C_2}(d_2),$$

$$\delta_{C_3 \rightarrow C_4}(d_3) = \sum_{z_3} \psi_{C_3}(z_3, d_3),$$

$$\delta_{C_4 \rightarrow C_5}(\phi_{3\rightarrow 1}) = \sum_{d_3} \psi_{C_4}(d_3, \phi_{3\rightarrow 1}) \times \delta_{C_3 \rightarrow C_4}(d_3),$$

$$\delta_{C_5 \rightarrow C_6}(a_1) = \sum_{\phi_{0\rightarrow 1}, \phi_{2\rightarrow 1}, \phi_{3\rightarrow 1}} \psi_{C_5}(\phi_{0\rightarrow 1}, \phi_{2\rightarrow 1}, \phi_{3\rightarrow 1}, a_1),$$

$$\times\, \delta_{C_2 \rightarrow C_5}(\phi_{2\rightarrow 1}) \times \delta_{C_4 \rightarrow C_5}(\phi_{3\rightarrow 1}),$$

$$\delta_{C_6 \rightarrow C_7}(z_1, d_1, z'_1) = \sum_{a_1, \phi_{1\rightarrow 1}} \psi_{C_6}(z_1, d_1, \phi_{1\rightarrow 1}, a_1, z'_1) \times \delta_{C_5 \rightarrow C_6}(a_1),$$

$$\delta_{C_7 \rightarrow C_8}(z_1, z'_1, k'_1) = \sum_{d_1} \psi_{C_7}(z_1, d_1, z'_1, k'_1) \times \delta_{C_6 \rightarrow C_7}(z_1, d_1, z'_1).$$

Proposition 14.1 implies that the sought pmf can be found via

$$\iota'(z_1', \ell_1') = \sum_{z_1, k_1'} \psi_{C_8}(z_1, z_1', k_1', \ell_1') \times \delta_{C_7 \to C_8}(z_1, z_1', k_1').$$

Flow Line with Fishbone Information Structure

We consider the flow line model with the fishbone information structure example from Sect. 13.5. We first consider the query $C' = \{Z_3', Z_4'\}$ and then the query $C' = \{Z_3', L_3'\}$.

Query $C' = \{Z_3', Z_4'\}$ The efficient refactored distributional program for element $C' = \{Z_3', Z_4'\}$ of the graphical QPLEX map is shown below, with the for-loops expanded. The variables in the distributional program are $z_2, z_3, z_4, d_2, d_3, d_4, z_3'$, and z_4'. The cliques associated with each instruction are shown on each line of the instruction.

1: **function** GRAPHICALQPLEXMAPELEMENT$_{\{Z_3', Z_4'\}}^{(t)}(\iota)$
2:     $(z_2, z_3) \sim \iota(Z_2 = \cdot, Z_3 = \cdot)$ $\triangleright \{z_2, z_3\}$
3:     $z_4 \sim \iota(Z_4 = \cdot | z_3)$ $\triangleright \{z_3, z_4\}$
4:     $d_2 \sim$ NUMBEROFCOMPLETIONS$_{\iota,2}^{(t)}(z_2)$ $\triangleright \{z_2, d_2\}$
5:     $d_3 \sim$ NUMBEROFCOMPLETIONS$_{\iota,3}^{(t)}(z_3)$ $\triangleright \{z_3, d_3\}$
6:     $d_4 \sim$ NUMBEROFCOMPLETIONS$_{\iota,4}^{(t)}(z_4)$ $\triangleright \{z_4, d_4\}$
7:     $z_3' \leftarrow z_3 - d_3 + d_2$ $\triangleright \{z_3, d_2, d_3, z_3'\}$
8:     $z_4' \leftarrow z_4 - d_4 + d_3$ $\triangleright \{z_4, d_3, d_4, z_4'\}$
9:     **return** (z_3', z_4') $\triangleright \{z_3', z_4'\}$

Query Graph The graph produced by planting cliques associated with each instruction is shown below.

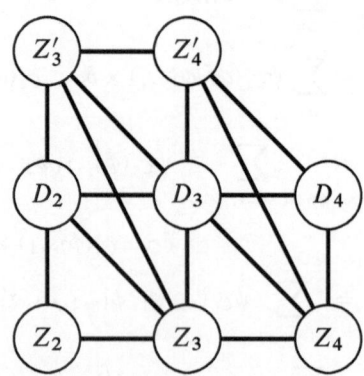

This graph is *not* triangulated, since the cycle $Z_3 - Z_4 - Z'_4 - Z'_3 - Z_3$ does not have a chord. We add the edge $Z'_3 - Z_4$ to triangulate this graph. The query graph is shown below, with the additional edge highlighted in blue.

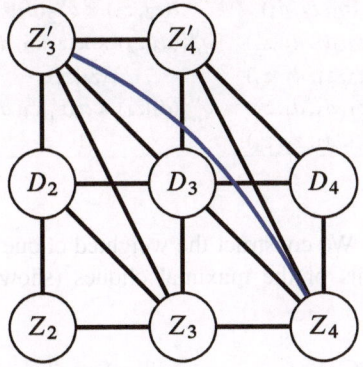

This graph is triangulated, since it is recursively simplicial with respect to the ordering $Z_2, D_2, D_4, Z_3, Z_4, D_3, Z'_3, Z'_4$.

Factors The maximal cliques in the query graph are:

$$C_1 = \{Z_2, Z_3, D_2\}, \qquad C_2 = \{Z_3, D_2, D_3, Z'_3\},$$
$$C_3 = \{Z_3, Z_4, D_3, Z'_3\}, \qquad C_4 = \{Z_4, D_3, D_4, Z'_4\},$$
$$C_5 = \{Z_4, D_3, Z'_3, Z'_4\}.$$

The next step is to assign each line number in the body of the program to a maximal clique. Lines 2 and 4 must be assigned to C_1; Line 7 must be assigned to C_2; Line 3 must be assigned to C_3; Lines 6 and 8 must be assigned to C_4; Line 9 must be assigned to C_5; and Line 5 can either be assigned to C_2 or C_3. So, one feasible assignment of line numbers pmfs to maximal cliques is given in the following table.

Maximal Clique	Line Number(s) Assigned
$C_1 = \{Z_2, Z_3, D_2\}$	2, 4
$C_2 = \{Z_3, D_2, D_3, Z'_3\}$	5, 7
$C_3 = \{Z_3, Z_4, D_3, Z'_3\}$	3
$C_4 = \{Z_4, D_3, D_4, Z'_4\}$	6, 8
$C_5 = \{Z_4, D_3, Z'_3, Z'_4\}$	–

After defining the pmfs, for $m = 2, 3$,

$$\pi(z'_m | z_m, d_m, d_{m-1}) = 1(z'_m = z_m - d_m + d_{m-1}),$$

we obtain the following factors associated with the query graph.

Factor	Definition
$\psi_{C_1}(z_2, z_3, d_2)$	$\iota(z_2, z_3) \times q_{\iota,2}^{(t)}(d_2\|z_2)$
$\psi_{C_2}(z_3, d_2, d_3, z_3')$	$q_{\iota,3}^{(t)}(d_3\|z_3) \times \pi(z_3'\|z_3, d_2, d_3)$
$\psi_{C_3}(z_3, z_4, d_3, z_3')$	$\iota(z_4\|z_3)$
$\psi_{C_4}(z_4, d_3, d_4, z_4')$	$q_{\iota,4}^{(t)}(d_4\|z_4) \times \pi(z_4'\|z_4, d_3, d_4)$
$\psi_{C_5}(z_4, d_3, z_3', z_4')$	1

Query Junction In-Tree   We construct the weighted clique intersection graph from the pairwise intersections of the maximal cliques (showing only the nonempty intersections):

$$C_1 \cap C_2 = \{Z_3, D_2\}, \qquad\qquad C_1 \cap C_3 = \{Z_3\},$$
$$C_2 \cap C_3 = \{Z_3, D_3, Z_3'\}, \qquad\qquad C_2 \cap C_4 = \{D_3\},$$
$$C_2 \cap C_5 = \{D_3, Z_3'\}, \qquad\qquad C_3 \cap C_4 = \{Z_4, D_3\},$$
$$C_3 \cap C_5 = \{Z_4, D_3, Z_3'\}, \qquad\qquad C_4 \cap C_5 = \{Z_4, D_3, Z_4'\}.$$

We need to find a maximum weight spanning tree in the weighted clique intersection graph and direct all edges toward $R_T = C_5$, since it is the unique maximal clique that contains $C' = \{Z_3', Z_4'\}$. This weighted clique intersection graph and the (unique) query junction in-tree are shown below.

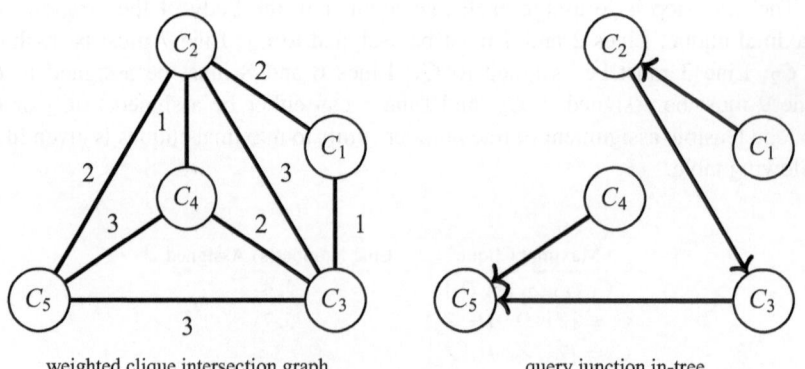

weighted clique intersection graph   query junction in-tree

Message Passing   The messages are shown below:

$$\delta_{C_1 \to C_2}(z_3, d_2) = \sum_{z_2} \psi_{C_1}(z_2, z_3, d_2),$$

$$\delta_{C_2 \to C_3}(z_3, d_3, z_3') = \sum_{d_2} \psi_{C_2}(z_3, d_2, d_3, z_3') \times \delta_{C_1 \to C_2}(z_3, d_2),$$

$$\delta_{C_3 \to C_5}(z_4, d_3, z_3') = \sum_{z_3} \psi_{C_3}(z_3, z_4, d_3, z_3') \times \delta_{C_2 \to C_3}(z_3, d_3, z_3'),$$

$$\delta_{C_4 \to C_5}(z_4, d_3, z_4') = \sum_{d_4} \psi_{C_4}(z_4, d_3, d_4, z_4').$$

(The last message does not depend on any previously calculated messages, so it can be calculated out of order.) Proposition 14.1 implies that the sought pmf can be found via

$$\iota'(z_3', z_4') = \sum_{z_4, d_3} \delta_{C_3 \to C_5}(z_4, d_3, z_3') \times \delta_{C_4 \to C_5}(z_4, d_3, z_4'),$$

where we have suppressed the factor $\psi_{C_5}(z_4, d_3, z_3', z_4')$ on the right-hand side as it is 1.

Query $C' = \{Z_3', L_3'\}$ The efficient refactored distributional program for element $C' = \{Z_3', L_3'\}$ of the graphical QPLEX map is shown below, with the for-loop expanded. The variables in the distributional program are z_2, z_3, d_2, d_3, z_3', k_3', and ℓ_3'. The cliques associated with each instruction are shown on each line of the instruction.

1: **function** GRAPHICALQPLEXMAPELEMENT$_{\{Z_3', L_3'\}}^{(t)}(\iota)$
2:     $(z_2, z_3) \sim \iota(Z_2 = \cdot, Z_3 = \cdot)$ $\triangleright \{z_2, z_3\}$
3:     $d_2 \sim$ NUMBEROFCOMPLETIONS$_{\iota, 2}^{(t)}(z_2)$ $\triangleright \{z_2, d_2\}$
4:     $d_3 \sim$ NUMBEROFCOMPLETIONS$_{\iota, 3}^{(t)}(z_3)$ $\triangleright \{z_3, d_3\}$
5:     $z_3' \leftarrow z_3 - d_3 + d_2$ $\triangleright \{z_3, d_2, d_3, z_3'\}$
6:     $k_3' \sim$ ENTITYTYPE$_3^{(t)}(z_3, d_3, z_3')$ $\triangleright \{z_3, d_3, z_3', k_3'\}$
7:     $\ell_3' \sim$ NEXTLABEL$_{\iota, 3}^{(t)}(z_3, k_3')$ $\triangleright \{z_3, k_3', \ell_3'\}$
8:     **return** (z_3', ℓ_3') $\triangleright \{z_3', \ell_3'\}$

Query Graph The graph produced by planting cliques associated with each instruction is shown below.

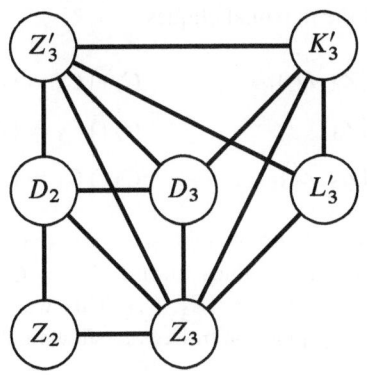

This graph is triangulated, since it is recursively simplicial with respect to the ordering Z_2, D_2, L_3', K_3', D_3, Z_3, Z_3'. Therefore, it is the query graph associated with the distributional program.

Factors The maximal cliques in this query graph are:

$$C_1 = \{Z_2, Z_3, D_2\}, \qquad\qquad C_2 = \{Z_3, D_2, D_3, Z_3'\},$$
$$C_3 = \{Z_3, D_3, Z_3', K_3'\}, \qquad\qquad C_4 = \{Z_3, Z_3', K_3', L_3'\}.$$

The next step is to assign each line number in the body of the program to a maximal clique. Lines 2 and 3 must be assigned to C_1; Line 5 must be assigned to C_2; Line 6 must be assigned to C_3; Line 7 must be assigned to C_4; and Line 4 can either be assigned to C_2 or C_3. So, one feasible assignment of line numbers to maximal cliques is given in the following table.

Maximal Clique	Line Number(s) Assigned
$C_1 = \{Z_2, Z_3, D_2\}$	2, 3
$C_2 = \{Z_3, D_2, D_3, Z_3'\}$	4, 5
$C_3 = \{Z_3, D_3, Z_3', K_3'\}$	6
$C_4 = \{Z_3, Z_3', K_3', L_3'\}$	7

We obtain the following factors associated with this query graph.

Factor	Definition
$\psi_{C_1}(z_2, z_3, d_2)$	$\iota(z_2, z_3) \times q_{\iota,2}^{(t)}(d_2\vert z_2)$
$\psi_{C_2}(z_3, d_2, d_3, z_3')$	$q_{\iota,3}^{(t)}(d_3\vert z_3) \times \pi(z_3'\vert z_3, d_2, d_3)$
$\psi_{C_3}(z_3, d_3, z_3', k_3')$	$q_3^{(t)}(k_3'\vert z_3, d_3, z_3')$
$\psi_{C_4}(z_3, z_3', k_3', \ell_3')$	$q_{\iota,3}^{(t)}(\ell_3'\vert z_3, k_3')$

Query Junction In-Tree We construct the weighted intersection graph from the pairwise intersections of the maximal cliques:

$$C_1 \cap C_2 = \{Z_3, D_2\}, \qquad\qquad C_1 \cap C_3 = \{Z_3\},$$
$$C_1 \cap C_4 = \{Z_3\}, \qquad\qquad C_2 \cap C_3 = \{Z_3, D_3, Z_3'\},$$
$$C_2 \cap C_4 = \{Z_3, Z_3'\}, \qquad\qquad C_3 \cap C_4 = \{Z_3, Z_3', K_3'\}.$$

We need to find a maximum weight spanning tree in the weighted clique intersection graph and direct all edges toward $R_T = C_4$, since it is the unique maximal clique that contains $C' = \{Z_3', L_3'\}$. This weighted clique intersection graph and the (unique) query junction in-tree are shown below.

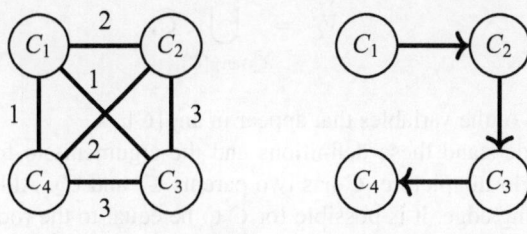

weighted clique intersection graph   query junction in-tree

Message Passing The messages are shown below:

$$\delta_{C_1 \to C_2}(z_3, d_2) = \sum_{z_2} \psi_{C_1}(z_2, z_3, d_2),$$

$$\delta_{C_2 \to C_3}(z_3, d_3, z_3') = \sum_{d_2} \psi_{C_2}(z_3, d_2, d_3, z_3') \times \delta_{C_1 \to C_2}(z_3, d_2),$$

$$\delta_{C_3 \to C_4}(z_3, z_3', k_3') = \sum_{d_3} \psi_{C_3}(z_3, d_3, z_3', k_3') \times \delta_{C_2 \to C_3}(z_3, d_3, z_3').$$

Proposition 14.1 implies that the sought pmf can be found via

$$\iota'(z_3', \ell_3') = \sum_{z_3, k_3'} \psi_{C_4}(z_3, z_3', k_3', \ell_3') \times \delta_{C_3 \to C_4}(z_3, z_3', k_3').$$

14.8   Proof of Proposition 14.1

We seek to efficiently calculate the function

$$\sum_{V \setminus R_T} \prod_{\hat{C} \in V_T} \psi_{\hat{C}}, \tag{14.5}$$

where $V = \bigcup_{C \in V_T} C$ is the set of all variables that appear in T. Note that this is a function with scope R_T.

The ideas underlying the proof of Proposition 14.1 are well-known; see, for example, Section 10.2 of [14]. The proof we present here uses the notation of this book. We begin by setting forth notation and definitions of certain function collections we require for the argument.

Notation For any $C \in V_T$, recall that $\text{pa}_T(C)$ denotes the set of parents of C in T. Let $\text{an}_T[C]$ denote the set of ancestors of C in T, including C, namely, the set of vertices of T with a directed path to C, possibly of zero length. We furthermore define the subset V_C^T of V via

$$V_C^T = \bigcup_{\widehat{C} \in \text{an}_T[C]} \widehat{C},$$

which consists of the variables that appear in $\text{an}_T[C]$.

To help understand these definitions and the arguments to follow, refer to the picture below. In the picture, C has two parents $\tilde{C}_1$ and $\tilde{C}_2$. Although C is shown with an outgoing edge, it is possible for C to be equal to the root R_T. The set V_C^T consists of all variables appearing in any of the maximal cliques associated with the vertices in $\text{an}_T[C]$. Clouds represent subtrees of T. The large cloud represents the subtree of T induced by the vertices in $\text{an}_T[C]$, and the small clouds represent the subtrees of T induced by the vertices in $\text{an}_T[\tilde{C}_1]$ and $\text{an}_T[\tilde{C}_2]$.

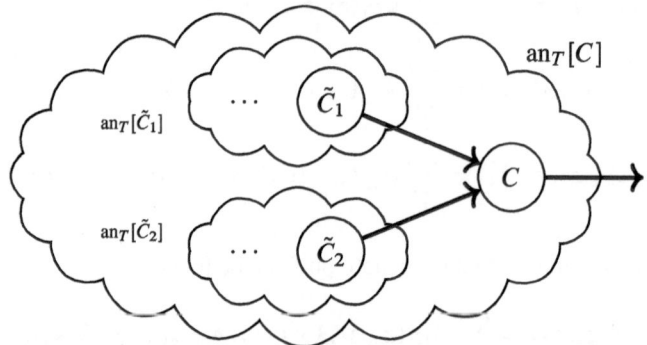

The key argument relies on decomposing the large cloud in terms of the smaller clouds. Indeed, the picture suggests that the two smaller clouds do not intersect. The following lemma establishes this fact and is required for arguments to follow. It applies the clique intersection property; see Sect. 11.3. Referring to the picture, the lemma states that the only variables that can appear in both clouds $\text{an}_T[\tilde{C}_1]$ and $\text{an}_T[\tilde{C}_2]$ must also belong to C. Indeed, if there were an element in the intersection of $V_{\tilde{C}_1}^T \setminus C$ and $V_{\tilde{C}_2}^T \setminus C$, then it would have to belong to C by the clique intersection property, since C is on the unique path from $\tilde{C}_1$ to $\tilde{C}_2$, an obvious contradiction.

Lemma 14.2 *For each $C \in \mathcal{C}(G)$, the collection of sets*

$$\{V_{\tilde{C}}^T \setminus C : \tilde{C} \in \text{pa}_T(C)\}$$

is a partition of $V_C^T \setminus C$.

Function Collections The strategy for calculating (14.5) is to calculate two collections of functions, and (14.5) is one of these functions. The first collection consists of functions we call *maximal clique functions*, and this collection is indexed by the set of vertices V_T of T. For all $C \in V_T$, these are defined via

$$\xi_C = \sum_{V_C^T \setminus C} \prod_{\widehat{C} \in \text{an}_T[C]} \psi_{\widehat{C}}.$$

The right-hand side is to be interpreted as ψ_C if C is a leaf. Note that ξ_{R_T} equals (14.5) since $V_{R_T}^T = V$.

The second collection of functions is indexed by the set of edges E_T of T. For each edge $(\tilde{C}, C) \in E_T$, we set

$$\hat{\delta}_{\tilde{C} \to C} = \sum_{V_{\tilde{C}}^T \setminus C} \prod_{\widehat{C} \in \mathrm{an}_T[\tilde{C}]} \psi_{\widehat{C}}.$$

Both the ξ_C and $\hat{\delta}_{\tilde{C} \to C}$ functions depend on the junction in-tree T, even though this is not explicit in the notation. Note that we have written $\delta_{\tilde{C} \to C}$ with a "hat" in this definition; we will establish momentarily that each $\hat{\delta}_{\tilde{C} \to C}$ equals the message $\delta_{\tilde{C} \to C}$ defined in Sect. 14.5. We apply the clique intersection property to verify that the scope of $\hat{\delta}_{\tilde{C} \to C}$ is $\tilde{C} \cap C$. Since the scope of the product $\prod_{\widehat{C} \in \mathrm{an}_T[\tilde{C}]} \psi_{\widehat{C}}$ in this definition is $V_{\tilde{C}}^T$, the scope of $\hat{\delta}_{\tilde{C} \to C}$ is $V_{\tilde{C}}^T \cap C$. By the clique intersection property, each element that belongs to both $V_{\tilde{C}}^T$ and C must also belong to $\tilde{C}$, and so $V_{\tilde{C}}^T \cap C = \tilde{C} \cap C$.

Recursive Calculation The following two lemmas show how the $\hat{\delta}_{\tilde{C} \to C}$ can be calculated from the maximal clique function $\xi_{\tilde{C}}$ and how the maximal clique function ξ_C can conversely be calculated from the $\hat{\delta}_{\tilde{C} \to C}$ associated with each of the incoming edges to C.

Lemma 14.3 *For each edge $(\tilde{C}, C) \in E_T$, we have that*

$$\hat{\delta}_{\tilde{C} \to C} = \sum_{\tilde{C} \setminus C} \xi_{\tilde{C}}. \tag{14.6}$$

Proof Fix some $(\tilde{C}, C) \in E_T$. We first note that the sets $V_{\tilde{C}}^T \setminus \tilde{C}$ and C do not intersect. Indeed, an element in their intersection would belong to some clique in $\mathrm{an}_T[\tilde{C}] \setminus \{\tilde{C}\}$ and $\tilde{C}$ by the clique intersection property, a contradiction. Since the set $V_{\tilde{C}}^T$ is the disjoint union of the sets $V_{\tilde{C}}^T \setminus \tilde{C}$ and $\tilde{C}$, the set $V_{\tilde{C}}^T \setminus C$ is the disjoint union of the sets $V_{\tilde{C}}^T \setminus \tilde{C}$ and $\tilde{C} \setminus C$. As a consequence of this observation, we obtain that

$$\hat{\delta}_{\tilde{C} \to C} = \sum_{V_{\tilde{C}}^T \setminus C} \prod_{\widehat{C} \in \mathrm{an}_T[\tilde{C}]} \psi_{\widehat{C}}$$

$$= \sum_{\tilde{C} \setminus C} \sum_{V_{\tilde{C}}^T \setminus \tilde{C}} \prod_{\widehat{C} \in \mathrm{an}_T[\tilde{C}]} \psi_{\widehat{C}}$$

$$= \sum_{\tilde{C} \setminus C} \xi_{\tilde{C}},$$

where the last equality follows from the definition of $\xi_{\tilde{C}}$.   □

Lemma 14.4 *For each $C \in \mathcal{C}(G)$, we have that*

$$\xi_C = \psi_C \times \prod_{\tilde{C} \in \mathrm{pa}_T(C)} \hat{\delta}_{\tilde{C} \to C}, \tag{14.7}$$

where an empty product should be interpreted as 1.

Proof Fix some $C \in \mathcal{C}(G)$. From the representation

$$\mathrm{an}_T[C] = \{C\} \cup \bigcup_{\tilde{C} \in \mathrm{pa}_T(C)} \mathrm{an}_T[\tilde{C}]$$

as a disjoint union, we obtain that

$$\xi_C = \psi_C \times \sum_{V_C^T \backslash C} \prod_{\tilde{C} \in \mathrm{pa}_T(C)} \left(\prod_{\widehat{C} \in \mathrm{an}_T[\tilde{C}]} \psi_{\widehat{C}} \right).$$

Consider the term in parentheses on the right-hand side. It defines a function for each $\tilde{C} \in \mathrm{pa}_T(C)$ with scope $V_{\tilde{C}}^T$. As a direct result of Lemma 14.2, each of the variables in $V_C^T \backslash C$ in the sum appears in *exactly one* of these functions. The distributive law for sums and products thus shows that

$$\xi_C = \psi_C \times \prod_{\tilde{C} \in \mathrm{pa}_T(C)} \sum_{V_{\tilde{C}}^T \backslash C} \left(\prod_{\widehat{C} \in \mathrm{an}_T[\tilde{C}]} \psi_{\widehat{C}} \right),$$

which proves the claim by definition of $\hat{\delta}_{\tilde{C} \to C}$.   □

Proof of Main Result Upon substituting (14.7) into (14.6), we obtain that

$$\hat{\delta}_{\tilde{C} \to C} = \sum_{\tilde{C} \backslash C} \left(\psi_{\tilde{C}} \times \prod_{\widehat{C} \in \mathrm{pa}_T(\tilde{C})} \hat{\delta}_{\widehat{C} \to \tilde{c}} \right), \tag{14.8}$$

which is the same recursion as in (14.2) for the messages. We show that $\hat{\delta}_{\tilde{C} \to C} = \delta_{\tilde{C} \to C}$ for all $(\tilde{C}, C) \in E_T$ by induction on the distance of $\tilde{C}$ to the root R_T.

Suppose then that the claim has been established for each edge $(\widehat{C}, \tilde{C})$ with $\tilde{C}$ at a distance at least $k \geq 1$ from the root. Fix some edge $(\tilde{C}, C) \in E_T$ with $\tilde{C}$ at a distance k from the root, so that C is at a distance $k - 1$ from the root. If $\tilde{C}$ is a leaf, then $\mathrm{pa}_T(\tilde{C})$ is empty. It then follows from their respective definitions that $\hat{\delta}_{\tilde{C} \to C} = \delta_{\tilde{C} \to C}$. Now suppose that $\tilde{C}$ is not a leaf. In this case, we have that $\hat{\delta}_{\widehat{C} \to \tilde{c}} = \delta_{\widehat{C} \to \tilde{c}}$

for all $\widehat{C} \in \mathrm{pa}_T(\tilde{C})$ by the induction hypothesis, and so $\hat{\delta}_{\tilde{C} \to C} = \delta_{\tilde{C} \to C}$ follows from (14.2) and (14.8). This completes the inductive argument, as we have now established the claim for every edge $(\tilde{C}, C) \in E_T$ with C at least at a distance $k - 1$ from the root. Proposition 14.1 can now be established by applying Lemma 14.4 to $C = R_T$.

Part III
Foundations

Chapter 15
Introduction to Foundations

In Part III, we provide the mathematical foundations of the QPLEX methodology. While the material presented in this part is self-contained from a mathematical perspective, it heavily uses the results presented in Chaps. 11 and 12.

Part III formally defines the kernel, compression, multinomial lift, QPLEX, and graphical QPLEX maps, which we previously introduced via distributional programs without discussing their (co)domains. We also establish various properties of these maps. Various compositions of these maps are shown to be projections (in the sense of Sect. 1.8) and play a central role in establishing what we call "period-by-period" optimality of the QPLEX iterates, the subject of Chap. 16, and the information iterates, the subject of Chap. 17. Some results in these chapters are inspired by existing results for dynamic Bayesian networks and probabilistic graphical models [14]. In Chap. 18, we present theoretical results that establish that the marginal pmfs of the counters for the QPLEX iterates are *identical* to the marginal pmfs of the counters for the kernel iterates under several special conditions. These conditions are precisely what should be expected based on the myriad of classical stochastic modeling results.

Throughout Part III, we fix some QPLEX chain $\{(\boldsymbol{Z}^{(t)}, \boldsymbol{H}^{(t)}) : t \geq 0\}$ with counter set Ω_Z, M activities, a collection of size function vectors $\{\boldsymbol{x}^{(t)} : t \geq 0\}$, and label sets $\Omega_1, \ldots, \Omega_M$. Nothing in Chaps. 16 and 17 requires a special structure of its transition probabilities $p^{(t)}(\boldsymbol{z}', \boldsymbol{h}'|\boldsymbol{z}, \boldsymbol{h})$, and in particular, we do not require simple or advanced transition dynamics. However, some results we derive in Chap. 18 are only valid under special assumptions on this QPLEX chain.

15.1   Foundations of QPLEX Calculus

This section provides an overview of the QPLEX calculus foundations. Before we are able to state the formal definitions of the various maps we use in Part III, we need to introduce several sets that will appear as (co)domains of these maps. We fix time t until we are ready to introduce the optimality properties of QPLEX iterates at the end of this section.

Recall that the state space for a QPLEX chain is

$$\Omega_{Z,H} = \Omega_Z \times \Omega_H$$

and that the definition of a QPLEX chain uses the set

$$\Omega_{Z,H}^{(t)} = \{(z, h) \in \Omega_{Z,H} : |h_m| = x_m^{(t)}(z) \text{ for all } m\}.$$

The random vector $(\boldsymbol{Z}^{(t)}, \boldsymbol{H}^{(t)})$ is an element of $\Omega_{Z,H}^{(t)}$ with probability 1 by virtue of $\{(\boldsymbol{Z}^{(t)}, \boldsymbol{H}^{(t)}) : t \geq 0\}$ being a QPLEX chain. We let $\mathcal{P}_{Z,H}$ denote the set of pmfs on $\Omega_{Z,H}$, and we let $\mathcal{P}_{Z,H}^{(t)}$ denote the subset of $\mathcal{P}_{Z,H}$ defined by

$$\mathcal{P}_{Z,H}^{(t)} = \{v \in \mathcal{P}_{Z,H} : v(\Omega_{Z,H}^{(t)}) = 1\}.$$

By the definition of a QPLEX chain, the marginal pmf of $(\boldsymbol{Z}^{(t)}, \boldsymbol{H}^{(t)})$ is an element of $\mathcal{P}_{Z,H}^{(t)}$.

We define the sets $\Omega_L = \Omega_1 \times \cdots \times \Omega_M$ and

$$\Omega_{Z,L} = \Omega_Z \times \Omega_L.$$

We also let $\mathcal{P}_{Z,L}$ denote the set of pmfs on $\Omega_{Z,L}$, and we let $\mathcal{P}_{Z,L}^{(t)}$ denote the subset of $\mathcal{P}_{Z,L}$ defined by

$$\mathcal{P}_{Z,L}^{(t)} = \{\mu \in \mathcal{P}_{Z,L} : \text{ for all } m, z, \text{ and } \ell_m,$$

$$\mu(z, \ell_m) = \mu(z) \times \theta_m^0(\ell_m) \text{ if } x_m^{(t)}(z) = 0\},$$

where we recall that θ_m^0 denotes a fixed "dummy" pmf on the label set Ω_m for each activity m. Even though these pmfs have played no significant role so far in this book, they are needed to formulate the projective properties of various maps.

Compression and Multinomial Lift Maps We have already used the distributional program representations of these maps in Sect. 3.2, but we require formal definitions for purposes of Part III. We will define the compression map $\mathcal{E}^{(t)}$ by setting, for all $v \in \mathcal{P}_{Z,H}^{(t)}$ and $(z, \ell) \in \Omega_{Z,L}$,

$$[\mathcal{E}^{(t)}(v)](z, \ell) = \sum_{h \in \Omega_H} \left(\prod_m \frac{h_m(\ell_m)}{|h_m|} \right) \times v(z, h),$$

where the ratio should be understood to mean $\theta_m^0(\ell_m)$ if $h_m = 0$. We will show that $\mathcal{P}_{Z,L}^{(t)}$ can be chosen as the codomain of $\mathcal{E}^{(t)}$.

We will define the multinomial lift map $\mathcal{L}^{(t)}$ by setting, for all $\mu \in \mathcal{P}_{Z,L}^{(t)}$ and $(z, h) \in \Omega_{Z,H}$,

$$[\mathcal{L}^{(t)}(\mu)](z, h) = \mu(z) \times \prod_m \Pr[\text{mult}(x_m^{(t)}(z), \mu(L_m = \cdot|z)) = h_m].$$

We will show that $\mathcal{P}_{Z,H}^{(t)}$ can be chosen as the codomain of $\mathcal{L}^{(t)}$. We will also show via an example why the map $\mathcal{L}^{(t)}$ is not linear.

Kernel and QPLEX Maps We revisit the diagram from Sect. 3.2 where we introduced the QPLEX map $\mathcal{Q}^{(t)}$. We can now annotate the two edges between the levels with the multinomial lift map $\mathcal{L}^{(t)}$ and the compression map $\mathcal{E}^{(t)}$. In view of the above (co)domains of these maps, the pmfs v and v' on the top level are elements of $\mathcal{P}_{Z,H}^{(t)}$ and $\mathcal{P}_{Z,H}^{(t+1)}$, respectively, while the pmfs μ and μ' on the bottom level are elements of $\mathcal{P}_{Z,L}^{(t)}$ and $\mathcal{P}_{Z,L}^{(t+1)}$, respectively.

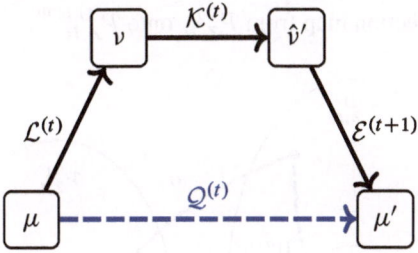

We will choose the domains and codomains of the kernel map $\mathcal{K}^{(t)}$ and the QPLEX map $\mathcal{Q}^{(t)}$ to match the structure of the pmfs in this diagram. In particular, we will formally define the kernel map $\mathcal{K}^{(t)}$ as a map from $\mathcal{P}_{Z,H}^{(t)}$ to $\mathcal{P}_{Z,H}^{(t+1)}$ and the QPLEX map $\mathcal{Q}^{(t)}$ as a map from $\mathcal{P}_{Z,L}^{(t)}$ to $\mathcal{P}_{Z,L}^{(t+1)}$. We previously introduced the kernel and QPLEX maps $\mathcal{K}^{(t)}$ and $\mathcal{Q}^{(t)}$ in terms of the transition probabilities $p^{(t)}(z', h'|z, h)$ without specifying their (co)domains; see (2.2) and Sect. 3.2, respectively.

Projection Maps Compositions of the compression and multinomial lift maps are projection maps and play a central role in Part III.

The *label conditional independence projection map* $\Pi^{(t), \text{lci}}$ is defined as the composition $\mathcal{E}^{(t)} \circ \mathcal{L}^{(t)}$. Let $\mathcal{P}_{Z,L}^{(t), \text{lci}}$ be the subset of $\mathcal{P}_{Z,L}^{(t)}$ defined by the property that the label variables are conditionally independent given the counter variables, i.e., each $\mu \in \mathcal{P}_{Z,L}^{(t),\text{lci}}$ satisfies $\mu(z, \ell) = \mu(z) \times \prod_m \mu(\ell_m | z)$ for all $(z, \ell) \in \Omega_{Z,L}$. As visualized in the diagram below, we will show that $\Pi^{(t),\text{lci}}$ is a projection map from $\mathcal{P}_{Z,L}^{(t)}$ onto $\mathcal{P}_{Z,L}^{(t),\text{lci}}$.

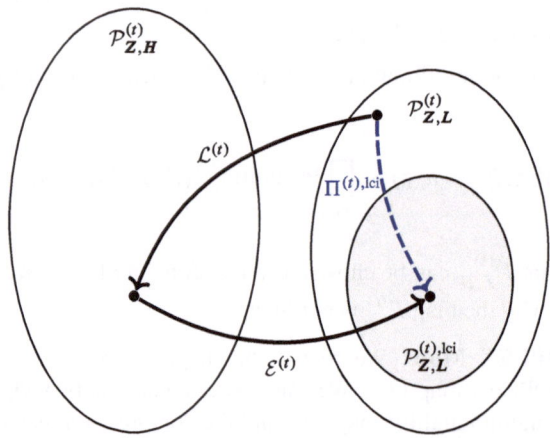

The *multinomial projection map* $\Pi^{(t),\text{mult}}$ is defined as the composition $\mathcal{L}^{(t)} \circ \mathcal{E}^{(t)}$. Let $\mathcal{P}_{Z,H}^{(t),\text{mult}}$ be the subset of $\mathcal{P}_{Z,H}^{(t)}$ consisting of $\nu \in \mathcal{P}_{Z,H}^{(t)}$ that can be written as $\nu = \mathcal{L}^{(t)}(\mu)$ for some $\mu \in \mathcal{P}_{Z,L}^{(t)}$. As visualized in the diagram below, we will show that $\Pi^{(t),\text{mult}}$ is a projection map from $\mathcal{P}_{Z,H}^{(t)}$ onto $\mathcal{P}_{Z,H}^{(t),\text{mult}}$.

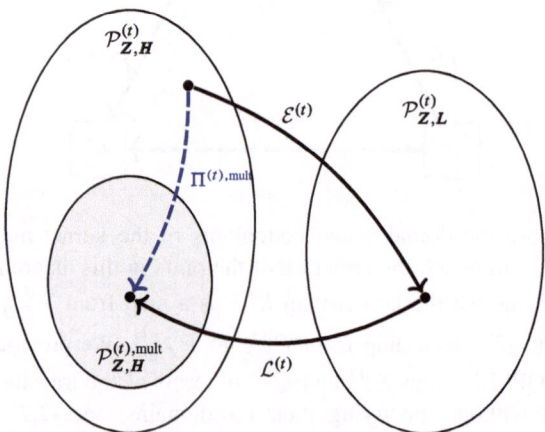

We will show that the map $\Pi^{(t),\text{mult}}$ possesses a fundamental optimality property that we will use to establish an optimality result for the QPLEX iterates. Fix some pmf $\hat{\nu} \in \mathcal{P}_{Z,H}^{(t)}$ with finite support. Suppose we seek to find the pmf $\nu \in \mathcal{P}_{Z,H}^{(t),\text{mult}}$ that minimizes the Kullback-Leibler divergence $D(\hat{\nu} \parallel \nu)$. (See Sect. 1.7 for the definition and properties of Kullback-Leibler divergence.) We show that the solution is unique and given by

$$\Pi^{(t),\mathrm{mult}}(\hat{v}) = \arg\inf_{v \in \mathcal{P}_{Z,H}^{(t),\mathrm{mult}}} D(\hat{v} \parallel v). \tag{15.1}$$

QPLEX Iterates and Period-by-Period Optimality Given some initial $\mu_Q^{(0)} \in \mathcal{P}_{Z,L}^{(0)}$, the QPLEX iterates are defined iteratively for $t \geq 0$ via

$$\mu_Q^{(t+1)} = \mathcal{Q}^{(t)}(\mu_Q^{(t)}).$$

Each QPLEX iterate $\mu_Q^{(t)}$ is a pmf of $(\boldsymbol{Z}, \boldsymbol{L})$ and lies in $\mathcal{P}_{Z,L}^{(t)}$.

The definition of QPLEX iterates should be compared to that of the kernel iterates, which start with some initial $v_{\mathrm{ker}}^{(0)} \in \mathcal{P}_{Z,H}^{(0)}$ and are defined iteratively for $t \geq 0$ via

$$v_{\mathrm{ker}}^{(t+1)} = \mathcal{K}^{(t)}(v_{\mathrm{ker}}^{(t)}).$$

Each kernel iterate $v_{\mathrm{ker}}^{(t)}$ is a pmf of $(\boldsymbol{Z}, \boldsymbol{H})$ and lies in $\mathcal{P}_{Z,H}^{(t)}$. The marginal pmfs of the counters for the QPLEX iterates approximate the corresponding marginal pmfs of the counters for the kernel iterates, but the QPLEX iterates have (much) smaller supports than the kernel iterates.

The QPLEX iterates can be embedded into our familiar diagram appropriately "glued" together, as shown below for two periods. We have added the multinomial projection maps $\Pi^{(1),\mathrm{mult}}$ and $\Pi^{(2),\mathrm{mult}}$ in the diagram. This diagram starts from some initial QPLEX iterate $\mu_Q^{(0)} \in \mathcal{P}_{Z,L}^{(0)}$ and shows how all pmfs in the diagram can be found by applying appropriate maps. For example, $\mu_Q^{(1)}$ can be calculated from $\mu_Q^{(0)}$ by considering either of the two directed paths from $\mu_Q^{(0)}$ to $\mu_Q^{(1)}$ shown in the diagram.

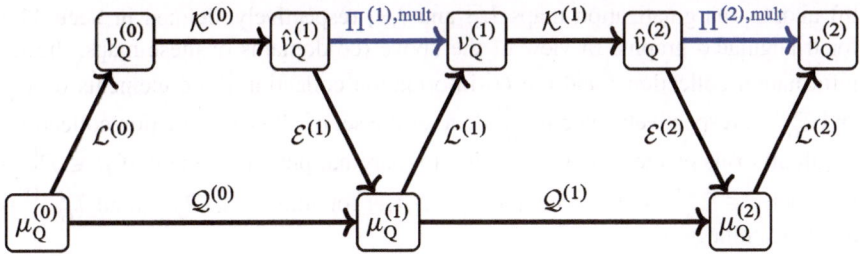

The maps $\Pi^{(t),\mathrm{mult}}$ (highlighted in blue) appear in this diagram as "bridges" between the single-period diagrams. It directly follows from (15.1) that

$$\Pi^{(t),\mathrm{mult}}(\hat{v}_Q^{(t)}) = \arg\inf_{v \in \mathcal{P}_{Z,H}^{(t),\mathrm{mult}}} D(\hat{v}_Q^{(t)} \parallel v).$$

As the diagram shows, we have that

$$\Pi^{(t),\text{mult}}(\hat{v}_Q^{(t)}) = v_Q^{(t)} = \mathcal{L}^{(t)}(\mu_Q^{(t)}).$$

Thus, for example, we see that the QPLEX iterate $\mu_Q^{(2)}$ possesses a period-by-period optimality property in the sense that the so-called lifted QPLEX iterate $\mathcal{L}^{(2)}(\mu_Q^{(2)})$ is the closest pmf (with respect to Kullback-Leibler divergence) to the pmf obtained by applying the kernel map $\mathcal{K}^{(1)}$ to the lifted QPLEX iterate $\mathcal{L}^{(1)}(\mu_Q^{(1)})$. This should be compared to the kernel iterate $v_{\text{ker}}^{(2)}$ being defined as the pmf obtained by applying the kernel map $\mathcal{K}^{(1)}$ to $v_{\text{ker}}^{(1)}$. The QPLEX iterates $\mu_Q^{(t)}$ are the best period-by-period choices if one accepts the QPLEX paradigm of (implicitly) working with "lower-dimensional" sets $\mathcal{P}_{Z,H}^{(t),\text{mult}}$ instead of $\mathcal{P}_{Z,H}^{(t)}$ to overcome the curse of dimensionality for histograms.

15.2   Foundations of Graphical QPLEX Calculus

This section provides an overview of the graphical QPLEX calculus foundations. Similarly to the last section, we define several projection maps that play a central role in the optimality properties of information iterates.

Graphical QPLEX Maps   We fix time t and arbitrary information structure graphs G and G'. As we noted in Chap. 13, it can be practically useful for G and G' to be different.

We revisit the diagram from Chap. 13 where we introduced the graphical QPLEX map $\mathcal{Q}_{G \to G'}^{(t)}$. We can now annotate the two edges between the second and third levels with the maps $L_G^{(t)}$ and $I_{G'}^{(t+1)}$, which are appropriate restrictions of the G-lift and G'-marginalization maps L_G and $I_{G'}$, respectively, defined in Sect. 11.5 for triangulated graphs. In view of the above (co)domains of these maps, the G-information collection ι and the G'-information collection ι' are elements of $\mathcal{I}_G^{(t)}$ and $\mathcal{I}_{G'}^{(t+1)}$, respectively, where $\mathcal{I}_G^{(t)}$ denotes the set of all G-information collections ι with all of its elements corresponding to marginal pmfs of some pmf $\mu \in \mathcal{P}_{Z,L}^{(t)}$, i.e., each $\iota \in \mathcal{I}_G^{(t)}$ can be written as $\iota = I_G(\mu)$ for some $\mu \in \mathcal{P}_{Z,L}^{(t)}$ (and $\mathcal{I}_{G'}^{(t+1)}$ is defined similarly).

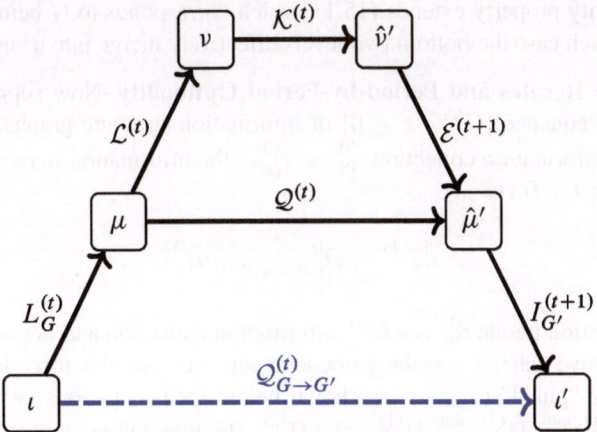

While the pmf $\mu = L_G^{(t)}(\iota)$ has the global Markov property with respect to G (see (11.8)), the pmf $\hat{\mu}' = \mathcal{Q}^{(t)}(\mu)$ will *not*, in general, possess the global Markov property with respect to G' since applying the QPLEX map $\mathcal{Q}^{(t)}$ destroys such structure. This is true even when $G = G'$ or when the kernel of the QPLEX chain uses simple transition dynamics. Similarly, the pmf $v = \mathcal{L}^{(t)}(\mu)$ has additional structure, because it is an element of $\mathcal{P}_{Z,H}^{(t),\mathrm{mult}}$, but applying the kernel map $\mathcal{K}^{(t)}$ destroys such structure. Thus, the pmfs on the top two levels of the "left-hand side" of the diagram have additional structure that is not present in the corresponding pmfs on the "right-hand side" of the diagram.

Projection Maps We again fix t and, this time, a single information structure graph G. Since the elements of the vertex set V of G correspond to the components of $\mathbf{Z}$ and $\mathbf{L}$, the set $\mathcal{P}_V$ from Chap. 11, defined as the set of possible pmfs of the random variables represented in V, is identical to $\mathcal{P}_{\mathbf{Z},\mathbf{L}}$. Thus, $\mathcal{M}_G$ is the set of all pmfs $\mu \in \mathcal{P}_{\mathbf{Z},\mathbf{L}}$ that possess the global Markov property with respect to G. We let $\mathcal{M}_G^{(t)} = \mathcal{P}_{\mathbf{Z},\mathbf{L}}^{(t)} \cap \mathcal{M}_G$ denote the subset of all $\mu \in \mathcal{P}_{\mathbf{Z},\mathbf{L}}^{(t)}$ that possess the global Markov property with respect to G. Let $\Pi_G^{(t)} = L_G^{(t)} \circ I_G^{(t)}$. It turns out that $\Pi_G^{(t)}$ is an appropriate restriction of the G-projection map Π_G from Sect. 11.5.

The *multinomial G-projection map* $\Pi_G^{(t),\mathrm{mult}}$ is defined as the composition $\mathcal{L}^{(t)} \circ \Pi_G^{(t)} \circ \mathcal{E}^{(t)}$. Let $\mathcal{M}_G^{(t),\mathrm{mult}}$ be the set of all $v \in \mathcal{P}_{\mathbf{Z},\mathbf{H}}^{(t)}$ that can be written as $v = \mathcal{L}^{(t)}(\mu)$ for some $\mu \in \mathcal{M}_G^{(t)}$. We will show that $\Pi_G^{(t),\mathrm{mult}}$ is a projection map from $\mathcal{P}_{\mathbf{Z},\mathbf{H}}^{(t)}$ onto $\mathcal{M}_G^{(t),\mathrm{mult}}$.

We also will show that $\Pi_G^{(t),\mathrm{mult}}$ possesses a fundamental optimality property. Fix some pmf $\hat{v} \in \mathcal{P}_{\mathbf{Z},\mathbf{H}}^{(t)}$ with finite support. Suppose we seek to find the pmf $v \in \mathcal{M}_G^{(t),\mathrm{mult}}$ that minimizes the Kullback-Leibler divergence $D(\hat{v} \parallel v)$. We show that the solution is unique and given by

$$\Pi_G^{(t),\mathrm{mult}}(\hat{v}) = \inf_{v \in \mathcal{M}_G^{(t),\mathrm{mult}}} D(\hat{v} \parallel v). \tag{15.2}$$

This optimality property extends (15.1), which corresponds to G being a complete graph, in which case the bottom two levels effectively merge into a single level.

Information Iterates and Period-by-Period Optimality   Now suppose we have an arbitrary sequence $\{G^{(t)} : t \geq 0\}$ of information structure graphs. Given some initial $G^{(0)}$-information collection $\iota_{IS}^{(0)} \in \mathcal{I}_{G^{(0)}}^{(0)}$, the information iterates are defined iteratively for $t \geq 0$ via

$$\iota_{IS}^{(t+1)} = \mathcal{Q}_{G^{(t)} \to G^{(t+1)}}^{(t)}(\iota_{IS}^{(t)}).$$

Each information iterate $\iota_{IS}^{(t)}$ is a $G^{(t)}$-information collection and lies in $\mathcal{I}_{G^{(t)}}^{(t)}$.

This iterative scheme can be embedded into our familiar three-level diagram appropriately "glued" together, as shown below for two periods. We have added the maps $\Pi_{G^{(1)}}^{(1),\text{mult}}$, $\Pi_{G^{(2)}}^{(2),\text{mult}}$, $\Pi_{G^{(1)}}^{(1)}$, and $\Pi_{G^{(2)}}^{(2)}$ (highlighted in blue) to this diagram. In this diagram, these maps act as "bridges" between the levels. This diagram starts from some initial information iterate $\iota_{IS}^{(0)} \in \mathcal{I}_{G^{(0)}}^{(0)}$ and shows how all pmfs in the diagram can be found by applying appropriate maps. Each information iterate $\iota_{IS}^{(t+1)}$ can be calculated from the information iterate $\iota_{IS}^{(t)}$ by considering any of the three directed paths from $\iota_{IS}^{(t)}$ to $\iota_{IS}^{(t+1)}$ shown in the diagram. Each pmf $\mu_{IS}^{(t+1)}$ can be calculated from the pmf $\mu_{IS}^{(t)}$ by considering any of the four directed paths from $\mu_{IS}^{(t)}$ to $\mu_{IS}^{(t+1)}$ shown in the diagram.

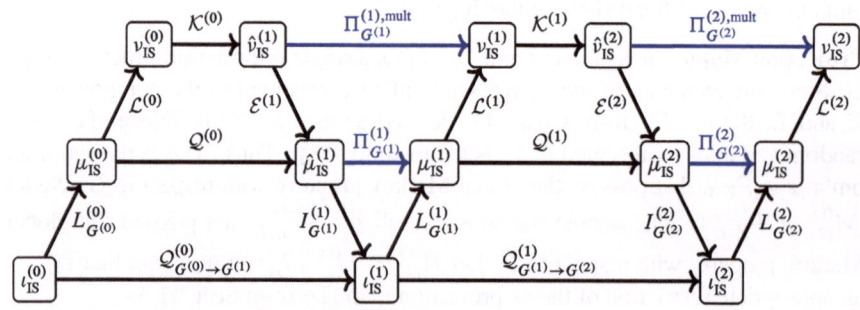

It directly follows from (15.2) that

$$\Pi_{G^{(t)}}^{(t),\text{mult}}(\hat{\nu}_{IS}^{(t)}) = \underset{\nu \in \mathcal{M}_{G^{(t)}}^{(t),\text{mult}}}{\arg\inf} \; D(\hat{\nu}_{IS}^{(t)} \parallel \nu).$$

As the diagram shows, we have that

$$\Pi_{G^{(t)}}^{(t),\text{mult}}(\hat{\nu}_{IS}^{(t)}) = \nu_{IS}^{(t)} = (\mathcal{L}^{(t)} \circ L_{G^{(t)}}^{(t)})(\iota_{IS}^{(t)}).$$

Thus, for example, we see that the information iterate $\iota_{\mathrm{IS}}^{(2)}$ possesses a period-by-period optimality property in the sense that the lifted QPLEX iterate $(\mathcal{L}^{(2)} \circ L_{G^{(2)}}^{(2)})(\iota_{\mathrm{IS}}^{(2)})$ is the closest pmf (with respect to Kullback-Leibler divergence) to the pmf obtained by applying the kernel map $\mathcal{K}^{(1)}$ to the lifted QPLEX iterate $(\mathcal{L}^{(1)} \circ L_{G^{(1)}}^{(1)})(\iota_{\mathrm{IS}}^{(1)})$. The information iterates $\iota_{\mathrm{IS}}^{(t)}$ are the best period-by-period choices if one accepts the QPLEX paradigm of (implicitly) working with lower-dimensional spaces $\mathcal{M}_{G^{(t)}}^{(t),\mathrm{mult}}$ instead of $\mathcal{P}_{Z,H}^{(t)}$ to overcome the curse of dimensionality for counters.

Performance Metrics Consider two information structure graphs $G = (V, E)$ and $\tilde{G} = (V, \tilde{E})$ such that $\tilde{E} \subseteq E$. Since $\mathcal{M}_{\tilde{G}} \subseteq \mathcal{M}_G$ (Lemma 11.4) and therefore $\mathcal{M}_{\tilde{G}}^{(t)} \subseteq \mathcal{M}_G^{(t)}$ and $\mathcal{M}_{\tilde{G}}^{(t),\mathrm{mult}} \subseteq \mathcal{M}_G^{(t),\mathrm{mult}}$, we have that

$$\inf_{\nu \in \mathcal{M}_G^{(t),\mathrm{mult}}} D(\hat{\nu} \parallel \nu) \leq \inf_{\nu \in \mathcal{M}_{\tilde{G}}^{(t),\mathrm{mult}}} D(\hat{\nu} \parallel \nu).$$

This shows that removing edges from an information structure graph can never lower the projection error measured via the Kullback-Leibler divergence. On the other hand, since removing edges breaks up maximal cliques into smaller ones, it could lead to lower computational requirements to calculate the information iterates.

Let $\mathrm{lci}(G)$ denote the information structure graph obtained from an information structure graph G by removing all edges between label variables. Each $\mu \in \mathcal{M}_{\mathrm{lci}(G)}$ has the property that the label variables are conditionally independent given the counter variables under μ.

For each element $G^{(t)}$ in the collection $\{G^{(t)} : t \geq 0\}$ of information structure graphs, we define $\tilde{G}^{(t)} = \mathrm{lci}(G^{(t)})$. The calculation of the information iterates $\tilde{\iota}_{\mathrm{IS}}^{(t)}$ generated via $\tilde{\iota}_{\mathrm{IS}}^{(t+1)} = \mathcal{Q}_{\tilde{G}^{(t)} \to \tilde{G}^{(t+1)}}^{(t)}(\tilde{\iota}_{\mathrm{IS}}^{(t)})$ may require less computational effort than calculating the information iterates $\iota_{\mathrm{IS}}^{(t)}$ via $\iota_{\mathrm{IS}}^{(t+1)} = \mathcal{Q}_{G^{(t)} \to G^{(t+1)}}^{(t)}(\iota_{\mathrm{IS}}^{(t)})$. We show that if each performance metric pmf tracked over time can be obtained from a marginal pmf associated with a subset of counter variables and *at most* one label variable, then the performance metric pmfs can all be calculated from the information iterates $\tilde{\iota}_{\mathrm{IS}}^{(t)}$, and so there is no need to work with the original information structure graphs $G^{(t)}$.

15.3 Exactness Results

Suppose that we are given initial pmfs $\mu_Q^{(0)} \in \mathcal{P}_{Z,L}^{(0)}$ and $\nu_{\mathrm{ker}}^{(0)} \in \mathcal{P}_{Z,H}^{(0)}$ satisfying $\nu_{\mathrm{ker}}^{(0)} = \mathcal{L}^{(0)}(\mu_Q^{(0)})$, which is immediately satisfied when no entities are undertaking any activities initially. Suppose also that each kernel iterate $\nu_{\mathrm{ker}}^{(t)}$ is an element of $\mathcal{P}_{Z,H}^{(t),\mathrm{mult}}$. Under these conditions, we show that each kernel iterate is the multinomial lift of the corresponding QPLEX iterate, i.e., $\nu_{\mathrm{ker}}^{(t)} = \mathcal{L}^{(t)}(\mu_Q^{(t)})$. We show that the marginal pmf of the counters of the QPLEX iterate is then the same as the marginal pmf of the counters of the kernel iterate.

We establish that the above conditions are satisfied in several cases with simple transition dynamics, which exactly correspond to what should be expected based on the myriad of classical stochastic modeling results. For example, these conditions are satisfied when the pmf of each activity duration is geometric, when the number of entities that can undertake any activity can only take on values 0 or 1, or when the routing pmf has an embedded Poisson property. As a final separate result, we consider single activity QPLEX chains with simple transition dynamics where the underlying model primitives are defined to be consistent with an activity duration that does not change over time. For this setting, we show that the pmf of the counter variables is the same whether the labels are interpreted as remaining activity duration or age.

Chapter 16
Optimality of QPLEX Iterates

This chapter rigorously defines and establishes properties of the compression, multinomial lift, kernel, and QPLEX maps, as well as two new maps, the label conditional independence and multinomial projection maps. We use these properties to establish what we call *period-by-period optimality* of the QPLEX iterates.

Throughout this chapter, unless stated otherwise, we fix time t, the counter set Ω_Z, the number of activities M, the label sets Ω_m, and the size function vectors $x^{(t)}$ and $x^{(t+1)}$.

16.1 Compression Maps

The compression map takes the information contained in a pmf $v \in \mathcal{P}_{Z,H}^{(t)}$ on $\Omega_{Z,H}^{(t)}$ and "compresses" it into a pmf μ with much smaller support. The definition of the compression map below formalizes this basic idea. We will show why it is possible to choose the stated codomain after establishing the requisite property of the compression map from which this follows. The sum in (16.1) is taken over Ω_H.

Definition 16.1 The *compression map* $\mathcal{E}^{(t)} : \mathcal{P}_{Z,H}^{(t)} \to \mathcal{P}_{Z,L}^{(t)}$ is defined by setting, for $(z, \ell) \in \Omega_{Z,L}$,

$$[\mathcal{E}^{(t)}(v)](z, \ell) = \sum_{h} \left(\prod_{m} \frac{h_m(\ell_m)}{|h_m|} \right) \times v(z, h), \qquad (16.1)$$

where the ratio should be understood to mean $\theta_m^0(\ell_m)$ if $h_m = 0$.

Here is a numerical example with two activities.

© The Author(s) 2025
A. B. Dieker, S. T. Hackman, *QPLEX: A Computational Modeling and Analysis Methodology for Stochastic Systems*, Springer Series in Operations Research and Financial Engineering, https://doi.org/10.1007/978-3-031-74870-7_16

Example 16.1 Fix some z, and suppose that $v \in \mathcal{P}^{(t)}_{Z,H}$ satisfies $v(z) = 1$; the particular value of z is not relevant. Suppose that $x_1^{(t)}(z) = 2$ and $x_2^{(t)}(z) = 1$ and that each entity's label can either be 0, 1, or 2. The set of *pairs* of histograms with the requisite sizes

$$\{\boldsymbol{h} = (h_1, h_2) : |h_1| = x_1^{(t)}(z) = 2, |h_2| = x_2^{(t)}(z) = 1\}$$

is the Cartesian product

$$\{(2, 0, 0), (0, 2, 0), (0, 0, 2), (1, 1, 0), (1, 0, 1), (0, 1, 1)\}$$
$$\times \{(1, 0, 0), (0, 1, 0), (0, 0, 1)\},$$

which contains 18 elements. Keep in mind that the first component of each histogram in vector form corresponds to label 0, etc. Suppose that $v(z, \boldsymbol{H} = \cdot)$ has three nonzero values, as displayed in the table below.

$\boldsymbol{h} = (h_1, h_2)$	$((1, 1, 0), (0, 0, 1))$	$((0, 2, 0), (1, 0, 0))$	$((1, 0, 1), (0, 1, 0))$
$v(z, \boldsymbol{h})$	0.2	0.3	0.5

Let $\mu = \mathcal{E}^{(t)}(v)$. In this two-activity example,

$$\mu(z, (\ell_1, \ell_2)) = \sum_{\boldsymbol{h}} \frac{h_1(\ell_1)}{|h_1|} \times \frac{h_2(\ell_2)}{|h_2|} \times v(z, \boldsymbol{h}).$$

The values of $\mu(z, \boldsymbol{L} = \cdot) = [\mathcal{E}^{(t)}(v)](z, \boldsymbol{L} = \cdot)$ are displayed in the table below.

$\ell_1 \backslash \ell_2$	0	1	2
0	0	0.25	0.1
1	0.3	0	0.1
2	0	0.25	0

For example, we have, for $\ell_1 = 1$, $\ell_2 = 2$, that

$$\mu(z, L_1 = 1, L_2 = 2) =$$

$$\frac{1}{2} \times \frac{1}{1} \times 0.2 + \frac{2}{2} \times \frac{0}{1} \times 0.3 + \frac{0}{2} \times \frac{0}{1} \times 0.5 = 0.1,$$

as marked in beige in the table.   △

Properties The compression map $\mathcal{E}^{(t)}$ has five key properties. Throughout our discussion of these properties, we fix some $\nu \in \mathcal{P}_{Z,H}^{(t)}$.

Counter Pmfs The first property is that

$$[\mathcal{E}^{(t)}(\nu)](z) = \nu(z), \tag{16.2}$$

i.e., the marginal pmfs $\nu(z)$ and $[\mathcal{E}^{(t)}(\nu)](z)$ coincide. Indeed, we have that

$$[\mathcal{E}^{(t)}(\nu)](z) = \sum_{\ell}[\mathcal{E}^{(t)}(\nu)](z, \ell)$$

$$= \sum_{h}\left(\prod_{m}\sum_{\ell_m}\frac{h_m(\ell_m)}{|h_m|}\right) \times \nu(z, h)$$

$$= \sum_{h} \nu(z, h).$$

This property, in conjunction with the definition of $\mathcal{E}^{(t)}$, has the immediate consequence that, whenever $\nu(z) > 0$,

$$[\mathcal{E}^{(t)}(\nu)](\ell|z) = \sum_{h}\left(\prod_{m}\frac{h_m(\ell_m)}{|h_m|}\right) \times \nu(h|z). \tag{16.3}$$

Label Pmf Given Counters The second property is that, as long as $\nu(z) > 0$, each conditional pmf $[\mathcal{E}^{(t)}(\nu)](\ell_m|z)$ can always be expressed as

$$[\mathcal{E}^{(t)}(\nu)](\ell_m|z) = \sum_{h_m}\frac{h_m(\ell_m)}{|h_m|} \times \nu(h_m|z), \tag{16.4}$$

where the ratio should again be interpreted as $\theta_m^0(\ell_m)$ when $h_m = 0$.

To verify that (16.4) holds, we note that by (16.3) and $|h_m| = \sum_{\ell_m} h_m(\ell_m)$, we have that

$$[\mathcal{E}^{(t)}(\nu)](\ell_m|z) = \sum_{(\ell_{\tilde{m}})_{\tilde{m} \neq m}} \sum_{h}\left(\prod_{\tilde{m}}\frac{h_{\tilde{m}}(\ell_{\tilde{m}})}{|h_{\tilde{m}}|}\right) \times \nu(h|z)$$

$$= \sum_{h}\left(\sum_{(\ell_{\tilde{m}})_{\tilde{m} \neq m}} \frac{h_m(\ell_m)}{|h_m|} \times \prod_{\tilde{m} \neq m}\frac{h_{\tilde{m}}(\ell_{\tilde{m}})}{|h_{\tilde{m}}|}\right) \times \nu(h|z)$$

$$= \sum_{h}\frac{h_m(\ell_m)}{|h_m|} \times \left(\prod_{\tilde{m} \neq m}\sum_{\ell_{\tilde{m}}}\frac{h_{\tilde{m}}(\ell_{\tilde{m}})}{|h_{\tilde{m}}|}\right) \times \nu(h|z)$$

$$= \sum_{h} \frac{h_m(\ell_m)}{|h_m|} \times \nu(\boldsymbol{h}|\boldsymbol{z})$$

$$= \sum_{h_m} \frac{h_m(\ell_m)}{|h_m|} \times \nu(h_m|\boldsymbol{z}).$$

This property implies that $\mathcal{E}^{(t)}(\nu) \in \mathcal{P}_{\boldsymbol{Z},\boldsymbol{L}}^{(t)}$, as can be seen as follows. For $\boldsymbol{z}$ with $x_m^{(t)}(\boldsymbol{z}) = 0$, the sum over h_m has only a single nonzero summand corresponding to the zero histogram as a consequence of $\nu \in \mathcal{P}_{\boldsymbol{Z},\boldsymbol{H}}^{(t)}$, and therefore the right-hand side of (16.4) is $\theta_m^0(\ell_m)$ in that case. The fact that $\mathcal{E}^{(t)}(\nu) \in \mathcal{P}_{\boldsymbol{Z},\boldsymbol{L}}^{(t)}$ ensures that the codomain of $\mathcal{E}^{(t)}$ can indeed be chosen to be $\mathcal{P}_{\boldsymbol{Z},\boldsymbol{L}}^{(t)}$.

Label Variables Not Conditionally Independent The third property relates to the conditional independence of the label variables given the counter variables under $\mathcal{E}^{(t)}(\nu)$. Specifically, even though the entities undertaking activities are chosen uniformly at random, independently across the activities and independently of the counter variables, the label variables need *not* be conditionally independent given the counter variables under $\mathcal{E}^{(t)}(\nu)$. That is, in general,

$$[\mathcal{E}^{(t)}(\nu)](\boldsymbol{\ell}|\boldsymbol{z}) \neq \prod_{m} [\mathcal{E}^{(t)}(\nu)](\ell_m|\boldsymbol{z})$$

since, in general, whenever $\nu(\boldsymbol{z}) > 0$,

$$\sum_{h} \left(\prod_{m} \frac{h_m(\ell_m)}{|h_m|} \right) \times \nu(\boldsymbol{h}|\boldsymbol{z}) \neq \prod_{m} \left(\sum_{h_m} \frac{h_m(\ell_m)}{|h_m|} \times \nu(h_m|\boldsymbol{z}) \right). \tag{16.5}$$

Here are two numerical examples to illustrate this point.

Example 16.2 Let $\mu = \mathcal{E}^{(t)}(\nu)$ for the choice of ν in Example 16.1. The conditional pmfs $\mu(\ell_m|\boldsymbol{z})$ for $m = 1, 2$ have been added in the table below.

ℓ_1 \ ℓ_2	0	1	2	
0	0	0.25	0.1	0.35
1	0.3	0	0.1	0.4
2	0	0.25	0	0.25
	0.3	0.5	0.2	

$\mu(\ell_1|\boldsymbol{z})$ labels the rightmost column; $\mu(\ell_2|\boldsymbol{z})$ labels the bottom row.

It shows that

$$\mu(L_1 = 1, L_2 = 2|z) = 0.1 \neq 0.4 \times 0.2 = \mu(L_1 = 1|z) \times \mu(L_2 = 2|z),$$

illustrating (16.5). △

Example 16.3 Suppose there are one counter and two activities, and that the labels can take the value 0 or 1. Fix some strictly positive integer z. Suppose that under $v \in \mathcal{P}^{(t)}_{Z,H}$, the number of entities at each activity always equals z, and there are two possible equally likely histogram vectors given by

$$\boldsymbol{h}^{(1)} = (h_1^{(1)}, h_2^{(1)}) = ((0, z), (z, 0))$$

$$\boldsymbol{h}^{(2)} = (h_1^{(2)}, h_2^{(2)}) = ((z, 0), (0, z)).$$

Let $\mu = \mathcal{E}^{(t)}(v)$. Since the labels can never be the same, we have that $\mu(L_1 = 0, L_2 = 0|z) = 0$. The pmfs $\mu(L_1 = \cdot|z)$ and $\mu(L_2 = \cdot|z)$ are, respectively, the weighted average of the first or second component of each histogram vector normalized by z. Since each histogram vector is equally likely, we have that $\mu(L_m = \cdot|z) = (0.5, 0.5)$ for $m = 1, 2$, as confirmed by using (16.4) for $m = 1$ via

$$\mu(L_1 = 0|z) = \frac{h_1^{(1)}(0)}{|h_1^{(1)}|} \times v(h_1^{(1)}|z) + \frac{h_1^{(2)}(0)}{|h_1^{(2)}|} \times v(h_1^{(2)}|z)$$

$$= \frac{0}{z} \times 0.5 + \frac{z}{z} \times 0.5 = 0.5.$$

Consequently, we obtain that

$$\mu(L_1 = 0|z) \times \mu(L_2 = 0|z) = 0.25 \neq \mu(L_1 = 0, L_2 = 0|z) = 0,$$

again illustrating (16.5). △

Conditional independence of the label variables given the counter variables *is* assured under an additional assumption that the histogram variables are conditionally independent given the counter variables under v, since $v(\boldsymbol{h}|z) = \prod_m v(h_m|z)$ implies that the sum and product can be interchanged. Indeed, we have that

$$[\mathcal{E}^{(t)}(v)](\boldsymbol{\ell}|z) = \sum_{\boldsymbol{h}} \prod_m \left(\frac{h_m(\ell_m)}{|h_m|} \times v(h_m|z) \right)$$

$$= \prod_m \left(\sum_{h_m} \frac{h_m(\ell_m)}{|h_m|} \times v(h_m|z) \right)$$

$$= \prod_m [\mathcal{E}^{(t)}(v)](\ell_m|z).$$

Linearity The fourth property is that the compression map $\mathcal{E}^{(t)}$ is "linear": for any finite collection of pmfs $\{v_k\}$ and nonnegative weights $\{\alpha_k\}$ with $\sum_k \alpha_k = 1$, we have that

$$\mathcal{E}^{(t)}\left(\sum_k \alpha_k v_k\right) = \sum_k \alpha_k \mathcal{E}^{(t)}(v_k).$$

This follows immediately from the definition of $\mathcal{E}^{(t)}$.

Many-to-One The fifth and final property is that the compression map is many-to-one. This is already apparent even in the single-activity case since whenever $v(z) > 0$, $[\mathcal{E}^{(t)}(v)](\ell|z)$ only depends on $v(h|z)$ through a "mean" in (16.4). As a result, uncountably many v compress to the same μ. The following example constructs two pmfs v_1 and v_2 such that the labels of the entities are perfectly positively correlated under v_1, while the labels of the entities are independent under v_2. Both pmfs, however, compress to the same pmf.

Example 16.4 Suppose there is a single activity and the labels can either be 0 or 1, so that relevant histograms and pmfs on this set can be represented as two-dimensional vectors. Fix $z \in \Omega_Z$, and suppose that two entities are undertaking the activity at time t. Let $v_1 \in \mathcal{P}_{Z,H}^{(t)}$ satisfy $v_1(z) = 1$ and

$$v_1(h|z) = 0.5 \times 1(h = (2, 0)) + 0.5 \times 1(h = (0, 2)).$$

This pmf encodes perfectly positively correlated labels of the two entities, as both entities have either label 0 or 1. Using (16.4), we obtain that

$$[\mathcal{E}^{(t)}(v_1)](L = \cdot|z) = \frac{(2, 0)}{2} \times 0.5 + \frac{(0, 2)}{2} \times 0.5 = (0.5, 0.5).$$

Let $v_2 \in \mathcal{P}_{Z,H}^{(t)}$ satisfy $v_2(z) = 1$ and

$$v_2(h|z) =$$
$$0.25 \times 1(h = (2, 0)) + 0.25 \times 1(h = (0, 2)) + 0.5 \times 1(h = (1, 1)).$$

Thus, $v_2(h|z)$ is the pmf of the histogram if the two labels of the entities undertaking the activity are independent. Again using (16.4), we obtain that

$$[\mathcal{E}^{(t)}(v_2)](\ell|z) = \frac{(2, 0)}{2} \times 0.25 + \frac{(0, 2)}{2} \times 0.25 + \frac{(1, 1)}{2} \times 0.5 = (0.5, 0.5),$$

which shows that $[\mathcal{E}^{(t)}(v_1)](\ell|z) = [\mathcal{E}^{(t)}(v_2)](\ell|z)$. Since $v_1(z) = v_2(z)$ for all z, we have that $[\mathcal{E}^{(t)}(v_1)](z) = [\mathcal{E}^{(t)}(v_2)](z)$ by (16.2) and therefore that $\mathcal{E}^{(t)}(v_1) = \mathcal{E}^{(t)}(v_2)$. $\triangle$

Here is another example.

Example 16.5 Again consider the single activity case with some fixed $z \in \Omega_Z$. The pmf $\tilde{\nu}$ below compresses to the pmf $\mu(z, L = \cdot) = (0.35, 0.4, 0.25)$.

h	$(2,0,0)$	$(0,2,0)$	$(0,0,2)$	$(1,1,0)$	$(1,0,1)$	$(0,1,1)$
$\tilde{\nu}(z,h)$	0.1225	0.16	0.0625	0.28	0.175	0.2

For example,

$$[\mathcal{E}(\tilde{\nu})](z, L = 1) = \frac{2}{2} \times 0.16 + \frac{1}{2} \times 0.28 + \frac{1}{2} \times 0.2 = 0.4.$$

The pmf $\mu(z, L = \cdot) = (0.35, 0.4, 0.25)$ is the same as in Example 2.2, where it is obtained as the compression of a different pmf than $\tilde{\nu}$. $\triangle$

16.2   Multinomial Lift Maps

We have shown how to "compress" an element of $\mathcal{P}_{Z,H}^{(t)}$ to an element of $\mathcal{P}_{Z,L}^{(t)}$. We now discuss how to "lift" an element of $\mathcal{P}_{Z,L}^{(t)}$ to an element of $\mathcal{P}_{Z,H}^{(t)}$. Below is a formal definition of the multinomial lift map.

Definition 16.2 The *multinomial lift map* $\mathcal{L}^{(t)} : \mathcal{P}_{Z,L}^{(t)} \to \mathcal{P}_{Z,H}^{(t)}$ is defined by setting, for $(z, h) \in \Omega_{Z,H}$,

$$[\mathcal{L}^{(t)}(\mu)](z, h) = \mu(z) \times \prod_m \Pr[\text{mult}(x_m^{(t)}(z), \mu(L_m = \cdot|z)) = h_m]. \qquad (16.6)$$

We rely on several of our conventions in this definition. First, if $x_m^{(t)}(z) = 0$, then the multinomial probability is simply $1(h_m = 0)$. Second, if $\mu(z) = 0$, then the second parameter of the multinomial distribution is undefined, but the right-hand side of (16.6) is taken to be 0 by convention. Third, the multiplicand on the right-hand side is 0 for "incompatible" histograms h_m with $|h_m| \neq x_m^{(t)}(z)$, implying that $\mathcal{L}^{(t)}$ is well-defined in the sense that its range is indeed included in its codomain.

Example 16.6 Here is a single activity example. Fix some $\mu \in \mathcal{P}_{Z,L}^{(t)}$ and some $z \in \Omega_Z$ with $\mu(z) = 1$. Suppose that the number of entities undertaking this one activity is $x(z) = 2$. We specifically take $\mu(z, L = \cdot) = (0.35, 0.4, 0.25)$ to be the same as in Example 16.5. Let $\nu = \mathcal{L}^{(t)}(\mu)$. The values of $\nu(z, H = \cdot)$ are given in the table below.

h	$(2,0,0)$	$(0,2,0)$	$(0,0,2)$	$(1,1,0)$	$(1,0,1)$	$(0,1,1)$
$v(z,h)$	0.1225	0.16	0.0625	0.28	0.175	0.2

For example,

$$v(z, H = (1,1,0)) = \binom{2}{1,1,0} \times (0.35)^1 \times (0.4)^1 \times (0.25)^0 = 0.28.$$

The pmf v of this example is the *same* as the pmf $\tilde{v}$ in Example 16.5. We argued in that example that, in the notation of the current example, $\mathcal{E}^{(t)}(v) = \mu$. Thus, for this particular choice of μ, we have that

$$\mu = \mathcal{E}^{(t)}(v) = \mathcal{E}^{(t)}(\mathcal{L}^{(t)}(\mu)) = [\mathcal{E}^{(t)} \circ \mathcal{L}^{(t)}](\mu),$$

which suggests that the composition map $\mathcal{E}^{(t)} \circ \mathcal{L}^{(t)}$ is the identity map on $\mathcal{P}_{Z,L}^{(t)}$. This is indeed the case, but *only* when there is a single activity, as in this example. More generally, this composition map will be shown to be a key projection map. See Sect. 16.4 for details.   △

Example 16.7 Here is an example with two activities. We use the setup of Example 16.2. The conditional pmfs $\mu(\ell_m|z)$ for $m = 1, 2$ are, respectively, $\mu(L_1 = \cdot|z) = (0.35, 0.4, 0.25)$ and $\mu(L_2 = \cdot|z) = (0.3, 0.5, 0.2)$. Set $v = \mathcal{L}^{(t)}(\mu)$. Since $\mu(z) = 1$, we have, for histograms $h_1 = (1,0,1)$ and $h_2 = (0,0,1)$, that

$$v(z, (h_1, h_2)) = v(z, h_1) \times v(z, h_2),$$

where

$$v(z, h_1) = \binom{2}{1,0,1} \times (0.35)^1 \times (0.4)^1 \times (0.25)^0,$$

$$v(z, h_2) = \binom{1}{0,0,1} \times (0.3)^0 \times (0.5)^0 \times (0.2)^1$$

by definition of $\mathcal{L}^{(t)}$.   △

Properties The multinomial lift map $\mathcal{L}^{(t)}$ has four key properties. Fix some $\mu \in \mathcal{P}_{Z,L}^{(t)}$.

Counter Pmfs The first property is that

$$[\mathcal{L}^{(t)}(\mu)](z) = \sum_h \mu(z) \times \prod_m \Pr[\text{mult}(x_m^{(t)}(z), \mu(L_m = \cdot|z)) = h_m] = \mu(z),$$

$$(16.7)$$

i.e., the marginal pmfs $\mu(z)$ and $[\mathcal{L}^{(t)}(\mu)](z)$ coincide. This follows from the definition of $\mathcal{L}^{(t)}$.

Histogram Pmf Given Counters The second property is that, as long as $\mu(z) > 0$, each conditional pmf $[\mathcal{L}^{(t)}(\mu)](h_m|z)$ is given by

$$[\mathcal{L}^{(t)}(\mu)](h_m|z) = \Pr[\text{mult}(x_m^{(t)}(z), \mu(L_m = \cdot|z)) = h_m]. \tag{16.8}$$

Indeed, by (16.6) and the first property, we have that

$$[\mathcal{L}^{(t)}(\mu)](h|z) = \prod_m \Pr[\text{mult}(x_m^{(t)}(z), \mu(L_m = \cdot|z)) = h_m] \tag{16.9}$$

and therefore that

$$
\begin{aligned}
[\mathcal{L}^{(t)}(\mu)](h_m|z) &= \sum_{(h_{\tilde{m}})_{\tilde{m} \neq m}} \prod_{\tilde{m}} \Pr[\text{mult}(x_{\tilde{m}}^{(t)}(z), \mu(L_{\tilde{m}} = \cdot|z)) = h_{\tilde{m}}] \\
&= \sum_{(h_{\tilde{m}})_{\tilde{m} \neq m}} \Pr[\text{mult}(x_m^{(t)}(z), \mu(L_m = \cdot|z)) = h_m] \\
&\qquad \times \prod_{\tilde{m} \neq m} \Pr[\text{mult}(x_{\tilde{m}}^{(t)}(z), \mu(L_{\tilde{m}} = \cdot|z)) = h_{\tilde{m}}] \\
&= \Pr[\text{mult}(x_m^{(t)}(z), \mu(L_m = \cdot|z)) = h_m] \\
&\qquad \times \prod_{\tilde{m} \neq m} \left(\sum_{h_{\tilde{m}}} \Pr[\text{mult}(x_{\tilde{m}}^{(t)}(z), \mu(L_{\tilde{m}} = \cdot|z)) = h_{\tilde{m}}] \right) \\
&= \Pr[\text{mult}(x_m^{(t)}(z), \mu(L_m = \cdot|z)) = h_m].
\end{aligned}
$$

Histogram Conditional Independence The third property is that the histogram variables are conditionally independent given the counter variables under $\mathcal{L}^{(t)}(\mu)$. Indeed, as long as $\mu(z) > 0$, we have by (16.8) and (16.9) that

$$
\begin{aligned}
\prod_m [\mathcal{L}^{(t)}(\mu)](h_m|z) &= \prod_m \Pr[\text{mult}(x_m^{(t)}(z), \mu(L_m = \cdot|z)) = h_m] \\
&= [\mathcal{L}^{(t)}(\mu)](h|z).
\end{aligned}
$$

No Linearity The fourth and final property is that the multinomial lift map is *not* "linear," namely, in general, for $\mu^1, \mu^2 \in \mathcal{P}_{Z,L}^{(t)}$ and $\alpha \in (0, 1)$,

$$\mathcal{L}^{(t)}(\alpha\mu^1 + (1 - \alpha)\mu^2) \neq \alpha\mathcal{L}^{(t)}(\mu^1) + (1 - \alpha)\mathcal{L}^{(t)}(\mu^2).$$

Here is a simple example.

Example 16.8 Suppose there is a single activity, and the label can only be 0 or 1. Fix some z such that $x^{(t)}(z) = 2$, and suppose that $\mu_1, \mu_2 \in \mathcal{P}_{Z,L}^{(t)}$ are such that $\mu_1(z) = \mu_2(z) = 1$. Also fix $h = (2, 0)$. If $\mathcal{L}^{(t)}$ were linear, then the following identity

$$[\mathcal{L}^{(t)}(\alpha \times \mu_1 + (1 - \alpha) \times \mu_2)](h|z)$$
$$= \alpha \times [\mathcal{L}^{(t)}(\mu_1)](h|z) + (1 - \alpha) \times [\mathcal{L}^{(t)}(\mu_2)](h|z)$$

holds for all $\alpha \in [0, 1]$. For this particular choice of h and by definition of $\mathcal{L}^{(t)}$, the identity

$$(\alpha \times \mu_1(L = 0|z) + (1 - \alpha) \times \mu_2(L = 0|z))^2$$
$$= \alpha \times \mu_1(L = 0|z)^2 + (1 - \alpha) \times \mu_2(L = 0|z)^2$$

would therefore hold for all $\alpha \in [0, 1]$, but it clearly does not hold when $\mu_1(L = 0|z) \neq \mu_2(L = 0|z)$.   △

16.3   Kernel and QPLEX Maps

Here are the formal definitions of the kernel and QPLEX maps. Assume we are given the transition probabilities $p^{(t)}(z', h'|z, h)$ for each z, h, z', and h'.

Definition 16.3 The *kernel map* $\mathcal{K}^{(t)}$: $\mathcal{P}_{Z,H}^{(t)} \rightarrow \mathcal{P}_{Z,H}^{(t+1)}$ associated with the transition probabilities $p^{(t)}(z', h'|z, h)$ is defined by setting, for $(z', h') \in \Omega_{Z,H}$,

$$[\mathcal{K}^{(t)}(v)](z', h') = \sum_{z,h} v(z, h) \times p^{(t)}(z', h'|z, h).$$

The sum in this definition is taken over $\Omega_{Z,H}$ but can also be viewed as a sum over $\Omega_{Z,H}^{(t)}$ because $v \in \mathcal{P}_{Z,H}^{(t)}$.

Definition 16.4 The *QPLEX map* $\mathcal{Q}^{(t)}$: $\mathcal{P}_{Z,L}^{(t)} \rightarrow \mathcal{P}_{Z,L}^{(t+1)}$ is defined as the composition

$$\mathcal{Q}^{(t)} = \mathcal{E}^{(t+1)} \circ \mathcal{K}^{(t)} \circ \mathcal{L}^{(t)}.$$

Of course, given some $\mu \in \mathcal{P}_{Z,L}^{(t)}$, applying the kernel map $\mathcal{K}^{(t)}$ to $\mathcal{L}^{(t)}(\mu)$ may be practically infeasible to calculate. Chapters 4 and 6 showed how to directly and efficiently calculate $\mathcal{Q}^{(t)}(\mu)$ *without* having to calculate $\mathcal{K}^{(t)}(\mathcal{L}^{(t)}(\mu))$ under certain assumptions on the transition probabilities.

16.4 Label Conditional Independence Projection Maps

This section defines the label conditional independence projection map $\Pi^{(t),\text{lci}}$ using the compression and multinomial lift maps. After a formal definition, we establish its properties, which will justify its name.

Definition 16.5 The *label conditional independence projection map* $\Pi^{(t),\text{lci}}$: $\mathcal{P}_{Z,L}^{(t)} \to \mathcal{P}_{Z,L}^{(t)}$ is defined as the composition

$$\Pi^{(t),\text{lci}} = \mathcal{E}^{(t)} \circ \mathcal{L}^{(t)}.$$

Basic Properties The map $\Pi^{(t),\text{lci}}$ has three basic properties that follow from the definitions and properties of $\mathcal{E}^{(t)}$ and $\mathcal{L}^{(t)}$ developed in Sects. 16.1 and 16.2. Fix some $\mu \in \mathcal{P}_{Z,L}^{(t)}$.

First, using (16.2) and (16.7), we find that

$$[\Pi^{(t),\text{lci}}(\mu)](z) = [\mathcal{E}^{(t)}(\mathcal{L}^{(t)}(\mu))](z) = [\mathcal{L}^{(t)}(\mu)](z)) = \mu(z). \tag{16.10}$$

Second, suppose that $\mu(z) > 0$ and therefore $[\Pi^{(t),\text{lci}}(\mu)](z) > 0$. We deduce from (16.4) and (16.8) that

$$[\Pi^{(t),\text{lci}}(\mu)](\ell_m | z)$$

$$= [\mathcal{E}^{(t)}(\mathcal{L}^{(t)}(\mu))](\ell_m | z)$$

$$= \sum_{h_m} \frac{h_m(\ell_m)}{|h_m|} \times \Pr[\text{mult}(x_m^{(t)}(z), \mu(L_m = \cdot | z)) = h_m]$$

$$= \begin{cases} \mu(\ell_m | z) & \text{if } x_m^{(t)}(z) > 0 \\ \theta_m^0(\ell_m) & \text{if } x_m^{(t)}(z) = 0 \end{cases}$$

$$= \mu(\ell_m | z), \tag{16.11}$$

where the second to last equality uses the fact that the sum can be taken over h_m with $|h_m| = x_m^{(t)}(z)$ if $x_m^{(t)}(z) > 0$, and the last equality holds because $\mu \in \mathcal{P}_{Z,L}^{(t)}$.

Third, as long as $[\Pi^{(t),\text{lci}}(\mu)](z) > 0$ or equivalently $\mu(z) > 0$, we have that

$$[\Pi^{(t),\text{lci}}(\mu)](z, \ell) = [\Pi^{(t),\text{lci}}(\mu)](z) \times \prod_m [\Pi^{(t),\text{lci}}(\mu)](\ell_m | z). \tag{16.12}$$

This property justifies naming this map using the modifier "label conditional independence." It follows from the fact that the histogram variables are conditionally independent given the counter variables under $\mathcal{L}^{(t)}(\mu)$ (see Sect. 16.2), in which case the label variables are conditionally independent given the counter variables under the compressed pmf $\mathcal{E}^{(t)}(\mathcal{L}^{(t)}(\mu)) = \Pi^{(t),\text{lci}}(\mu)$ as noted in Sect. 16.1.

Projection Property The following proposition shows that $\Pi^{(t),\text{lci}}$ is a projection map. Let $\mathcal{P}_{Z,L}^{(t),\text{lci}}$ denote the set of all $\mu \in \mathcal{P}_{Z,L}^{(t)}$ satisfying

$$\mu(z, \ell) = \mu(z) \times \prod_m \mu(\ell_m | z).$$

Proposition 16.6 $\Pi^{(t),\text{lci}}$ *is a projection map from* $\mathcal{P}_{Z,L}^{(t)}$ *onto* $\mathcal{P}_{Z,L}^{(t),\text{lci}}$.

Before we can see why this proposition holds, we require the following lemma. It immediately follows upon substituting (16.10) and (16.11) into the right-hand side of (16.12).

Lemma 16.7 *For any* $\mu \in \mathcal{P}_{Z,L}^{(t)}$ *and* $(z, \ell) \in \Omega_{Z,L}$, *we have that*

$$[\Pi^{(t),\text{lci}}(\mu)](z, \ell) = \mu(z) \times \prod_m \mu(\ell_m | z). \qquad (16.13)$$

This lemma could be used as an alternative definition of $\Pi^{(t),\text{lci}}$. It shows that $\Pi^{(t),\text{lci}}$ only depends on t only through its domain and codomain. An immediate consequence is the following lemma.

Lemma 16.8 $\Pi^{(t),\text{lci}}$ *is the identity map on* $\mathcal{P}_{Z,L}^{(t)}$ *if there is only one activity, i.e.,* $M = 1$.

We are now ready to establish Proposition 16.6. We verify the two defining properties of a projection map in Definition 1.2. It directly follows from (16.12) and the definition of $\mathcal{P}_{Z,L}^{(t),\text{lci}}$ that $\Pi^{(t),\text{lci}}(\mu) \in \mathcal{P}_{Z,L}^{(t),\text{lci}}$ for all $\mu \in \mathcal{P}_{Z,L}^{(t)}$. If $\mu \in \mathcal{P}_{Z,L}^{(t),\text{lci}}$, then the right-hand side of (16.13) simply equals $\mu(z, \ell)$ and therefore $\Pi^{(t),\text{lci}}(\mu) = \mu$.

Compositions For $\mu \in \mathcal{P}_{Z,L}^{(t)}$, Lemma 16.8 implies that $\Pi^{(t),\text{lci}}(\mu) = \mu$ in the single-activity setting. This identity does not hold in general, but the next proposition shows that applying the multinomial lift map $\mathcal{L}^{(t)}$ to both $\Pi^{(t),\text{lci}}(\mu)$ and μ results in *exactly* the same pmf.

Proposition 16.9 *We have that* $\mathcal{L}^{(t)} \circ \Pi^{(t),\text{lci}} = \mathcal{L}^{(t)}$.

To verify this claim, we note that, for any $\mu \in \mathcal{P}_{Z,L}^{(t)}$ and $(z, h) \in \Omega_{Z,H}$,

$$[\mathcal{L}^{(t)}(\Pi^{(t),\text{lci}}(\mu))](z, h)$$

$$= [\Pi^{(t),\text{lci}}(\mu)](z) \times \prod_m \Pr[\text{mult}(x_m^{(t)}(z), [\Pi^{(t),\text{lci}}(\mu)](L_m = \cdot | z)) = h_m]$$

$$= \mu(z) \times \prod_m \Pr[\text{mult}(x_m^{(t)}(z), \mu(L_m = \cdot | z)) = h_m],$$

where the second equality follows from (16.10) and (16.11). The right-hand side is $[\mathcal{L}^{(t)}(\mu)](z, h)$ by definition of $\mathcal{L}^{(t)}$.

An immediate consequence is a similar property for the QPLEX map $\mathcal{Q}^{(t)}$ given by

$$\mathcal{Q}^{(t)} \circ \Pi^{(t),\mathrm{lci}} = \mathcal{E}^{(t+1)} \circ \mathcal{K}^{(t)} \circ \mathcal{L}^{(t)} \circ \Pi^{(t),\mathrm{lci}} = \mathcal{E}^{(t+1)} \circ \mathcal{K}^{(t)} \circ \mathcal{L}^{(t)} = \mathcal{Q}^{(t)}.$$

We summarize this identity in the following corollary, as we will frequently use it.

Corollary 16.10 *We have that $\mathcal{Q}^{(t)} \circ \Pi^{(t),\mathrm{lci}} = \mathcal{Q}^{(t)}$.*

16.5 Multinomial Projection Maps

This section defines the multinomial projection map $\Pi^{(t),\mathrm{mult}}$ using the compression and multinomial lift maps. After a formal definition, we establish its properties, which will justify its name.

Definition 16.11 The *multinomial projection map* $\Pi^{(t),\mathrm{mult}} : \mathcal{P}_{Z,H}^{(t)} \to \mathcal{P}_{Z,H}^{(t)}$ is defined as the composition

$$\Pi^{(t),\mathrm{mult}} = \mathcal{L}^{(t)} \circ \mathcal{E}^{(t)}.$$

Basic Properties The map $\Pi^{(t),\mathrm{mult}}$ has three basic properties, which follow from the definitions and properties of $\mathcal{E}^{(t)}$ and $\mathcal{L}^{(t)}$ developed in Sects. 16.1 and 16.2. Fix some $\nu \in \mathcal{P}_{Z,H}^{(t)}$.

First, using (16.2) and (16.7), we find that

$$[\Pi^{(t),\mathrm{mult}}(\nu)](z) = [\mathcal{L}^{(t)}(\mathcal{E}^{(t)}(\nu))](z) = [\mathcal{E}^{(t)}(\nu)](z)) = \nu(z).$$

Second, suppose that $\nu(z) > 0$ and therefore $[\Pi^{(t),\mathrm{mult}}(\nu)](z) > 0$. We deduce from (16.4) and (16.8) that

$$[\Pi^{(t),\mathrm{mult}}(\nu)](h_m|z)$$

$$= [\mathcal{L}^{(t)}(\mathcal{E}^{(t)}(\nu))](h_m|z)$$

$$= \Pr[\mathrm{mult}(x_m^{(t)}(z), [\mathcal{E}^{(t)}(\nu)](L_m = \cdot|z)) = h_m]$$

$$= \Pr\left[\mathrm{mult}\left(x_m^{(t)}(z), \sum_{\tilde{h}_m} \frac{\tilde{h}_m(L_m = \cdot)}{|\tilde{h}_m|} \times \nu(\tilde{h}_m|z) \right) = h_m \right],$$

where the ratio (and thus the sum over $\tilde{h}_m$) in the last expression should be interpreted as θ_m^0 if $x_m^{(t)}(z) = 0$.

Third, the histogram variables are conditionally independent given the counter variables under $\Pi^{(t),\text{mult}}(\nu)$ by the corresponding property of $\mathcal{L}^{(t)}$ (see Sect. 16.2), i.e., we have that

$$[\Pi^{(t),\text{mult}}(\nu)](\boldsymbol{h}|\boldsymbol{z}) = \prod_m [\Pi^{(t),\text{mult}}(\nu)](h_m|\boldsymbol{z}).$$

Projection Property Let $\mathcal{P}_{\boldsymbol{Z},\boldsymbol{H}}^{(t),\text{mult}}$ denote the image of $\mathcal{P}_{\boldsymbol{Z},\boldsymbol{L}}^{(t)}$ under $\mathcal{L}^{(t)}$, i.e.,

$$\mathcal{P}_{\boldsymbol{Z},\boldsymbol{H}}^{(t),\text{mult}} = \{\nu \in \mathcal{P}_{\boldsymbol{Z},\boldsymbol{H}}^{(t)} : \nu = \mathcal{L}^{(t)}(\mu) \text{ for some } \mu \in \mathcal{P}_{\boldsymbol{Z},\boldsymbol{L}}^{(t)}\}.$$

Thus, from the definition of $\mathcal{L}^{(t)}$, any $\nu \in \mathcal{P}_{\boldsymbol{Z},\boldsymbol{H}}^{(t),\text{mult}}$ can be written as

$$\nu(\boldsymbol{z}, \boldsymbol{h}) = \mu(\boldsymbol{z}) \times \prod_m \Pr[\text{mult}(x_m^{(t)}(\boldsymbol{z}), \mu(L_m = \cdot|\boldsymbol{z})) = h_m]$$

for some $\mu \in \mathcal{P}_{\boldsymbol{Z},\boldsymbol{L}}^{(t)}$.

The following proposition justifies our name for $\Pi^{(t),\text{mult}}$.

Proposition 16.12 $\Pi^{(t),\text{mult}}$ *is a projection map from* $\mathcal{P}_{\boldsymbol{Z},\boldsymbol{H}}^{(t)}$ *onto* $\mathcal{P}_{\boldsymbol{Z},\boldsymbol{H}}^{(t),\text{mult}}$.

To see why this proposition holds, we verify the two defining properties of a projection map in Definition 1.2. The definition of $\mathcal{P}_{\boldsymbol{Z},\boldsymbol{H}}^{(t),\text{mult}}$ and the identity $\Pi^{(t),\text{mult}} = \mathcal{L}^{(t)} \circ \mathcal{E}^{(t)}$ imply that

$$\Pi^{(t),\text{mult}}(\mathcal{P}_{\boldsymbol{Z},\boldsymbol{H}}^{(t)}) \subseteq \mathcal{P}_{\boldsymbol{Z},\boldsymbol{H}}^{(t),\text{mult}},$$

establishing the first property. For any $\nu \in \mathcal{P}_{\boldsymbol{Z},\boldsymbol{H}}^{(t),\text{mult}}$, we may pick some $\mu \in \mathcal{P}_{\boldsymbol{Z},\boldsymbol{L}}^{(t)}$ such that $\nu = \mathcal{L}^{(t)}(\mu)$. The definitions of $\Pi^{(t),\text{mult}}$, and $\Pi^{(t),\text{lci}}$ together with Proposition 16.9 imply that

$$\Pi^{(t),\text{mult}}(\nu) = (\Pi^{(t),\text{mult}} \circ \mathcal{L}^{(t)})(\mu)$$
$$= (\mathcal{L}^{(t)} \circ \mathcal{E}^{(t)} \circ \mathcal{L}^{(t)})(\mu)$$
$$= (\mathcal{L}^{(t)} \circ \Pi^{(t),\text{lci}})(\mu)$$
$$= \mathcal{L}^{(t)}(\mu) = \nu,$$

which establishes the second property.

Propositions 16.6 and 16.12 together establish that the maps $\mathcal{E}^{(t)}$ and $\mathcal{L}^{(t)}$, appropriately restricted, are bijections and each other's inverse, as shown in the diagram below. (See Sect. 1.1 for the definition of bijection.)

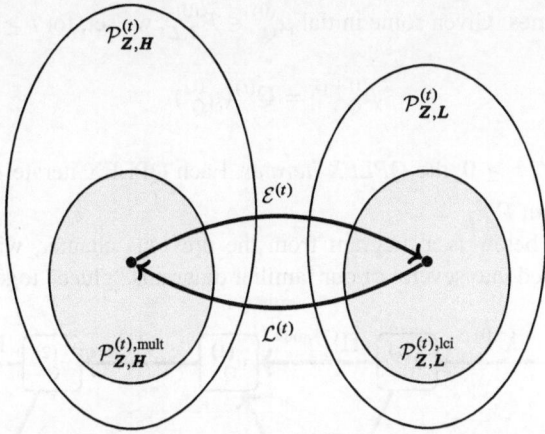

Proposition 16.13 *The* $(\mathcal{P}_{Z,L}^{(t),\mathrm{lci}}, \mathcal{P}_{Z,H}^{(t),\mathrm{mult}})$-*restriction of* $\mathcal{L}^{(t)}$ *and the* $(\mathcal{P}_{Z,H}^{(t),\mathrm{mult}},$ $\mathcal{P}_{Z,L}^{(t),\mathrm{lci}})$-*restriction of* $\mathcal{E}^{(t)}$ *are bijections and each other's inverse.*

The restrictions of $\mathcal{L}^{(t)}$ and $\mathcal{E}^{(t)}$ in this proposition are well-defined since $\mathcal{L}^{(t)}(\mathcal{P}_{Z,L}^{(t),\mathrm{lci}}) \subseteq \mathcal{L}^{(t)}(\mathcal{P}_{Z,L}^{(t)}) = \mathcal{P}_{Z,L}^{(t),\mathrm{mult}}$ and

$$\mathcal{E}^{(t)}(\mathcal{P}_{Z,H}^{(t),\mathrm{mult}}) = (\mathcal{E}^{(t)} \circ \mathcal{L}^{(t)})(\mathcal{P}_{Z,L}^{(t)}) = \Pi^{(t),\mathrm{lci}}(\mathcal{P}_{Z,L}^{(t)}) = \mathcal{P}_{Z,L}^{(t),\mathrm{lci}},$$

where the last equality follows from Proposition 16.6.

The proof of this proposition rests on the definition of a bijection in Sect. 1.1 in conjunction with Propositions 16.6 and 16.12. Indeed, Proposition 16.6 implies that $(\mathcal{E}^{(t)} \circ \mathcal{L}^{(t)})(\mu) = \mu$ for $\mu \in \mathcal{P}_{Z,L}^{(t),\mathrm{lci}}$, while Proposition 16.12 implies that $(\mathcal{L}^{(t)} \circ \mathcal{E}^{(t)})(\nu) = \nu$ for $\nu \in \mathcal{P}_{Z,H}^{(t),\mathrm{mult}}$.

We close this section with an optimality result for the map $\Pi^{(t),\mathrm{mult}}$. This proposition is a special case of Proposition 17.14 in Chap. 17. For completeness, we give a self-contained proof in Sect. 16.7.

Proposition 16.14 *Fix some pmf* $\hat{\nu} \in \mathcal{P}_{Z,H}^{(t)}$ *with finite support. The solution to the minimization problem* $\inf_{\nu \in \mathcal{P}_{Z,H}^{(t),\mathrm{mult}}} D(\hat{\nu} \parallel \nu)$ *is unique and corresponds to* $\nu^* = \Pi^{(t),\mathrm{mult}}(\hat{\nu})$.

16.6 Iterative Schemes and Period-by-Period Optimality

In this section, we consider iterative schemes, and so we no longer consider a fixed time epoch. For every $t \geq 0$, we assume we are given, in addition to the counter and label sets, a size function vector $\boldsymbol{x}^{(t)}$ and a kernel map $\mathcal{K}^{(t)}$, and thus a QPLEX map $\mathcal{Q}^{(t)}$ as well.

Iterative Schemes Given some initial $\mu_Q^{(0)} \in \mathcal{P}_{Z,L}^{(0)}$, we set, for $t \geq 0$,

$$\mu_Q^{(t+1)} = \mathcal{Q}^{(t)}(\mu_Q^{(t)}).$$

We call the $\mu_Q^{(t)}$, $t \geq 0$, the *QPLEX iterates*. Each QPLEX iterate $\mu_Q^{(t)}$ is a pmf of (Z, L) and lies in $\mathcal{P}_{Z,L}^{(t)}$.

Reproduced below is a diagram from the previous chapter, with the QPLEX iterates embedded into several of our familiar diagrams "glued" together.

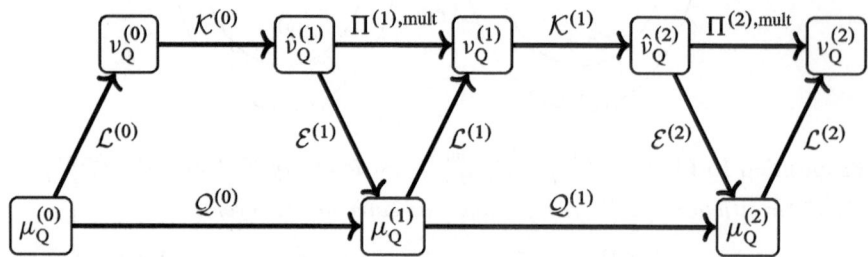

Given some initial $v_{\text{ker}}^{(0)} \in \mathcal{P}_{Z,H}^{(0)}$, we set, for $t \geq 0$,

$$v_{\text{ker}}^{(t+1)} = \mathcal{K}^{(t)}(v_{\text{ker}}^{(t)}). \tag{16.14}$$

We call the $v_{\text{ker}}^{(t)}$, $t \geq 0$, the *kernel iterates*. Each kernel iterate $v_{\text{ker}}^{(t)}$ is a pmf of (Z, H) and lies in $\mathcal{P}_{Z,H}^{(t)}$.

The kernel iterates $v_{\text{ker}}^{(t)}$ are *not* the same as the *lifted QPLEX iterates*

$$v_Q^{(t)} = \mathcal{L}^{(t)}(\mu_Q^{(t)}).$$

Although $v_{\text{ker}}^{(0)} = \mathcal{L}^{(0)}(\mu_Q^{(0)})$ implies that $v_{\text{ker}}^{(1)} = \hat{v}_Q^{(1)}$, subsequent kernel iterates $v_{\text{ker}}^{(t)}$ are (in general) not the same as $\hat{v}_Q^{(t)}$ or $v_Q^{(t)}$ for $t \geq 1$. Indeed, the kernel iterates are found by successive applications of the kernel maps $\mathcal{K}^{(t)}$, whereas the lifted QPLEX iterates $v_Q^{(t)}$ are found by successive applications of these kernel maps alternated with the projection maps $\Pi^{(t),\text{mult}}$.

Calculating these lifted QPLEX iterates from the corresponding QPLEX iterates $\mu_Q^{(t)}$ via the multinomial lift map is almost always practically infeasible. Still, they can be thought of as the QPLEX approximation to the kernel iterates. Since the marginal pmfs of the counters for a QPLEX iterate equal the corresponding marginal pmfs of the counters for a lifted QPLEX iterate by (16.7), the QPLEX methodology allows for approximating the corresponding marginal pmfs of the counters for the kernel iterates using lower-dimensional pmfs than the kernel iterates.

Period-by-Period Optimality We now show that the QPLEX iterates $\mu_Q^{(t)}$ lead to "optimal approximations" of the kernel iterates $v_{ker}^{(t)}$ if one accepts the QPLEX paradigm of (implicitly) working with the much "smaller" set $\mathcal{P}_{Z,H}^{(t),\text{mult}}$ instead of $\mathcal{P}_{Z,H}^{(t)}$ to overcome the curse of dimensionality for histograms. The optimality properties arise from the iterative schemes as opposed to the iterates, and as such they should be interpreted on a period-by-period basis.

Period-by-Period Optimality for the $v_Q^{(t)}$ The period-by-period optimality property of the lifted QPLEX iterates is that

$$v_Q^{(t+1)} = \underset{v \in \mathcal{P}_{Z,H}^{(t+1),\text{mult}}}{\arg\inf} D(\mathcal{K}^{(t)}(v_Q^{(t)}) \parallel v)$$

assuming $\mathcal{K}^{(t)}(v_Q^{(t)})$ has finite support.

This follows from

$$\Pi^{(t+1),\text{mult}}(\mathcal{K}^{(t)}(v_Q^{(t)})) = (\Pi^{(t+1),\text{mult}} \circ \mathcal{K}^{(t)})(v_Q^{(t)}) = v_Q^{(t+1)}$$

and a direct application of Proposition 16.14 with t replaced by $t + 1$ and with $\hat{v} = \mathcal{K}^{(t)}(v_Q^{(t)})$. Note that the resulting minimization problem has a unique solution, so the equality is warranted.

Period-by-Period Optimality for the $\mu_Q^{(t)}$ The period-by-period optimality property of the QPLEX iterates is that

$$\mu_Q^{(t+1)} \in \underset{\mu \in \mathcal{P}_{Z,L}^{(t+1)}}{\arg\inf} D(\mathcal{K}^{(t)}(\mathcal{L}^{(t)}(\mu_Q^{(t)})) \parallel \mathcal{L}^{(t+1)}(\mu)) \tag{16.15}$$

assuming $\mathcal{K}^{(t)}(\mathcal{L}^{(t)}(\mu_Q^{(t)}))$ has finite support.

We thus see that the QPLEX iterate $\mu_Q^{(t+1)}$ at time $t + 1$ is a pmf in $\mathcal{P}_{Z,L}^{(t+1)}$ with corresponding lifted QPLEX iterate $\mathcal{L}^{(t+1)}(\mu_Q^{(t+1)})$ as close as possible to the pmf obtained by applying the kernel map $\mathcal{K}^{(t)}$ to the current lifted QPLEX iterate $\mathcal{L}^{(t)}(\mu_Q^{(t)})$. This contrasts with the kernel iterative scheme, where $\mathcal{K}^{(t)}$ is applied to $v_{ker}^{(t)}$ directly in order to obtain $v_{ker}^{(t+1)}$; see (16.14).

To see why (16.15) holds, we invoke Proposition 16.14 to establish that the set of solutions to

$$\underset{\mu \in \mathcal{P}_{Z,L}^{(t)}}{\inf} D(\hat{v} \parallel \mathcal{L}^{(t)}(\mu)) \tag{16.16}$$

is

$$\{\mu \in \mathcal{P}_{Z,L}^{(t)} : \mathcal{L}^{(t)}(\mu) = \Pi^{(t),\text{mult}}(\hat{v})\}. \tag{16.17}$$

Since $\mathcal{E}^{(t)}(\hat{v})$ lies in this set by definition of $\Pi^{(t),\text{mult}}$, we have that

$$\mathcal{E}^{(t)}(\hat{v}) \in \arg\inf_{\mu \in \mathcal{P}_{Z,L}^{(t)}} D(\hat{v} \parallel \mathcal{L}^{(t)}(\mu)). \tag{16.18}$$

Substituting $\mathcal{K}^{(t)}(\mathcal{L}^{(t)}(\mu_Q^{(t)}))$ for $\hat{v}$ and $t + 1$ instead of t into (16.18) shows that

$$\mathcal{E}^{(t+1)}(\mathcal{K}^{(t)}(\mathcal{L}^{(t)}(\mu_Q^{(t)}))) = \mathcal{Q}^{(t)}(\mu_Q^{(t)}) = \mu_Q^{(t+1)}$$

is indeed a solution to the right-hand side of (16.15).

In the single activity case ($M = 1$), the set of solutions in (16.17) is the singleton $\{\mathcal{E}^{(t)}(\hat{v})\}$ in view of the bijection results in Proposition 16.13. In general, however, this is not the case. For instance, $\Pi^{(t),\text{lci}}(\mathcal{E}^{(t)}(\hat{v}))$ is also a solution, since $(\mathcal{L}^{(t)} \circ \Pi^{(t),\text{lci}})(\mathcal{E}^{(t)}(\hat{v})) = \mathcal{L}^{(t)}(\mathcal{E}^{(t)}(\hat{v})) = \Pi^{(t),\text{mult}}(\hat{v})$, where the first equality follows from Proposition 16.9. If the feasible set in (16.16) is restricted to $\mathcal{P}_{Z,L}^{(t),\text{lci}}$, then Proposition 16.13 implies that $\Pi^{(t),\text{lci}}(\mathcal{E}^{(t)}(\hat{v}))$ is the unique solution.

16.7 Proof of Proposition 16.14

We begin by establishing a lemma. Given some pmf $\hat{v}$ of an appropriate histogram variable H, it shows that the pmf θ^* of a label variable L that gives rise to the smallest Kullback-Leibler divergence among multinomial pmfs uses the "mean" histogram under $\hat{v}$. This result can be viewed as a special case of an M-projection, which projects distributions onto an exponential family by matching moments; see, for example, Section 8.5 of [14]. The lemma holds for some general (countably infinite) sets Ω_L. We let Ω_H denote the set of all histograms on Ω_L. We also let $\mathcal{P}_H$ and $\mathcal{P}_L$ denote the set of pmfs on Ω_H and Ω_L, respectively. Throughout this section, for notational convenience, we use $\text{mult}(n, p)$ interchangeably as a random variable and the pmf associated with that random variable.

Lemma 16.15 *Fix some pmf $\hat{v} \in \mathcal{P}_H$ with support in $\{h \in \Omega_H : |h| = x\}$ for some $x \geq 1$. The infimum $\inf_{\theta \in \mathcal{P}_L} D(\hat{v} \parallel \text{mult}(x, \theta))$ is uniquely attained by $\hat{\theta}$ defined via $\hat{\theta}(\ell) = \sum_h (h(\ell)/|h|) \times \hat{v}(h)$.*

Proof Fix $\hat{v}$ and define $\hat{\theta}$ as in the statement of the lemma. We first note that, for any $\theta \in \mathcal{P}_L$, $D(\hat{v} \parallel \text{mult}(x, \theta)) = \infty$ if there exists some $\ell \in \Omega_L$ with $\theta(\ell) = 0$ and $\hat{\theta}(\ell) > 0$. Indeed, the latter implies that there must exist some h with $h(\ell) > 0$ and $\hat{v}(h) > 0$, while the former shows that such h cannot lie in the support of $\text{mult}(x, \theta)$. Thus, we can assume, without loss of generality, that the support of θ includes the support of $\hat{\theta}$ and that the support of $\text{mult}(x, \theta)$ includes the support of $\hat{v}$.

We will show that, for any such $\theta \in \mathcal{P}_L$,

$$D(\hat{\nu} \| \text{mult}(x, \theta)) - D(\hat{\nu} \| \text{mult}(x, \hat{\theta})) = x \times D(\hat{\theta} \| \theta). \tag{16.19}$$

The claim then follows as the right-hand side is nonnegative and equal to zero if and only if $\theta = \hat{\theta}$. To establish (16.19), we use the fact that, for h in the support of $\hat{\nu}$,

$$\log\left(\frac{\Pr[\text{mult}(x, \hat{\theta}) = h]}{\Pr[\text{mult}(x, \theta) = h]}\right) = \sum_{\ell} h(\ell) \times \log\left(\frac{\hat{\theta}(\ell)}{\theta(\ell)}\right). \tag{16.20}$$

The sum on the right-hand side is taken over $\ell \in \Omega_L$ with $h(\ell) > 0$, so that $\hat{\theta}(\ell) > 0$ and then also $\theta(\ell) > 0$. Note that the denominator on the left-hand side is nonzero, since h must also lie in the support of $\text{mult}(x, \theta)$.

It follows from the definition of Kullback-Leibler divergence that

$$D(\hat{\nu} \| \text{mult}(x, \theta)) - D(\hat{\nu} \| \text{mult}(x, \hat{\theta}))$$

$$= \sum_{h} \hat{\nu}(h) \times \log\left(\frac{\Pr[\text{mult}(x, \hat{\theta}) = h]}{\Pr[\text{mult}(x, \theta) = h]}\right), \tag{16.21}$$

with the sum taken over the support of $\hat{\nu}$. Using (16.20), the right-hand side of (16.21) may be expressed as

$$\sum_{h} \hat{\nu}(h) \times \left(\sum_{\ell} h(\ell) \times \log\left(\frac{\hat{\theta}(\ell)}{\theta(\ell)}\right)\right)$$

$$= \sum_{\ell}\left[\sum_{h} \hat{\nu}(h) \times h(\ell)\right] \times \log\left(\frac{\hat{\theta}(\ell)}{\theta(\ell)}\right)$$

$$= \sum_{\ell}[x \times \hat{\theta}(\ell)] \times \log\left(\frac{\hat{\theta}(\ell)}{\theta(\ell)}\right),$$

which equals $x \times D(\hat{\theta} \| \theta)$, as required.   □

We now turn to the proof of Proposition 16.14. Fix some $\hat{\nu} \in \mathcal{P}^{(t)}_{Z,H}$ with finite support and set $\nu^* = \Pi^{(t),\text{mult}}(\hat{\nu}) \in \mathcal{P}^{(t),\text{mult}}_{Z,H}$. Consider an arbitrary fixed $\nu \in \mathcal{P}^{(t),\text{mult}}_{Z,H}$. We can assume, without loss of generality, that $D(\hat{\nu} \| \nu) < \infty$, which implies that the support of ν includes the support of $\hat{\nu}$.

By definition of $\mathcal{P}^{(t),\text{mult}}_{Z,H}$, we can pick some $\mu \in \mathcal{P}^{(t)}_{Z,L}$ satisfying $\nu = \mathcal{L}^{(t)}(\mu)$. Define $\mu^* = \mathcal{E}^{(t)}(\hat{\nu})$, and note that

$$\nu^* = \Pi^{(t),\text{mult}}(\hat{\nu}) = (\mathcal{L}^{(t)} \circ \mathcal{E}^{(t)})(\hat{\nu}) = \mathcal{L}^{(t)}(\mu^*).$$

It thus follows from these definitions that for $(z, h) \in \Omega_{Z, H}$

$$v(z, h) = \mu(z) \times \prod_m \Pr[\text{mult}(x_m^{(t)}(z), \mu(L_m = \cdot|z)) = h_m] \tag{16.22}$$

$$v^*(z, h) = \mu^*(z) \times \prod_m \Pr[\text{mult}(x_m^{(t)}(z), \mu^*(L_m = \cdot|z)) = h_m]. \tag{16.23}$$

Recall that by convention, the right-hand sides of (16.22) and (16.23) are zero if $\mu(z) = 0$ or $\mu^*(z) = 0$, respectively, even though the multiplicands are undefined. We need to show that $D(\hat{v} \parallel v) \geq D(\hat{v} \parallel v^*)$ and that this is an equality if and only if $v = v^*$.

From the definition of the Kullback-Leibler divergence, we deduce that

$$D(\hat{v} \parallel v) - D(\hat{v} \parallel v^*) = \sum_{z, h} \hat{v}(z, h) \log \left(\frac{v^*(z, h)}{v(z, h)} \right),$$

where the right-hand side should be interpreted as an extended real number; it can be $-\infty$ but not $+\infty$. In view of (16.22) and (16.23), it suffices to show that

$$\sum_z \hat{v}(z) \times \log \left(\frac{\mu^*(z)}{\mu(z)} \right) \geq 0 \tag{16.24}$$

and, for every m,

$$\sum_{z, h_m} \hat{v}(z, h_m) \times \log \left(\frac{\Pr[\text{mult}(x_m^{(t)}(z), \mu^*(L_m = \cdot|z)) = h_m]}{\Pr[\text{mult}(x_m^{(t)}(z), \mu(L_m = \cdot|z)) = h_m]} \right) \geq 0, \tag{16.25}$$

with equalities if and only if $v = v^*$.

We begin with (16.24). By definition of μ^* and (16.2),

$$\mu^*(z) = [\mathcal{E}^{(t)}(\hat{v})](z) = \hat{v}(z), \tag{16.26}$$

so the left-hand side of (16.24) is simply $D(\mu^*(Z = \cdot) \parallel \mu(Z = \cdot))$, which is nonnegative and nonzero unless $\mu(Z = \cdot) = \mu^*(Z = \cdot)$.

We now turn to (16.25). We first discuss why the left-hand side is well-defined. Fix some z and h_m with $\hat{v}(z, h_m) > 0$. It immediately follows from (16.26) that $\mu^*(z) > 0$. Moreover, by (16.22) and the assumption that the support of v contains the support of $\hat{v}$ made at the beginning of this proof, we also have $\mu(z) = v(z) \geq v(z, h) > 0$. Therefore, the conditional pmfs in this expression are well-defined. Next, fix some z with $\hat{v}(z) > 0$. The quantity

$$\sum_{h_m} \hat{v}(h_m|z) \times \log \left(\frac{\Pr[\text{mult}(x_m^{(t)}(z), \mu^*(L_m = \cdot|z)) = h_m]}{\Pr[\text{mult}(x_m^{(t)}(z), \mu(L_m = \cdot|z)) = h_m]} \right) \tag{16.27}$$

can be expressed as

$$D(\hat{v}(H_m = \cdot|z) \parallel \text{mult}(x_m^{(t)}(z), \mu(L_m = \cdot|z)))$$
$$- D(\hat{v}(H_m = \cdot|z) \parallel \text{mult}(x_m^{(t)}(z), \mu^*(L_m = \cdot|z))). \tag{16.28}$$

By Lemma 16.15, the first term on the right-hand side of (16.28) is at least as large as

$$D(\hat{v}(H_m = \cdot|z) \parallel \text{mult}(x_m^{(t)}(z), [\mathcal{E}^{(t)}(\hat{v})](L_m = \cdot|z))),$$

which is the second term on the right-hand side of (16.28) since $\mu^* = \mathcal{E}^{(t)}(\hat{v})$. Since the left-hand side of (16.25) is a weighted average over $\hat{v}(z)$ of nonnegative quantities, it is nonnegative, too. This argument also shows that (16.27) is zero only if $\mu(\ell_m|z) = \mu^*(\ell_m|z)$ for all ℓ_m.

We have now established that v^* is a solution to the minimization problem in the statement of the proposition, and it remains to show that it is the unique solution. Suppose v chosen at the beginning of this proof is another solution. The preceding arguments show that the left-hand sides of (16.24) and (16.25) must then be zero. This implies that $\mu(z) = \mu^*(z)$ for all z and $\mu(\ell_m|z) = \mu^*(\ell_m|z)$ for all m, ℓ_m, and all z with $\hat{v}(z) > 0$. Since $\hat{v}(z) = \mu^*(z)$, the condition $\hat{v}(z) > 0$ is equivalent to $\mu^*(z) > 0$. Thus, we must have $\mu(z, \ell_m) = \mu^*(z, \ell_m)$ for all m, z, and ℓ_m. In view of (16.22) and (16.23), this means that $v = v^*$. This establishes the uniqueness, as claimed.

Chapter 17
Optimality of Information Iterates

This chapter defines and establishes properties of the graphical QPLEX map and several projection maps. We use these properties to establish what we call *period-by-period optimality* of the information iterates. We investigate how the choice of information structure graph affects performance metrics.

As in the previous chapter, unless stated otherwise, we fix time t, the counter set Ω_Z, the number of activities M, the label sets Ω_m, and the size function vectors $x^{(t)}$ and $x^{(t+1)}$. We let the vertex set V consist of all counter and label variables.

17.1 Graphical QPLEX Maps

Throughout this section, we fix arbitrary information structure graphs $G = (V, E)$ and $G' = (V, E')$ compatible with $x^{(t)}$ and $x^{(t+1)}$, respectively. Since the set $\mathcal{P}_V$ defined in Chap. 11 corresponds to $\mathcal{P}_{Z,L}$, the G-lift map L_G from Sect. 11.5 has domain $\mathcal{I}_G$ and codomain $\mathcal{P}_{Z,L}$, and the G-marginalization map I_G from Sect. 11.5 has domain $\mathcal{P}_{Z,L}$ and codomain $\mathcal{I}_G$. We define $\mathcal{I}_G^{(t)}$ as the image of $\mathcal{P}_{Z,L}^{(t)}$ under I_G, i.e.,

$$\mathcal{I}_G^{(t)} = \left\{ \iota \in \mathcal{I}_G : \iota = I_G(\mu) \text{ for some } \mu \in \mathcal{P}_{Z,L}^{(t)} \right\}. \tag{17.1}$$

Thus, each $\iota \in \mathcal{I}_G^{(t)}$ has the property that there exists some $\mu \in \mathcal{P}_{Z,L}^{(t)}$ such that each of the pmfs in the G-information collection ι is a marginal pmf of $\mu \in \mathcal{P}_{Z,L}^{(t)}$.

The maps $L_G^{(t)}$ and $I_G^{(t)}$ defined below are suitable restrictions of the maps L_G and I_G, respectively.

Definition 17.1 The *restricted G-lift map $L_G^{(t)}$* is the $(\mathcal{I}_G^{(t)}, \mathcal{P}_{Z,L}^{(t)})$-restriction of the G-lift map L_G.

© The Author(s) 2025
A. B. Dieker, S. T. Hackman, *QPLEX: A Computational Modeling and Analysis Methodology for Stochastic Systems*, Springer Series in Operations Research and Financial Engineering, https://doi.org/10.1007/978-3-031-74870-7_17

Definition 17.2 The *restricted G-marginalization map* $I_G^{(t)}$ is the $(\mathcal{P}_{Z,L}^{(t)}, \mathcal{I}_G^{(t)})$-restriction of the G-marginalization map I_G.

The restricted G-marginalization map $I_G^{(t)}$ is well-defined in view of the definition of $\mathcal{I}_G^{(t)}$. The following lemma, proved in Sect. 17.7, establishes that $L_G^{(t)}$ is in fact well-defined.

Lemma 17.3 *We have that* $L_G(\mathcal{I}_G^{(t)}) \subseteq \mathcal{P}_{Z,L}^{(t)}$.

Here is a formal definition of the graphical QPLEX map.

Definition 17.4 The *graphical QPLEX map* $\mathcal{Q}_{G \to G'}^{(t)} : \mathcal{I}_G^{(t)} \to \mathcal{I}_{G'}^{(t+1)}$ is defined as the composition

$$\mathcal{Q}_{G \to G'}^{(t)} = I_{G'}^{(t+1)} \circ \mathcal{Q}^{(t)} \circ L_G^{(t)}.$$

Of course, given some $\iota \in \mathcal{I}_G^{(t)}$, it may be computationally infeasible to evaluate $\mathcal{Q}^{(t)}(L_G^{(t)}(\iota))$. Chapters 13 and 14 showed how to directly and efficiently calculate $\mathcal{Q}_{G \to G'}^{(t)}(\iota)$ *without* having to calculate $\mathcal{Q}^{(t)}(L_G^{(t)}(\iota))$ under structural assumptions on $\mathcal{Q}^{(t)}$.

17.2   Projection Maps

Throughout this section, we fix some arbitrary information structure graph G compatible with $x^{(t)}$. We formally define three projection maps—the restricted G-projection map, the restricted G-label conditional independence map, and the multinomial G-projection map—and establish requisite properties that will be used in subsequent sections to establish a "period-by-period" optimality property of appropriate iterates.

Restricted G-Projection Maps In the context of the current chapter, the G-projection map $\Pi_G = L_G \circ I_G$ (see Sect. 11.5) has domain and codomain $\mathcal{P}_{Z,L}$. The restricted G-projection map $\Pi_G^{(t)}$ in the definition below is also the $(\mathcal{P}_{Z,L}^{(t)}, \mathcal{P}_{Z,L}^{(t)})$-restriction of Π_G by definition of $L_G^{(t)}$ and $I_G^{(t)}$.

Definition 17.5 The *restricted G-projection map* $\Pi_G^{(t)} : \mathcal{P}_{Z,L}^{(t)} \to \mathcal{P}_{Z,L}^{(t)}$ is defined as the composition

$$\Pi_G^{(t)} = L_G^{(t)} \circ I_G^{(t)}.$$

Recall that $\mathcal{M}_G$ is the set of all pmfs $\mu \in \mathcal{P}_{Z,L}$ that possess the global Markov property with respect to G. We define the subset $\mathcal{M}_G^{(t)}$ of $\mathcal{P}_{Z,L}^{(t)}$ as the set of pmfs in $\mathcal{P}_{Z,L}^{(t)}$ that possess the global Markov property with respect to G, i.e.,

$$\mathcal{M}_G^{(t)} = \mathcal{P}_{Z,L}^{(t)} \cap \mathcal{M}_G.$$

The following two propositions are consequences of Definition 17.5 and previously established results.

Proposition 17.6 $\Pi_G^{(t)}$ *is a projection map from* $\mathcal{P}_{Z,L}^{(t)}$ *onto* $\mathcal{M}_G^{(t)}$.

We verify this proposition by establishing the two requisite properties of Definition 1.2. Since Π_G is a projection map from $\mathcal{P}_{Z,L}$ onto $\mathcal{M}_G$ (Proposition 11.14), the image of $\mathcal{P}_{Z,L}^{(t)}$ under $\Pi_G^{(t)}$ is included in $\mathcal{M}_G^{(t)}$, which establishes the first property, and $\Pi_G^{(t)}(\mu) = \mu$ if $\mu \in \mathcal{M}_G^{(t)} \subseteq \mathcal{M}_G$, which establishes the second property.

Proposition 17.7 *For each pmf* $\hat{\mu} \in \mathcal{P}_{Z,L}^{(t)}$ *with finite support, the solution to the minimization problem* $\inf_{\mu \in \mathcal{M}_G^{(t)}} D(\hat{\mu} \parallel \mu)$ *is unique and corresponds to* $\mu^* = \Pi_G^{(t)}(\hat{\mu})$.

To see why this proposition holds, since $\hat{\mu} \in \mathcal{P}_{Z,L}^{(t)}$, it follows from Proposition 17.6 that $\mu^* = \Pi_G^{(t)}(\hat{\mu})$ lies in $\mathcal{M}_G^{(t)}$, and so μ^* is a feasible solution to the minimization problem. The definition of $\Pi_G^{(t)}$ implies that $\mu^* = \Pi_G^{(t)}(\hat{\mu}) = \Pi_G(\hat{\mu})$. The claim now follows from Proposition 11.18, which establishes that $\Pi_G(\hat{\mu})$ is the unique minimizer of the Kullback-Leibler divergence between $\hat{\mu}$ and the set $\mathcal{M}_G$, which contains $\mathcal{M}_G^{(t)}$.

Restricted G-Label Conditional Independence Maps We start with a formal definition. Recall that the label conditional independence map $\Pi^{(t),\text{lci}}$ is defined as the composition $\mathcal{E}^{(t)} \circ \mathcal{L}^{(t)}$; see Sect. 16.4.

Definition 17.8 The *restricted G-label conditional independence map* $\Pi_G^{(t),\text{lci}}$: $\mathcal{M}_G^{(t)} \to \mathcal{M}_G^{(t)}$ is defined as the $(\mathcal{M}_G^{(t)}, \mathcal{M}_G^{(t)})$-restriction of $\Pi^{(t),\text{lci}}$.

As a consequence of Proposition 17.10, we have that $\Pi_G^{(t),\text{lci}}(\mathcal{M}_G^{(t)}) \subseteq \mathcal{M}_{\text{lci}(G)}^{(t)}$. Since $\mathcal{M}_{\text{lci}(G)}^{(t)} \subseteq \mathcal{M}_G^{(t)}$, it therefore follows that $\Pi_G^{(t),\text{lci}}$ is well-defined.

The following proposition will be used to derive the projection property of $\Pi_G^{(t),\text{lci}}$. Recall that $\text{lci}(G)$ is the graph obtained from G by removing all edges between label variables. We know that, for any $\mu \in \mathcal{P}_{Z,L}^{(t)}$, $\Pi_G^{(t)}(\mu)$ has the global Markov property with respect to G. We also know from Sect. 16.4 that the label variables are conditionally independent given the counter variables under $\tilde{\mu} = \Pi^{(t),\text{lci}}(\Pi_G^{(t)}(\mu))$. The proposition shows that $\tilde{\mu}$ also has the global Markov property with respect to $\text{lci}(G)$. Its proof may be found in Sect. 17.7.

Proposition 17.9 *We have that* $\Pi^{(t),\text{lci}} \circ \Pi_G^{(t)} = \Pi_{\text{lci}(G)}^{(t)}$.

The next two propositions are consequences of Proposition 17.9. The first proposition should be compared with Proposition 16.6, which states that $\Pi^{(t),\text{lci}}$ is a projection map from $\mathcal{P}_{Z,L}^{(t)}$ onto $\mathcal{P}_{Z,L}^{(t),\text{lci}}$.

Proposition 17.10 $\Pi_G^{(t),\mathrm{lci}}$ *is a projection map from* $\mathcal{M}_G^{(t)}$ *onto* $\mathcal{M}_{\mathrm{lci}(G)}^{(t)}$.

To see why this proposition holds, we verify the two requisite properties of Definition 1.2. We first note that, for $\mu \in \mathcal{M}_G^{(t)}$, by Proposition 17.9 and the fact that $\Pi_G^{(t)}(\mu) = \mu$ by Proposition 17.6,

$$\Pi_G^{(t),\mathrm{lci}}(\mu) = \Pi^{(t),\mathrm{lci}}(\mu) = (\Pi^{(t),\mathrm{lci}} \circ \Pi_G^{(t)})(\mu) = \Pi_{\mathrm{lci}(G)}^{(t)}(\mu). \tag{17.2}$$

Since the right-hand side is an element of $\mathcal{M}_{\mathrm{lci}(G)}^{(t)}$ (again by Proposition 17.6), the image of $\mathcal{M}_G^{(t)}$ under $\Pi_G^{(t),\mathrm{lci}}$ is indeed a subset of $\mathcal{M}_{\mathrm{lci}(G)}^{(t)}$, as required for the first property. The right-hand side equals μ if μ is an element of the subset $\mathcal{M}_{\mathrm{lci}(G)}^{(t)}$ of $\mathcal{M}_G^{(t)}$, as required for the second property.

The second proposition is a corresponding Kullback-Leibler divergence optimality result.

Proposition 17.11 *For each pmf* $\hat{\mu} \in \mathcal{M}_G^{(t)}$ *with finite support, the solution to the minimization problem* $\inf_{\mu \in \mathcal{M}_{\mathrm{lci}(G)}^{(t)}} D(\hat{\mu} \parallel \mu)$ *is unique and corresponds to* $\mu^* = \Pi^{(t),\mathrm{lci}}(\hat{\mu})$.

This proposition holds since Proposition 17.7 establishes that $\Pi_{\mathrm{lci}(G)}^{(t)}(\hat{\mu})$ is the unique solution to this minimization problem and (17.2) shows that $\mu^* = \Pi_{\mathrm{lci}(G)}^{(t)}(\hat{\mu})$.

Multinomial G-Projection Maps We again start with a formal definition.

Definition 17.12 The *multinomial G-projection map* $\Pi_G^{(t),\mathrm{mult}} : \mathcal{P}_{Z,H}^{(t)} \to \mathcal{P}_{Z,H}^{(t)}$ is defined as the composition

$$\Pi_G^{(t),\mathrm{mult}} = \mathcal{L}^{(t)} \circ \Pi_G^{(t)} \circ \mathcal{E}^{(t)}.$$

Define the set $\mathcal{M}_G^{(t),\mathrm{mult}}$ to be the image of $\mathcal{M}_G^{(t)}$ under the multinomial lift map $\mathcal{L}^{(t)}$, i.e.,

$$\mathcal{M}_G^{(t),\mathrm{mult}} = \{\nu \in \mathcal{P}_{Z,H}^{(t)} : \nu = \mathcal{L}^{(t)}(\mu) \text{ for some } \mu \in \mathcal{M}_G^{(t)}\}. \tag{17.3}$$

Since $\mathcal{M}_G^{(t)} \subseteq \mathcal{P}_{Z,L}^{(t)}$, we have that $\mathcal{M}_G^{(t),\mathrm{mult}} \subseteq \mathcal{P}_{Z,H}^{(t),\mathrm{mult}}$.

The following proposition justifies calling $\Pi_G^{(t),\mathrm{mult}}$ a projection map.

Proposition 17.13 $\Pi_G^{(t),\mathrm{mult}}$ *is a projection map from* $\mathcal{P}_{Z,H}^{(t)}$ *onto* $\mathcal{M}_G^{(t),\mathrm{mult}}$.

To see why this proposition holds, we verify the two requisite properties of Definition 1.2. Fix some $\nu \in \mathcal{P}_{Z,H}^{(t)}$. We have that

$$\Pi_G^{(t),\mathrm{mult}}(\nu) = (\mathcal{L}^{(t)} \circ \Pi_G^{(t)} \circ \mathcal{E}^{(t)})(\nu) = \mathcal{L}^{(t)}(\Pi_G^{(t)}(\mathcal{E}^{(t)}(\nu))),$$

and since $\Pi_G^{(t)}(\mathcal{E}^{(t)}(\nu)) \in \mathcal{M}_G^{(t)}$, it follows that $\Pi_G^{(t),\mathrm{mult}}(\nu) \in \mathcal{M}_G^{(t),\mathrm{mult}}$, which establishes that the image of $\mathcal{P}_{\mathbf{Z},\mathbf{H}}^{(t)}$ under $\Pi_G^{(t),\mathrm{mult}}$ is indeed a subset of $\mathcal{M}_G^{(t),\mathrm{mult}}$. To establish the second property, fix some $\nu \in \mathcal{M}_G^{(t),\mathrm{mult}}$. We must show that $\Pi_G^{(t),\mathrm{mult}}(\nu) = \nu$. To this end, we may write $\nu = \mathcal{L}^{(t)}(\mu)$ for some $\mu \in \mathcal{M}_G^{(t)}$. The following identities establish the claim:

$$\Pi_G^{(t),\mathrm{mult}}(\nu) = \left(\mathcal{L}^{(t)} \circ \Pi_G^{(t)} \circ \mathcal{E}^{(t)} \circ \mathcal{L}^{(t)} \right)(\mu)$$

$$= \left(\mathcal{L}^{(t)} \circ \Pi_G^{(t)} \circ \Pi^{(t),\mathrm{lci}} \right)(\mu)$$

$$= \left(\mathcal{L}^{(t)} \circ \Pi^{(t),\mathrm{lci}} \right)(\mu)$$

$$= \mathcal{L}^{(t)}(\mu) = \nu.$$

The first equality follows from the definition of $\Pi_G^{(t),\mathrm{mult}}$ and μ. The second equality follows from the definition of $\Pi^{(t),\mathrm{lci}}$. The third equality uses that $\Pi^{(t),\mathrm{lci}}$ projects $\mathcal{M}_G^{(t)}$ onto $\mathcal{M}_{\mathrm{lci}(G)}^{(t)}$ (Proposition 17.10) in conjunction with $\mathcal{M}_{\mathrm{lci}(G)}^{(t)} \subseteq \mathcal{M}_G^{(t)}$ and the fact that $\Pi_G^{(t)}$ is a projection map onto $\mathcal{M}_G^{(t)}$ (Proposition 17.6). The fourth equality follows from Proposition 16.9, and the last equality follows from the definition of μ.

The next proposition establishes an optimality result for $\Pi_G^{(t),\mathrm{mult}}$, namely, that $\Pi_G^{(t),\mathrm{mult}}(\hat{\nu})$ minimizes the Kullback-Leibler divergence between $\hat{\nu} \in \mathcal{P}_{\mathbf{Z},\mathbf{H}}^{(t)}$ and the set $\mathcal{M}_G^{(t),\mathrm{mult}}$. This proposition extends Proposition 16.14, which corresponds to G being a complete graph. Its proof is given in Sect. 17.7.

Proposition 17.14 *For each pmf $\hat{\nu} \in \mathcal{P}_{\mathbf{Z},\mathbf{H}}^{(t)}$ with finite support, the solution to the minimization problem* $\inf_{\nu \in \mathcal{M}_G^{(t),\mathrm{mult}}} D(\hat{\nu} \parallel \nu)$ *is unique and corresponds to* $\nu^* = \Pi_G^{(t),\mathrm{mult}}(\hat{\nu})$.

17.3 Iterative Schemes and Period-by-Period Optimality

In this section, we consider iterative schemes, and so we no longer consider a fixed time epoch. For every $t \geq 0$, we assume we are given, in addition to the counter and label sets, a size function vector $\mathbf{x}^{(t)}$, a kernel map $\mathcal{K}^{(t)}$, and an information structure graph $G^{(t)}$ compatible with $\mathbf{x}^{(t)}$, and thus a graphical QPLEX map $\mathcal{Q}_{G^{(t)} \to G^{(t+1)}}^{(t)}$ as well.

Iterative Schemes The information iterates are defined via the following iterative scheme. Given some initial $\iota_{\mathrm{IS}}^{(0)} \in \mathcal{I}_{G^{(0)}}^{(0)}$, we set, for $t \geq 0$,

$$\iota_{\mathrm{IS}}^{(t+1)} = \mathcal{Q}_{G^{(t)} \to G^{(t+1)}}^{(t)}(\iota_{\mathrm{IS}}^{(t)}). \tag{17.4}$$

We call the $G^{(t)}$-information collections $\iota_{\mathrm{IS}}^{(t)}$, $t \geq 0$, the *information iterates*. Each information iterate $\iota_{\mathrm{IS}}^{(t)}$ is an $G^{(t)}$-information collection and lies in $\mathcal{I}_{G^{(t)}}^{(t)}$.

For convenience, we have reproduced our familiar three-level diagram below, appropriately "glued" together.

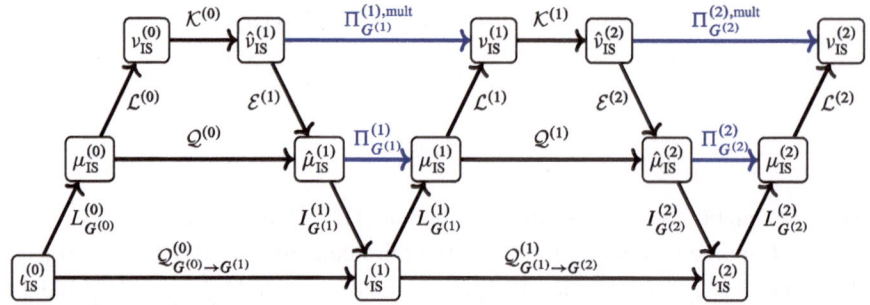

By considering the other two directed paths in this diagram, we obtain two alternative versions of (17.4) via

$$\iota_{\mathrm{IS}}^{(t+1)} = \left(I_{G^{(t+1)}}^{(t+1)} \circ \mathcal{E}^{(t+1)} \circ \mathcal{K}^{(t)} \circ \mathcal{L}^{(t)} \circ L_{G^{(t)}}^{(t)} \right) \left(\iota_{\mathrm{IS}}^{(t)} \right)$$

$$= \left(I_{G^{(t+1)}}^{(t+1)} \circ \mathcal{Q}^{(t)} \circ L_{G^{(t)}}^{(t)} \right) \left(\iota_{\mathrm{IS}}^{(t)} \right).$$

The QPLEX iterates $\mu_{\mathrm{Q}}^{(t)}$ discussed in the previous chapter are *not* the same as the $\mu_{\mathrm{IS}}^{(t)}$ in this diagram. (In fact, the QPLEX iterates are not visible in this diagram.) Indeed, the QPLEX iterates are found by successive applications of the QPLEX maps $\mathcal{Q}^{(t)}$, whereas as shown in the diagram, the $\mu_{\mathrm{IS}}^{(t)}$ are found by successive applications of these QPLEX maps *alternated* with the projection maps $\Pi_{G^{(t)}}^{(t)}$. The *lifted information iterate*

$$\mu_{\mathrm{IS}}^{(t)} = L_{G^{(t)}}^{(t)} \left(\iota_{\mathrm{IS}}^{(t)} \right)$$

approximates the QPLEX iterate $\mu_{\mathrm{Q}}^{(t)}$.

Similarly, the kernel iterates $\nu_{\mathrm{ker}}^{(t)}$ are defined by successive applications of kernel maps $\mathcal{K}^{(t)}$ and are also not visible in this diagram. They are not the same as the $\nu_{\mathrm{IS}}^{(t)}$ in this diagram, which are calculated by alternately applying the kernel maps $\mathcal{K}^{(t)}$ and the projection maps $\Pi_{G^{(t)}}^{(t),\mathrm{mult}}$. The *twice-lifted information iterate*

$$v_{\text{IS}}^{(t)} = \left(\mathcal{L}^{(t)} \circ L_{G^{(t)}}^{(t)}\right)\left(\iota_{\text{IS}}^{(t)}\right)$$

approximates $v_{\text{ker}}^{(t)}$. Of course, it is generally computationally infeasible to calculate the former, but the pmf of many performance metrics can be calculated from $\iota_{\text{IS}}^{(t)}$ without having to calculate $v_{\text{IS}}^{(t)}$.

Period-by-Period Optimality We now establish a period-by-period optimality property for the pmfs appearing on each of the three levels of our diagram. Throughout, we fix t as we consider an arbitrary single period, and we make the requisite finite support assumptions.

Period-by-Period Optimality for the $v_{\text{IS}}^{(t)}$ The period-by-period optimality property of the twice-lifted information iterates is that

$$v_{\text{IS}}^{(t+1)} = \underset{v \in \mathcal{M}_{G^{(t+1)}}^{(t+1),\text{mult}}}{\arg\inf}\ D(\mathcal{K}^{(t)}(v_{\text{IS}}^{(t)}) \parallel v).$$

This follows from

$$\Pi_{G^{(t+1)}}^{(t+1),\text{mult}}\left(\mathcal{K}^{(t)}(v_{\text{IS}}^{(t)})\right) = v_{\text{IS}}^{(t+1)}$$

and a direct application of Proposition 17.14 with t replaced by $t + 1$, G by $G^{(t+1)}$, and with $\hat{v} = \mathcal{K}^{(t)}(v_{\text{IS}}^{(t)})$. Note that the resulting minimization problem has a unique solution, so the equality is warranted.

Period-by-Period Optimality for the $\mu_{\text{IS}}^{(t)}$ The period-by-period optimality property of the lifted information iterates is that

$$\mu_{\text{IS}}^{(t+1)} \in \underset{\mu \in \mathcal{M}_{G^{(t+1)}}^{(t+1)}}{\arg\inf}\ D(\mathcal{K}^{(t)}(\mathcal{L}^{(t)}(\mu_{\text{IS}}^{(t)})) \parallel \mathcal{L}^{(t+1)}(\mu)). \tag{17.5}$$

To see why this holds, we invoke Proposition 17.14 once again, in conjunction with (17.3), to establish that the set of solutions to

$$\underset{\mu \in \mathcal{M}_G^{(t)}}{\inf}\ D(\hat{v} \parallel \mathcal{L}^{(t)}(\mu))$$

is

$$\left\{\mu \in \mathcal{M}_G^{(t)} : \mathcal{L}^{(t)}(\mu) = \Pi_G^{(t),\text{mult}}(\hat{v})\right\}. \tag{17.6}$$

Since $(\Pi_G^{(t)} \circ \mathcal{E}^{(t)})(\hat{v})$ lies in this set by definition of $\Pi_G^{(t),\text{mult}} = \mathcal{L}^{(t)} \circ \Pi_G^{(t)} \circ \mathcal{E}^{(t)}$, we have that

$$(\Pi_G^{(t)} \circ \mathcal{E}^{(t)})(\hat{v}) \in \arg\inf_{\mu \in \mathcal{M}_G^{(t)}} D(\hat{v} \parallel \mathcal{L}^{(t)}(\mu)). \tag{17.7}$$

Substituting $\mathcal{K}^{(t)}(\mathcal{L}^{(t)}(\mu_{\mathrm{IS}}^{(t)}))$ for $\hat{v}$, $G^{(t+1)}$ for G, and $t+1$ for t into (17.7) shows that

$$\left(\Pi_{G^{(t+1)}}^{(t+1)} \circ \mathcal{E}^{(t+1)}\right)\left(\mathcal{K}^{(t)}\left(\mathcal{L}^{(t)}\left(\mu_{\mathrm{IS}}^{(t)}\right)\right)\right) = (\Pi_{G^{(t+1)}}^{(t+1)} \circ \mathcal{Q}^{(t)})(\mu_{\mathrm{IS}}^{(t)}) = \mu_{\mathrm{IS}}^{(t+1)}$$

is indeed a solution to the right-hand side of (17.5).

Period-by-Period Optimality for the $\iota_{\mathrm{IS}}^{(t)}$ The period-by-period optimality property of the information iterates is that

$$\iota_{\mathrm{IS}}^{(t+1)} \in \arg\inf_{\iota \in \mathcal{I}_{G^{(t+1)}}^{(t+1)}} D((\mathcal{K}^{(t)} \circ \mathcal{L}^{(t)} \circ L_{G^{(t)}}^{(t)})(\iota_{\mathrm{IS}}^{(t)}) \parallel (\mathcal{L}^{(t+1)} \circ L_{G^{(t+1)}}^{(t+1)})(\iota)). \tag{17.8}$$

We thus see that the next information iterate $\iota_{\mathrm{IS}}^{(t+1)}$ is a $G^{(t+1)}$-information collection with corresponding twice-lifted information iterate $v_{\mathrm{IS}}^{(t+1)} = (\mathcal{L}^{(t+1)} \circ L_{G^{(t+1)}}^{(t+1)})(\iota_{\mathrm{IS}}^{(t+1)})$ as close as possible to the pmf obtained by applying the kernel map $\mathcal{K}^{(t)}$ to the current twice-lifted information iterate $v_{\mathrm{IS}}^{(t)} = (\mathcal{L}^{(t)} \circ L_{G^{(t)}}^{(t)})(\iota_{\mathrm{IS}}^{(t)})$. This contrasts with the kernel iterative scheme, where $\mathcal{K}^{(t)}$ is applied to $v_{\mathrm{ker}}^{(t)}$ directly in order to obtain $v_{\mathrm{ker}}^{(t+1)}$; see (16.14).

To see why this holds, we begin by establishing that

$$L_G^{(t)}\left(\mathcal{I}_G^{(t)}\right) = \mathcal{M}_G^{(t)}. \tag{17.9}$$

Indeed, from (17.1) and the definition of $I_G^{(t)}$, we find that $\mathcal{I}_G^{(t)} = I_G^{(t)}(\mathcal{P}_{Z,L}^{(t)})$, and by Proposition 17.6, we thus obtain that $(L_G^{(t)} \circ I_G^{(t)})(\mathcal{P}_{Z,L}^{(t)}) = \mathcal{M}_G^{(t)}$, which establishes the claim.

Next, we use $L_G^{(t)}(\mathcal{I}_G^{(t)}) = \mathcal{M}_G^{(t)}$ to conclude that $(\mathcal{L}^{(t)} \circ L_G^{(t)})(\mathcal{I}_G^{(t)}) = \mathcal{M}_G^{(t),\mathrm{mult}}$ by (17.3). Invoking Proposition 17.14 once again, the set of solutions to

$$\inf_{\iota \in \mathcal{I}_G^{(t)}} D(\hat{v} \parallel (\mathcal{L}^{(t)} \circ L_G^{(t)})(\iota))$$

is

$$\left\{ \iota \in \mathcal{I}_G^{(t)} : \left(\mathcal{L}^{(t)} \circ L_G^{(t)}\right)(\iota) = \Pi_G^{(t),\mathrm{mult}}(\hat{v}) \right\}. \tag{17.10}$$

Since $(I_G^{(t)} \circ \mathcal{E}^{(t)})(\hat{v})$ lies in this set by definition of $\Pi_G^{(t),\mathrm{mult}} = \mathcal{L}^{(t)} \circ L_G^{(t)} \circ I_G^{(t)} \circ \mathcal{E}^{(t)}$, we have that

$$\left(I_G^{(t)} \circ \mathcal{E}^{(t)} \right)(\hat{v}) \in \arg\inf_{\mu \in \mathcal{M}_G^{(t)}} D(\hat{v} \parallel \mathcal{L}^{(t)}(\mu)). \tag{17.11}$$

Substituting $(\mathcal{K}^{(t)} \circ \mathcal{L}^{(t)} \circ L_{G^{(t)}}^{(t)})(\iota_{\mathrm{IS}}^{(t)})$ for $\hat{v}$, $G^{(t+1)}$ for G, and $t+1$ for t in (17.11) shows that

$$\left(I_{G^{(t+1)}}^{(t+1)} \circ \mathcal{E}^{(t+1)} \right) \left(\left(\mathcal{K}^{(t)} \circ \mathcal{L}^{(t)} \circ L_{G^{(t)}}^{(t)} \right) \left(\iota_{\mathrm{IS}}^{(t)} \right) \right)$$
$$= \left(I_{G^{(t+1)}}^{(t+1)} \circ \mathcal{Q}^{(t)} \circ L_{G^{(t)}}^{(t)} \right) \left(\iota_{\mathrm{IS}}^{(t)} \right) = \iota_{\mathrm{IS}}^{(t+1)}$$

is indeed a solution to the right-hand side of (17.8).

In Sect. 17.5, we study a condition on the information structure graphs $G^{(t)}$ that implies that the sets (17.6) and (17.10) are singletons, but in general they are not.

17.4 Bijections

This section discusses the sets $\mathcal{P}_{Z,H}^{(t)}$, $\mathcal{P}_{Z,L}^{(t)}$, and $\mathcal{I}_G^{(t)}$, and bijections between various previously introduced subsets. Here, we again fix time t and an arbitrary information structure graph G compatible with $\boldsymbol{x}^{(t)}$. The results from this section will be used in the next section to study information iterates generated using label conditional independence graphs.

We start with the sets $\mathcal{P}_{Z,H}^{(t)}$ and $\mathcal{P}_{Z,L}^{(t)}$. The Venn diagram below summarizes the eight sets of pmfs we have introduced as well as their relationships, which are established in this section. A set with the same size, shape, shade, and position on the left-hand side and right-hand side in this diagram signifies that there is a bijection between the sets.

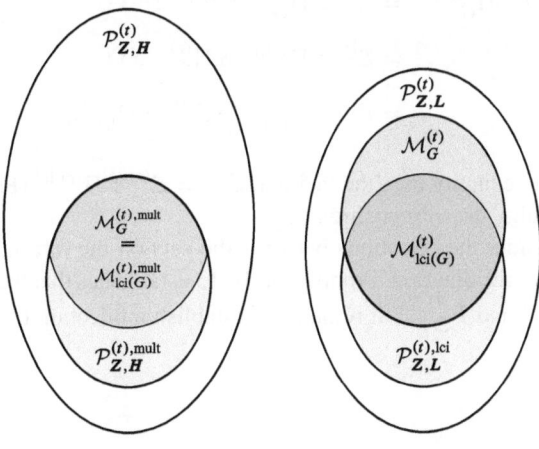

We first consider the sets shown on the right-hand side of this diagram. Recall that $\mathcal{P}_{Z,L}^{(t),\text{lci}}$ denotes the set of all $\mu \in \mathcal{P}_{Z,L}^{(t)}$ satisfying

$$\mu(z, \ell) = \mu(z) \times \prod_m \mu(\ell_m | z).$$

Lemma 17.15 *We have that* $\mathcal{M}_{\text{lci}(G)}^{(t)} = \mathcal{M}_G^{(t)} \cap \mathcal{P}_{Z,L}^{(t),\text{lci}}$.

To see why this lemma holds, fix some $\mu \in \mathcal{M}_{\text{lci}(G)}^{(t)}$. Since $\mu \in \mathcal{M}_{\text{lci}(G)}^{(t)} = \mathcal{M}_{\text{lci}(G)} \cap \mathcal{P}_{Z,L}^{(t)}$, the pmf μ is an element of $\mathcal{P}_{Z,L}^{(t),\text{lci}}$. Since $\mathcal{M}_{\text{lci}(G)} \subseteq \mathcal{M}_G$, we also have that $\mu \in \mathcal{M}_G \cap \mathcal{P}_{Z,L}^{(t)} = \mathcal{M}_G^{(t)}$. Consequently, $\mathcal{M}_{\text{lci}(G)}^{(t)} \subseteq \mathcal{M}_G^{(t)} \cap \mathcal{P}_{Z,L}^{(t),\text{lci}}$.

To show the reverse inclusion, fix some $\mu \in \mathcal{M}_G^{(t)} \cap \mathcal{P}_{Z,L}^{(t),\text{lci}}$. We have that

$$\Pi_{\text{lci}(G)}^{(t)}(\mu) = \left(\Pi^{(t),\text{lci}} \circ \Pi_G^{(t)} \right)(\mu) = \Pi^{(t),\text{lci}}(\mu) = \mu.$$

The first equality uses Proposition 17.9, the second equality uses that $\Pi_G^{(t)}$ is a projection map onto $\mathcal{M}_G^{(t)}$ (Proposition 17.6), and the last equality uses that $\Pi^{(t),\text{lci}}$ is a projection map onto $\mathcal{P}_{Z,L}^{(t),\text{lci}}$ (Proposition 16.6). Since $\Pi_{\text{lci}(G)}^{(t)}$ is a projection map onto $\mathcal{M}_{\text{lci}(G)}^{(t)}$ (Proposition 17.10), we may conclude that $\mu \in \mathcal{M}_{\text{lci}(G)}^{(t)}$, as required.

Next, consider the sets shown on the left-hand side of the diagram. We establish the following lemma.

Lemma 17.16 *We have that* $\mathcal{M}_G^{(t),\text{mult}} = \mathcal{M}_{\text{lci}(G)}^{(t),\text{mult}}$.

To see why this lemma holds, it suffices to establish that $\Pi_G^{(t),\text{mult}} = \Pi_{\text{lci}(G)}^{(t),\text{mult}}$ by Proposition 17.13. The definitions of $\Pi_G^{(t),\text{mult}}$ and $\Pi_{\text{lci}(G)}^{(t),\text{mult}}$ imply that

$$\Pi_G^{(t),\text{mult}} = \mathcal{L}^{(t)} \circ \Pi_G^{(t)} \circ \mathcal{E}^{(t)}$$

$$= \mathcal{L}^{(t)} \circ \Pi^{(t),\text{lci}} \circ \Pi_G^{(t)} \circ \mathcal{E}^{(t)}$$

$$= \mathcal{L}^{(t)} \circ \Pi_{\text{lci}(G)}^{(t)} \circ \mathcal{E}^{(t)} = \Pi_{\text{lci}(G)}^{(t),\text{mult}},$$

where the second equality uses the fact that $\mathcal{L}^{(t)} = \mathcal{L}^{(t)} \circ \Pi^{(t),\text{lci}}$ (Proposition 16.9) and the last equality uses Proposition 17.9.

We now consider the bijections between the sets on the left-hand side and the right-hand side of the diagram. Proposition 16.13 establishes that there is a bijection between $\mathcal{P}_{Z,H}^{(t),\text{mult}}$ and $\mathcal{P}_{Z,L}^{(t),\text{lci}}$. It remains to establish a bijection between $\mathcal{M}_G^{(t),\text{mult}}$ and $\mathcal{M}_{\text{lci}(G)}^{(t)}$.

Proposition 17.17 *The $(\mathcal{M}_G^{(t),\mathrm{mult}}, \mathcal{M}_{\mathrm{lci}(G)}^{(t)})$-restriction of $\mathcal{E}^{(t)}$ and the $(\mathcal{M}_{\mathrm{lci}(G)}^{(t)}, \mathcal{M}_G^{(t),\mathrm{mult}})$-restriction of $\mathcal{L}^{(t)}$ are bijections and each other's inverse.*

The restrictions in this proposition are well-defined. Indeed, by definition of $\Pi_G^{(t),\mathrm{lci}}$ and Proposition 17.10, we have that

$$\mathcal{E}^{(t)}\left(\mathcal{M}_G^{(t),\mathrm{mult}}\right) = \left(\mathcal{E}^{(t)} \circ \mathcal{L}^{(t)}\right)\left(\mathcal{M}_G^{(t)}\right) = \Pi^{(t),\mathrm{lci}}\left(\mathcal{M}_G^{(t)}\right) = \Pi_G^{(t),\mathrm{lci}}\left(\mathcal{M}_G^{(t)}\right) = \mathcal{M}_{\mathrm{lci}(G)}^{(t)},$$

whereas $\mathcal{M}_{\mathrm{lci}(G)}^{(t)} \subseteq \mathcal{M}_G^{(t)}$ implies by definition of $\mathcal{M}_G^{(t),\mathrm{mult}}$ that

$$\mathcal{L}^{(t)}\left(\mathcal{M}_{\mathrm{lci}(G)}^{(t)}\right) \subseteq \mathcal{L}^{(t)}\left(\mathcal{M}_G^{(t)}\right) = \mathcal{M}_G^{(t),\mathrm{mult}}.$$

To see why this proposition holds, we verify the two requirements from Sect. 1.1. Indeed, for $\mu \in \mathcal{M}_{\mathrm{lci}(G)}^{(t)} = \mathcal{M}_G^{(t)} \cap \mathcal{P}_{\mathbf{Z},L}^{(t),\mathrm{lci}}$ (Lemma 17.15), we have that

$$(\mathcal{E}^{(t)} \circ \mathcal{L}^{(t)})(\mu) = \Pi^{(t),\mathrm{lci}}(\mu) = \mu,$$

where the last equality uses that $\Pi^{(t),\mathrm{lci}}$ is a projection map onto $\mathcal{P}_{\mathbf{Z},L}^{(t),\mathrm{lci}}$ (Proposition 16.6). Moreover, for $\nu \in \mathcal{M}_G^{(t),\mathrm{mult}} = \mathcal{M}_{\mathrm{lci}(G)}^{(t),\mathrm{mult}}$, we can select some $\mu \in \mathcal{M}_{\mathrm{lci}(G)}^{(t)}$ such that $\nu = \mathcal{L}^{(t)}(\mu)$, so that

$$(\mathcal{L}^{(t)} \circ \mathcal{E}^{(t)})(\nu) = (\mathcal{L}^{(t)} \circ \mathcal{E}^{(t)} \circ \mathcal{L}^{(t)})(\mu) = \mathcal{L}^{(t)}(\Pi^{(t),\mathrm{lci}}(\mu)) = \mathcal{L}^{(t)}(\mu) = \nu,$$

where the second to last equality uses Proposition 16.9.

Now that we have established the results underlying the above Venn diagram of $\mathcal{P}_{\mathbf{Z},H}^{(t)}, \mathcal{P}_{\mathbf{Z},L}^{(t)}$, and various subsets, we consider their relationships with the sets $\mathcal{I}_G^{(t)}$ and $\mathcal{I}_{\mathrm{lci}(G)}^{(t)}$. The next result is a bijection between $\mathcal{M}_G^{(t)}$ and $\mathcal{I}_G^{(t)}$, visualized below.

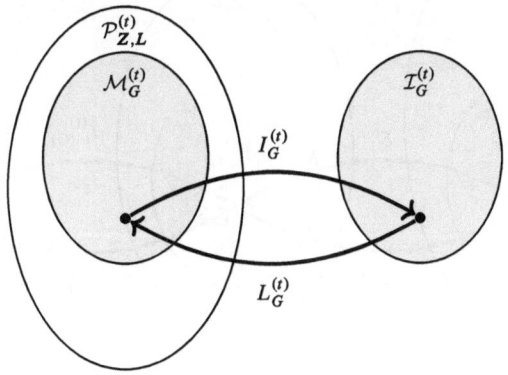

Proposition 17.18 *The* $(\mathcal{M}_G^{(t)}, \mathcal{I}_G^{(t)})$-*restriction of* $I_G^{(t)}$ *and the* $(\mathcal{I}_G^{(t)}, \mathcal{M}_G^{(t)})$-*restriction of* $L_G^{(t)}$ *are bijections and each other's inverse.*

The restrictions in this proposition are well-defined. Indeed, $L_G^{(t)}$ is well-defined by (17.9), and $I_G^{(t)}$ is well-defined, since the definition of $\mathcal{I}_G^{(t)}$ implies that

$$I_G^{(t)}\left(\mathcal{M}_G^{(t)}\right) \subseteq I_G^{(t)}\left(\mathcal{P}_{Z,L}^{(t)}\right) = \mathcal{I}_G^{(t)}.$$

To see why this proposition holds, we verify the two requirements from Sect. 1.1. For $\mu \in \mathcal{M}_G^{(t)}$, we have that

$$L_G^{(t)}\left(I_G^{(t)}(\mu)\right) = \Pi_G^{(t)}(\mu) = \mu,$$

where the last equality is a consequence of the fact that $\Pi_G^{(t)}$ is a projection map from $\mathcal{P}_{Z,L}^{(t)}$ onto $\mathcal{M}_G^{(t)}$ (Proposition 17.6). Moreover, for $\iota \in \mathcal{I}_G^{(t)}$, we trivially have that

$$I_G^{(t)}\left(L_G^{(t)}(\iota)\right) = \iota,$$

since the composition $I_G \circ L_G$ is the identity map on $\mathcal{I}_G$ (see Sect. 11.5) and therefore $I_G^{(t)} \circ L_G^{(t)}$ is the identity map on $\mathcal{I}_G^{(t)}$.

Propositions 17.17 and 17.18 imply that there are bijections between each of the sets $\mathcal{M}_G^{(t),\mathrm{mult}}$, $\mathcal{M}_{\mathrm{lci}(G)}^{(t)}$, and $\mathcal{I}_{\mathrm{lci}(G)}^{(t)}$. (Recall that $\mathrm{lci}(G)$ is an information structure graph when G is an information structure graph.) This is illustrated in the diagram below. These propositions also lead to the corollary stated below.

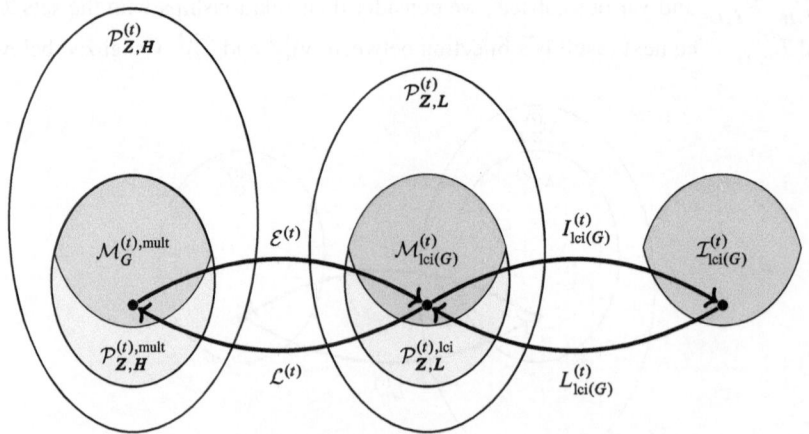

Corollary 17.19 *The* $(\mathcal{M}_G^{(t),\mathrm{mult}}, \mathcal{I}_{\mathrm{lci}(G)}^{(t)})$*-restriction of* $I_{\mathrm{lci}(G)}^{(t)} \circ \mathcal{E}^{(t)}$ *and the* $(\mathcal{I}_{\mathrm{lci}(G)}^{(t)}, \mathcal{M}_G^{(t),\mathrm{mult}})$*-restriction of* $\mathcal{L}^{(t)} \circ L_{\mathrm{lci}(G)}^{(t)}$ *are bijections and each other's inverse.*

17.5 Label Conditional Independence Graphs

In this section, we develop several consequences of the previous section. For every $t \geq 0$, we are given a size function vector $\boldsymbol{x}^{(t)}$ and a kernel map $\mathcal{K}^{(t)}$. We are also given a label conditional independence graph G_{lci}, which has no edges between label vertices (see Definition 12.5) and is assumed to be compatible with each of the size function vectors $\boldsymbol{x}^{(t)}$.

For notational convenience in the presentation to follow, we also suppose in this section that the size function vectors are identical across time, so the sets $\mathcal{P}_{Z,H}^{(t)}$, $\mathcal{P}_{Z,L}^{(t)}$, $\mathcal{M}_{G_{\mathrm{lci}}}^{(t),\mathrm{mult}}$, $\mathcal{M}_{G_{\mathrm{lci}}}^{(t)}$, and $\mathcal{I}_{G_{\mathrm{lci}}}^{(t)}$, and the maps $\mathcal{E}^{(t)}$, $\mathcal{L}^{(t)}$, $I_{G_{\mathrm{lci}}}^{(t)}$, $L_{G_{\mathrm{lci}}}^{(t)}$, $\Pi_{G_{\mathrm{lci}}}^{(t),\mathrm{mult}}$, and $\Pi_{G_{\mathrm{lci}}}^{(t)}$ do not change over time. Therefore, we use the superscript "(0)" in these sets and maps. Of course, the information structure graph G_{lci} needs to be compatible with this size function vector.

Bijections in Glued Diagram Recall that $\mu_{\mathrm{IS}}^{(t)} = L_{G_{\mathrm{lci}}}^{(t)}(\iota_{\mathrm{IS}}^{(t)})$ and $\nu_{\mathrm{IS}}^{(t)} = \mathcal{L}^{(t)}(\mu_{\mathrm{IS}}^{(t)})$ for $t \geq 0$. We know from Propositions 17.17 and 17.18 that these identities can be "reversed," i.e., $\iota_{\mathrm{IS}}^{(t)} = I_{G_{\mathrm{lci}}}^{(t)}(\mu_{\mathrm{IS}}^{(t)})$ and $\mu_{\mathrm{IS}}^{(t)} = \mathcal{E}^{(t)}(\nu_{\mathrm{IS}}^{(t)})$. The following diagram stresses this fact using arrows in both directions. It is a "generic" section of the three-level diagram glued together; see also Sect. 17.3.

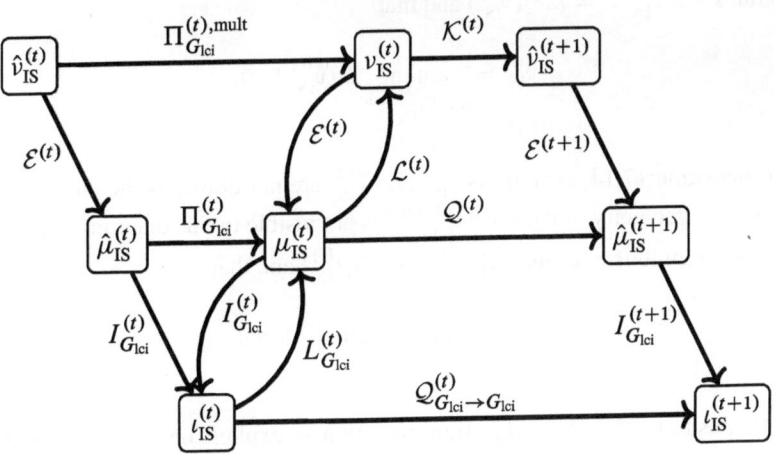

Period-by-Period Optimality Due to the bijection results from Propositions 17.17 and 17.18, the information iterates $\iota_{\mathrm{IS}}^{(t)}$ may be viewed as low-dimensional representations of the twice-lifted information iterates $v_{IS}^{(t)} \in \mathcal{P}_{Z,H}^{(0)}$ (or lifted information iterates $\mu_{IS}^{(t)} \in \mathcal{P}_{Z,L}^{(0)}$) that are repeatedly "optimally" projected onto $\mathcal{M}_{G_{lci}}^{(0),\text{mult}}$ (or $\mathcal{M}_{G_{lci}}^{(0)}$) after first applying a kernel (or QPLEX) map. (Not shown in the diagram is that $v_{IS}^{(t+1)} = \Pi_{G_{lci}}^{(t+1),\text{mult}}(\hat{v}_{IS}^{(t+1)})$ and that $\mu_{IS}^{(t+1)} = \Pi_{G_{lci}}^{(t+1)}(\hat{\mu}_{IS}^{(t+1)})$.)

The left-hand side, center, and right-hand side of the diagram below correspond, respectively, to the top, middle, and bottom levels of the above diagram. We show the first few iterates on each level.

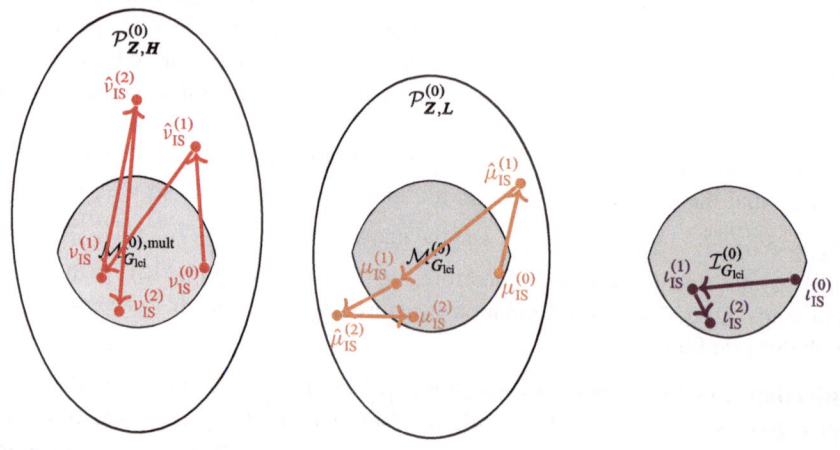

In view of the bijection, we made the relative positions of the dots in the gray sets identical in the three parts of the diagram. The left-most part of the diagram illustrates that $\hat{v}_{IS}^{(t+1)} = \mathcal{K}^{(t)}(v_{IS}^{(t)})$ and that

$$v_{IS}^{(t)} = \underset{v \in \mathcal{M}_{G_{lci}}^{(0),\text{mult}}}{\arg\inf} D(\hat{v}_{IS}^{(t)} \| v),$$

see Proposition 17.14. (The pmfs $v_{IS}^{(1)}$ and $v_{IS}^{(2)}$ are not drawn as the closest points, because the geometry of the set $\mathcal{M}_{G_{lci}}^{(0),\text{mult}}$ is not visible in this diagram.) The center of the diagram illustrates that $\hat{\mu}_{IS}^{(t+1)} = \mathcal{Q}^{(t)}(\mu_{IS}^{(t)})$ and that

$$\mu_{IS}^{(t)} = \underset{\mu \in \mathcal{M}_{G_{lci}}^{(0)}}{\arg\inf} D(\hat{\mu}_{IS}^{(t)} \| \mu),$$

see Proposition 17.7. No projection operation is explicit on the right-hand side of the diagram, yet the information iterates incorporate the projection operations implicitly.

17.6 Performance Metrics

Removing edges from an information structure graph has several consequences. On the one hand, it can never lower the projection error measured via the Kullback-Leibler divergence; on the other hand, it may decrease the computational requirements. The purpose of this section is to show that working with $\tilde{G} = \mathrm{lci}(G)$ instead of G can be a preferred choice.

We again fix time t and arbitrary information structure graphs G and G' compatible with $x^{(t)}$ and $x^{(t+1)}$ and define the graphs $\tilde{G} = \mathrm{lci}(G)$ and $\tilde{G}' = \mathrm{lci}(G')$, respectively, by removing their label vertices. We say that a $\tilde{G}$-information collection $\tilde{\iota}$ is *consistent* with a G-information collection ι if each element of $\tilde{\iota}$ is a marginal pmf of an element of ι. The following proposition states that consistency is preserved under appropriate graphical QPLEX maps.

Proposition 17.20 *If* $\tilde{\iota} \in \mathcal{I}_{\tilde{G}}^{(t)}$ *is consistent with* $\iota \in \mathcal{I}_{G}^{(t)}$, *then* $\mathcal{Q}_{\tilde{G} \to \tilde{G}'}^{(t)}(\tilde{\iota}) \in \mathcal{I}_{\tilde{G}'}^{(t+1)}$ *is consistent with* $\mathcal{Q}_{G \to G'}^{(t)}(\iota) \in \mathcal{I}_{G'}^{(t+1)}$.

To understand the significance of this proposition, let the information graphs $G^{(t)}$ all be equal to G for $t \geq 0$, and consider one or more performance metrics that are tracked over time. Suppose the pmfs of these performance metrics can be obtained from marginal pmfs associated with one or more cliques in $\tilde{G}$. In any of our queueing network models, for example, the distributions of the number of customers at each station and the virtual waiting time at each station would be of prime interest. We can calculate their pmfs over time from the information iterates $\tilde{\iota}_{\mathrm{IS}}^{(t)}$ without calculating the information iterates $\iota_{\mathrm{IS}}^{(t)}$, which are computationally more expensive. The condition that the performance metrics are associated with cliques in $\tilde{G}$ can only be satisfied if each such clique contains *at most* one label variable, which is the case in the above examples. For example, in the multiserver queueing model with abandonment, the removal of the edge between label variables L_B and L_S does not affect the sequence of marginal pmfs of the counter variable Z.

The remainder of this section is devoted to showing why this proposition holds. We establish three lemmas, which together immediately yield the proposition. Each lemma corresponds to going up, over, and down, respectively, in the diagram below. The topmost part of the diagram is our familiar three-level diagram defining $\mathcal{Q}_{G \to G'}^{(t)}$ as a composition. The bottommost part of the diagram is the analog for the graphical QPLEX map $\mathcal{Q}_{\tilde{G} \to \tilde{G}'}^{(t)}$. Given an arbitrary G-information collection $\iota \in \mathcal{I}_{G}^{(t)}$, as can be seen in the diagram, we define $\mu = L_G^{(t)}(\iota)$, $\hat{\mu}' = \mathcal{Q}^{(t)}(\mu)$, and $\iota' = I_{G'}^{(t+1)}(\hat{\nu})$. Similarly, we define $\tilde{\mu} = L_{\tilde{G}}^{(t)}(\tilde{\iota})$, $\hat{\tilde{\mu}}' = \mathcal{Q}^{(t)}(\tilde{\mu})$, and $\tilde{\iota}' = I_{\tilde{G}'}^{(t+1)}(\hat{\tilde{\mu}}')$. We use these (collections of) pmfs freely in the statements of the lemmas. The red arrow and rectangle illustrate the statements of two of the lemmas, as detailed below.

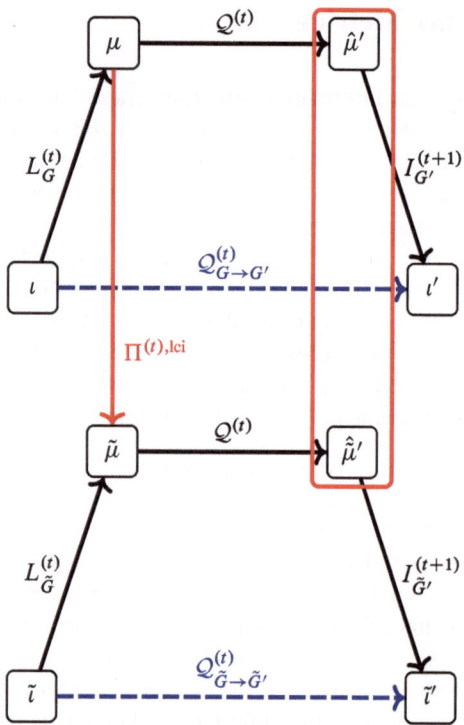

The first lemma states that if $\tilde{\iota}$ is consistent with ι, then we can obtain $\tilde{\mu}$ from μ by applying $\Pi^{(t),\text{lci}}$ as illustrated with the red arrow.

Lemma 17.21 *If $\tilde{\iota}$ is consistent with ι, then we have that $\Pi^{(t),\text{lci}}(\mu) = \tilde{\mu}$.*

To see why this lemma is true, we first note that Proposition 11.9 implies that $\mu_C = \iota_C$ for each maximal clique C in G. Since each maximal clique in $\tilde{G}$ is a subset of a maximal clique in G and $\tilde{\iota}$ is consistent with ι by assumption, it thus follows that $I_{\tilde{G}}^{(t)}(\mu) = \tilde{\iota}$. We thus find that

$$\Pi^{(t),\text{lci}}(\mu) = \left(\Pi^{(t),\text{lci}} \circ \Pi_G^{(t)}\right)(\mu) = \Pi_{\tilde{G}}^{(t)}(\mu) = \left(L_{\tilde{G}}^{(t)} \circ I_{\tilde{G}}^{(t)}\right)(\mu) = L_{\tilde{G}}^{(t)}(\tilde{\iota}) = \tilde{\mu},$$

where the first equality uses that $\mu \in \mathcal{M}_G^{(t)}$ (Corollary 11.12) and Proposition 17.6, the second uses Proposition 17.9, the third uses the definition of $\Pi_{\tilde{G}}^{(t)}$, the fourth uses the aforementioned fact that $I_{\tilde{G}}^{(t)}(\mu) = \tilde{\iota}$ under the stated assumption, and the last equality uses the definition of $\tilde{\mu}$.

The next lemma takes the conclusion of the previous lemma as an assumption and establishes that the pmfs enclosed in the red rectangle are *identical*.

Lemma 17.22 *If $\Pi^{(t),\text{lci}}(\mu) = \tilde{\mu}$, then we have that $\hat{\mu}' = \hat{\tilde{\mu}}'$.*

To see why this lemma is true, we note that

$$\hat{\mu}' = Q^{(t)}(\mu) = (Q^{(t)} \circ \Pi^{(t),\text{lci}})(\mu) = Q^{(t)}(\tilde{\mu}) = \hat{\tilde{\mu}}',$$

where the first equality uses the definition of $\hat{\mu}'$, the second uses the fact that $Q^{(t)} \circ \Pi^{(t),\text{lci}} = Q^{(t)}$ (Corollary 16.10), the third uses the assumption of the lemma, and the last uses the definition of $\hat{\tilde{\mu}}'$.

The next lemma takes the conclusion of the previous lemma as an assumption and establishes that $\tilde{\iota}'$ is consistent with ι'. This lemma immediately follows from the fact that $I_{\tilde{G}'}^{(t+1)}(\mu)$ and $I_{G'}^{(t+1)}(\mu)$ must be consistent for any $\mu \in \mathcal{P}_{Z,L}^{(t)}$, since any maximal clique in $\tilde{G}'$ is a clique in G'.

Lemma 17.23 *If $\hat{\mu}' = \hat{\tilde{\mu}}'$, then $\tilde{\iota}'$ is consistent with ι'.*

17.7 Proofs

Throughout these proofs, we fix time t, some size function vector $\boldsymbol{x}^{(t)}$, and some information structure graph G compatible with $\boldsymbol{x}^{(t)}$. This means that we have also fixed N_m, the set of counter vertices that are adjacent in G to the label vertex L_m, and the scope S_m of $x_m^{(t)}$. We use these sets of counter vertices in all subsequent proofs.

Proof of Lemma 17.3

Given some $\iota \in \mathcal{I}_G^{(t)}$, we know there is some $\mu \in \mathcal{P}_{Z,L}^{(t)}$ such that $\iota = I_G(\mu)$. We need to show that $L_G(\iota) \in \mathcal{P}_{Z,L}^{(t)}$. Since $\Pi_G = L_G \circ I_G$, it is sufficient to show that $\Pi_G(\mu) \in \mathcal{P}_{Z,L}^{(t)}$.

Fix some $(z, \ell) \in \Omega_{Z,L}$. Writing $\tilde{\mu} = \Pi_G(\mu)$, we need to show that for every m, whenever $x_m^{(t)}(z) = 0$,

$$\tilde{\mu}(z, \ell_m) = \tilde{\mu}(z) \times \theta_m^0(\ell_m). \tag{17.12}$$

We obtain that

$$\tilde{\mu}(z, \ell_m) = \tilde{\mu}(z_{N_m}, \ell_m) \times \tilde{\mu}(z | z_{N_m}, \ell_m) = \tilde{\mu}(z_{N_m}, \ell_m) \times \tilde{\mu}(z | z_{N_m}),$$

where the first equality uses the chain rule and the second equality follows from the facts that $\tilde{\mu}$ has the global Markov property with respect to G and that N_m separates the set of other counter vertices and $\{L_m\}$ (Lemma 12.2(b)). The product on the right-hand side should be interpreted as 0 if $\tilde{\mu}(z_{N_m}) = 0$. Evidently, we also have that

$$\tilde{\mu}(z) = \tilde{\mu}(z_{N_m}) \times \tilde{\mu}(z|z_{N_m}).$$

The condition $x_m^{(t)}(z) = 0$ can be ascertained from z_{N_m}, since the scope S_m of $x_m^{(t)}$ is a subset of N_m by (R2) in the definition of information structure graph (see Definition 12.1). Substituting the identities in the preceding two displays into (17.12) yields that it suffices to show that for every m, z_{N_m}, ℓ_m, whenever $x_m^{(t)}(z_{S_m}) = 0$,

$$\tilde{\mu}(z_{N_m}, \ell_m) = \tilde{\mu}(z_{N_m}) \times \theta_m^0(\ell_m). \tag{17.13}$$

Since $N_m \cup \{L_m\}$ is a clique in G and the marginal pmfs of the maximal cliques in G are equal under μ and $\tilde{\mu}$ (Corollary 11.15), (17.13) is equivalent to

$$\mu(z_{N_m}, \ell_m) = \mu(z_{N_m}) \times \theta_m^0(\ell_m). \tag{17.14}$$

Since $\mu \in \mathcal{P}_{Z,L}^{(t)}$, we have, for every m, z, and ℓ_m, whenever $x_m^{(t)}(z_{S_m}) = 0$, that

$$\mu(z, \ell_m) = \mu(z) \times \theta_m^0(\ell_m).$$

Taking an appropriate sum over z yields (17.14), as desired.

Proof of Proposition 17.9

Fix some $\mu \in \mathcal{P}_{Z,L}^{(t)}$ and set $\tilde{\mu} = \Pi_G^{(t)}(\mu)$ and $\hat{\mu} = \Pi_{\mathrm{lci}(G)}^{(t)}(\mu)$. We must show that $\Pi^{(t),\mathrm{lci}}(\tilde{\mu}) = \hat{\mu}$.

We start by showing that $\hat{\mu}(z) = \tilde{\mu}(z)$ for all $z \in \Omega_Z$. In view of Lemma 12.3, it is sufficient to show that

$$\Pi_{G_Z}(\mu(Z = \cdot)) = \Pi_{\mathrm{lci}(G)_Z}(\mu(Z = \cdot)),$$

where we recall that G_Z and $\mathrm{lci}(G)_Z$ denote the respective subgraphs of G and $\mathrm{lci}(G)$ induced by the counter vertices. These two subgraphs on counter vertices are identical, since no edge between any two counter vertices is removed when obtaining $\mathrm{lci}(G)$ from G. Evidently, this implies that $\Pi_{G_Z} = \Pi_{\mathrm{lci}(G)_Z}$ and therefore $\hat{\mu}(z) = \tilde{\mu}(z)$.

We next fix some $(z, \ell) \in \Omega_{Z,L}$ with $\hat{\mu}(z) > 0$, or equivalently with $\tilde{\mu}(z) > 0$. Since $\mathrm{lci}(G)$ is a label conditional independence graph and $\hat{\mu} \in \mathcal{M}_{\mathrm{lci}(G)}^{(t)} \subseteq \mathcal{M}_{\mathrm{lci}(G)}$, Proposition 12.6 shows that

$$\hat{\mu}(z, \ell) = \hat{\mu}(z) \times \prod_m \hat{\mu}(\ell_m | z_{N_m}),$$

where we use the fact that the sets N_m are identical in lci(G) and G. Note that the conditional pmf on the right-hand side is well-defined, since $\hat{\mu}(z_{N_m}) \geq \hat{\mu}(z) > 0$. Since $N_m \cup \{L_m\}$ is a (maximal) clique in lci(G), Corollary 11.15 implies that $\hat{\mu}(z_{N_m}, \ell_m) = \mu(z_{N_m}, \ell_m)$. With $\hat{\mu}(z) = \tilde{\mu}(z)$, we thus obtain that

$$\hat{\mu}(z, \ell) = \tilde{\mu}(z) \times \prod_m \mu(\ell_m | z_{N_m}). \tag{17.15}$$

We finish the proof by showing that $[\Pi^{(t),\text{lci}}(\tilde{\mu})](z, \ell)$ equals the right-hand side in this equation. Using $\tilde{\mu} \in \mathcal{P}_{Z,L}^{(t)}$ and Lemma 16.7, we find that

$$[\Pi^{(t),\text{lci}}(\tilde{\mu})](z, \ell) = \tilde{\mu}(z) \times \prod_m \tilde{\mu}(\ell_m | z).$$

Moreover, from $\tilde{\mu} \in \mathcal{M}_G$, we know that $\tilde{\mu}(\ell_m | z) = \tilde{\mu}(\ell_m | z_{N_m})$ by Proposition 12.4(b). We thus have that

$$[\Pi^{(t),\text{lci}}(\tilde{\mu})](z, \ell) = \tilde{\mu}(z) \times \prod_m \tilde{\mu}(\ell_m | z_{N_m}). \tag{17.16}$$

The set $N_m \cup \{L_m\}$ is a clique in G, so again with Corollary 11.15, we find that $\tilde{\mu}(z_{N_m}, \ell_m) = \mu(z_{N_m}, \ell_m)$. The right-hand side of (17.16) therefore equals the right-hand side of (17.15).

Proof of Proposition 17.14

Fix some $\hat{\nu} \in \mathcal{P}_{Z,H}^{(t)}$ with finite support, and set $\nu^* = \Pi_G^{(t),\text{mult}}(\hat{\nu}) \in \mathcal{M}_G^{(t),\text{mult}}$. Consider an arbitrary fixed $\nu \in \mathcal{M}_G^{(t),\text{mult}}$. We can assume, without loss of generality, that $D(\hat{\nu} \| \nu) < \infty$, which implies that the support of ν includes the support of $\hat{\nu}$. It is readily seen that this implies that the support of $\nu(Z = \cdot)$ includes the support of $\hat{\nu}(Z = \cdot)$.

Define $\mu^* = (\Pi_G^{(t)} \circ \mathcal{E}^{(t)})(\hat{\nu})$, and note that

$$\nu^* = \Pi_G^{(t),\text{mult}}(\hat{\nu}) = (\mathcal{L}^{(t)} \circ \Pi_G^{(t)} \circ \mathcal{E}^{(t)})(\hat{\nu}) = \mathcal{L}^{(t)}(\mu^*).$$

It thus follows from Proposition 12.4(c) that, for $(z, h) \in \Omega_{Z,H}$,

$$\nu^*(z, h) = \mu^*(z) \times \prod_m \Pr[\text{mult}(x_m^{(t)}(z_{S_m}), \mu^*(L_m = \cdot | z_{N_m})) = h_m]. \tag{17.17}$$

Similarly, we set $\mu = (\Pi_G^{(t)} \circ \mathcal{E}^{(t)})(\nu)$, and using the fact that

$$\mathcal{L}^{(t)}(\mu) = (\mathcal{L}^{(t)} \circ \Pi_G^{(t)} \circ \mathcal{E}^{(t)})(\nu) = \Pi_G^{(t),\text{mult}}(\nu) = \nu,$$

we find that

$$\nu(z, h) = \mu(z) \times \prod_m \Pr[\text{mult}(x_m^{(t)}(z_{S_m}), \mu(L_m = \cdot | z_{N_m})) = h_m]. \tag{17.18}$$

To establish the claim, we will show that $D(\hat{\nu} \| \nu) \geq D(\hat{\nu} \| \nu^*)$ and that this is an equality if and only if $\nu = \nu^*$.

In view of Lemma 11.20, (17.17), and (17.18), it suffices to show that

$$\sum_z \hat{\nu}(z) \times \log\left(\frac{\mu^*(z)}{\mu(z)}\right) \geq 0 \tag{17.19}$$

and, for every m,

$$\sum_{z_{N_m}, h_m} \hat{\nu}(z_{N_m}, h_m) \times \log\left(\frac{\Pr[\text{mult}(x_m^{(t)}(z_{S_m}), \mu^*(L_m = \cdot | z_{N_m})) = h_m]}{\Pr[\text{mult}(x_m^{(t)}(z_{S_m}), \mu(L_m = \cdot | z_{N_m})) = h_m]}\right) \geq 0, \tag{17.20}$$

with equalities if and only if $\nu = \nu^*$. As part of the argument, we will show that both left-hand sides are well-defined under the convention that the sums are to be taken over the supports of $\hat{\nu}(Z = \cdot)$ and $\hat{\nu}(Z_{N_m} = \cdot, H_m = \cdot)$, respectively.

We begin with (17.19), which we establish by proving a suitable lower bound on $D(\hat{\nu}(Z = \cdot) \| \mu(Z = \cdot))$. Since $\mu(Z = \cdot) \in \mathcal{M}_{G_Z}^{(t)}$, Proposition 11.18 shows that

$$D(\hat{\nu}(Z = \cdot) \| \mu(Z = \cdot)) \geq D(\hat{\nu}(Z = \cdot) \| \Pi_{G_Z}^{(t)}(\hat{\nu}(Z = \cdot))).$$

We next observe that

$$\Pi_{G_Z}^{(t)}(\hat{\nu}(Z = \cdot)) = \Pi_{G_Z}^{(t)}([\mathcal{E}^{(t)}(\hat{\nu})](Z = \cdot))$$

$$= [\Pi_G^{(t)}(\mathcal{E}^{(t)}(\hat{\nu}))](Z = \cdot)$$

$$= \mu^*(Z = \cdot), \tag{17.21}$$

where the first equality follows from (16.2) with $\nu = \hat{\nu}$ and the second equality follows upon applying Lemma 12.3 with $\mu = \mathcal{E}^{(t)}(\hat{\nu})$. The established inequality

$$D(\hat{\nu}(Z = \cdot) \| \mu(Z = \cdot)) \geq D(\hat{\nu}(Z = \cdot) \| \mu^*(Z = \cdot))$$

implies (17.19). This argument also shows that (17.19) is a strict inequality unless $\mu(\mathbf{Z} = \cdot) = \mu^*(\mathbf{Z} = \cdot)$.

We now turn to (17.20). Assume that $\hat{v}(z_{N_m}) > 0$. We first argue that

$$\sum_{h_m} \hat{v}(h_m | z_{N_m}) \times \log\left(\frac{\Pr[\text{mult}(x_m^{(t)}(z_{S_m}), \mu^*(L_m = \cdot | z_{N_m})) = h_m]}{\Pr[\text{mult}(x_m^{(t)}(z_{S_m}), \mu(L_m = \cdot | z_{N_m})) = h_m]}\right) \tag{17.22}$$

is well-defined. In view of Corollary 11.15 and (17.21), since N_m is a clique in G, we have that $\hat{v}(z_{N_m}) = \mu^*(z_{N_m})$, and so the conditional pmf in the numerator is well-defined. To see that the conditional pmf in the denominator is also well-defined, we note that $\hat{v}(z_{N_m}) > 0$ implies that there exists some z and h_m with $\hat{v}(z, h_m) > 0$, and so the assumption that the support of v contains the support of $\hat{v}$ made at the beginning of this proof in turn implies that $v(z_{N_m}) \geq v(z, h_m) > 0$.

Equation (17.22) can be expressed as

$$D(\hat{v}(H_m = \cdot | z_{N_m}) \parallel \text{mult}(x_m^{(t)}(z_{S_m}), \mu(L_m = \cdot | z_{N_m})))$$

$$- D(\hat{v}(H_m = \cdot | z_{N_m}) \parallel \text{mult}(x_m^{(t)}(z_{S_m}), \mu^*(L_m = \cdot | z_{N_m}))). \tag{17.23}$$

The first term on the right-hand side of (17.23) is at least as large as

$$D(\hat{v}(H_m = \cdot | z_{N_m}) \parallel \text{mult}(x_m^{(t)}(z_{S_m}), [\mathcal{E}^{(t)}(\hat{v})](L_m = \cdot | z_{N_m})))$$

by Lemma 16.15. It follows from Corollary 11.15 that the pmfs $\mu^* = (\Pi_G^{(t)} \circ \mathcal{E}^{(t)})(\hat{v})$ and $\mathcal{E}^{(t)}(\hat{v})$ agree on the maximal cliques in G. Since $N_m \cup \{L_m\}$ is a clique in G for each m, we obtain that

$$\mu^*(z_{N_m}, \ell_m) = [\mathcal{E}^{(t)}(\hat{v})](z_{N_m}, \ell_m), \tag{17.24}$$

and therefore (17.22) is nonnegative whenever $\hat{v}(z_{N_m}) > 0$. Since the left-hand side of (17.20) is a weighted average over $\hat{v}(z_{N_m})$ of nonnegative quantities, it is nonnegative, too.

We have now established that v^* is a solution to the minimization problem stated in the proposition, and it remains to show that it is the unique solution. Suppose v picked at the beginning of this proof is a different solution. The preceding arguments show that (17.19) and (17.20) imply that $\mu(z) = \mu^*(z)$ for all z and $\mu(\ell_m | z_{N_m}) = [\mathcal{E}^{(t)}(\hat{v})](\ell_m | z_{N_m})$ for all m, ℓ_m, and z_{N_m} with $\hat{v}(z_{N_m}) > 0$. In view of (17.24), the condition $\mu(\ell_m | z_{N_m}) = [\mathcal{E}^{(t)}(\hat{v})](\ell_m | z_{N_m})$ is equivalent to $\mu(\ell_m | z_{N_m}) = \mu^*(\ell_m | z_{N_m})$. Moreover, it follows from (17.24) and $[\mathcal{E}^{(t)}(\hat{v})](z) = \hat{v}(z)$ that $\hat{v}(z_{N_m}) > 0$ is equivalent to $\mu^*(z_{N_m}) > 0$. Therefore, μ must satisfy (1) $\mu(z) = \mu^*(z)$ for all z and (2) $\mu(\ell_m | z_{N_m}) = \mu^*(\ell_m | z_{N_m})$ for all m, ℓ_m, and z_{N_m} with $\mu^*(z_{N_m}) > 0$, and so we must have that $\mu(z_{N_m}, \ell_m) = \mu^*(z_{N_m}, \ell_m)$ for all z_{N_m} and ℓ_m. In view of (17.17) and (17.18), we conclude that $v = v^*$, which establishes the uniqueness, as claimed.

Chapter 18
Exactness Results

This chapter shows that, under various assumptions, the kernel iterates are equal to the lifted QPLEX iterates, i.e., QPLEX iterates after applying a multinomial lift map. We call such a result an *exactness result*, and we establish such results in three specific settings. In the first two settings, we consider QPLEX chains that encompass as special cases the multiserver queueing model from Chap. 2 in which (1) the service duration distribution is geometric and does not vary over time and (2) the number of servers is infinite. The third setting considers an *arbitrary* single activity QPLEX chain in which the size functions only take on values 0 or 1.

We close this chapter by showing that the marginal pmfs of the QPLEX iterates for the counters are unaffected by the choice of label interpretation (age or remaining activity duration) for single activity QPLEX chains with simple transition dynamics.

18.1 Conditions for Exactness

This section lays the foundation for the exactness results presented in this chapter. It formulates conditions under which the kernel iterates are the same as the lifted QPLEX iterates. The next three sections verify these conditions in the three settings described above.

The diagram below illustrates our first result. Fix time t and start from $\mu \in \mathcal{P}_{Z,L}^{(t)}$ in the lower left corner. Under the assumption that $\nu' \in \mathcal{P}_{Z,H}^{(t+1),\text{mult}}$, the following proposition states that we can obtain ν' in the upper right corner by following the path defined by the familiar black arrows or by following the path defined by the red arrows.

© The Author(s) 2025
A. B. Dieker, S. T. Hackman, *QPLEX: A Computational Modeling and Analysis Methodology for Stochastic Systems*, Springer Series in Operations Research and Financial Engineering, https://doi.org/10.1007/978-3-031-74870-7_18

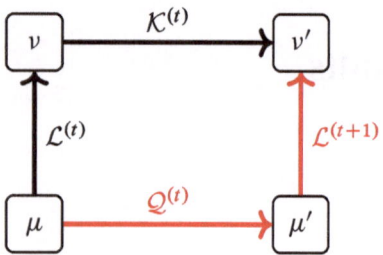

Proposition 18.1 *Let $\mu \in \mathcal{P}_{Z,L}^{(t)}$ be arbitrary, and set $\mu' = \mathcal{Q}^{(t)}(\mu)$, $\nu = \mathcal{L}^{(t)}(\mu)$, and $\nu' = \mathcal{K}^{(t)}(\nu)$. If $\nu' \in \mathcal{P}_{Z,H}^{(t+1),\mathrm{mult}}$, then we have that $\nu' = \mathcal{L}^{(t+1)}(\mu')$.*

To see why this proposition is true, note that $\nu' \in \mathcal{P}_{Z,H}^{(t+1),\mathrm{mult}}$ implies that $\nu' = \Pi^{(t+1),\mathrm{mult}}(\nu')$ as $\Pi^{(t+1),\mathrm{mult}}$ is a projection map onto $\mathcal{P}_{Z,H}^{(t+1),\mathrm{mult}}$ (Proposition 16.12). We thus find that

$$\nu' = \Pi^{(t+1),\mathrm{mult}}(\nu') = (\Pi^{(t+1),\mathrm{mult}} \circ \mathcal{K}^{(t)})(\nu) = (\Pi^{(t+1),\mathrm{mult}} \circ \mathcal{K}^{(t)} \circ \mathcal{L}^{(t)})(\mu).$$

By definition of $\Pi^{(t+1),\mathrm{mult}}$ and the QPLEX map $\mathcal{Q}^{(t)}$, the map on the right-hand side can be written as

$$\Pi^{(t+1),\mathrm{mult}} \circ \mathcal{K}^{(t)} \circ \mathcal{L}^{(t)} = \mathcal{L}^{(t+1)} \circ \mathcal{E}^{(t+1)} \circ \mathcal{K}^{(t)} \circ \mathcal{L}^{(t)} = \mathcal{L}^{(t+1)} \circ \mathcal{Q}^{(t)}.$$

We conclude that

$$\nu' = (\mathcal{L}^{(t+1)} \circ \mathcal{Q}^{(t)})(\mu) = \mathcal{L}^{(t+1)}(\mu'),$$

as claimed.

Given some $\nu_{\mathrm{ker}}^{(0)} \in \mathcal{P}_{Z,H}^{(0)}$ and $\mu_{Q}^{(0)} \in \mathcal{P}_{Z,L}^{(0)}$, recall that the iterative schemes for the kernel and the QPLEX iterates are given by $\nu_{\mathrm{ker}}^{(t+1)} = \mathcal{K}^{(t)}(\nu_{\mathrm{ker}}^{(t)})$ and $\mu_{Q}^{(t+1)} = \mathcal{Q}^{(t)}(\mu_{Q}^{(t)})$, respectively, for $t \geq 0$. The corollary below states that if the kernel iterates and QPLEX iterates are initially "linked," they remain linked under a condition on the kernel iterates.

Corollary 18.2 (Exactness of QPLEX Iterates) *Suppose that $\nu_{\mathrm{ker}}^{(t)} \in \mathcal{P}_{Z,H}^{(t),\mathrm{mult}}$ for all $t \geq 1$. If $\nu_{\mathrm{ker}}^{(0)} = \mathcal{L}^{(0)}(\mu_{Q}^{(0)})$, then we have that $\nu_{\mathrm{ker}}^{(t)} = \mathcal{L}^{(t)}(\mu_{Q}^{(t)})$ for all $t \geq 1$.*

We make three remarks about this corollary. First, the condition $\nu_{\mathrm{ker}}^{(0)} = \mathcal{L}^{(0)}(\mu_{Q}^{(0)})$ in this corollary is satisfied when initially no entities are undertaking any activities, i.e., $\{z : x^{(0)}(z) = 0\}$ is a singleton and $\mu_{Q}^{(0)}(z) = 1(x^{(0)}(z) = 0)$, in which case $\nu_{\mathrm{ker}}^{(0)}(z) = 1(x^{(0)}(z) = 0)$ as well.

Second, the conclusion of this corollary implies with (16.7) that $\nu_{\mathrm{ker}}^{(t)}(z) = \mu_{Q}^{(t)}(z)$ for all $t \geq 1$, i.e., the marginal pmf of the counters for a QPLEX iterate is the same as the marginal pmf of the counters for the corresponding kernel iterate. This can be viewed as the pmfs of the counters for the QPLEX iterates being exact.

Third, more generally, given functions $r^{(t)}$ on $\mathcal{P}_{Z,H}^{(t)}$, applying $r^{(t)}$ to the kernel iterate $v_{\text{ker}}^{(t)}$ yields the same result as applying $r^{(t)} \circ \mathcal{L}^{(t)}$ to the QPLEX iterate $\mu_Q^{(t)}$. For instance, $r^{(t)}(v_{\text{ker}}^{(t)})$ can represent the pmf of the virtual waiting time as defined in Appendix A.4. Thus, specialized to this example, the conditions of this corollary also imply that the virtual waiting time pmf is exact.

18.2 Geometric Activity Durations

In this section, we consider QPLEX chains with simple transition dynamics where the underlying model primitives are defined to be consistent, with each activity having a geometric activity duration that does not change over time. Throughout this section, we assume that we are given a success probability $\rho_m \in (0, 1)$ for each activity m, which is the parameter of the geometric activity duration.

The result we establish holds for both the remaining activity duration and age label interpretations. (Section 18.5 shows that the QPLEX iterates under the two interpretations are, in a sense, equivalent even if the activity duration is not geometric.) We use the remaining activity duration label interpretation, as it is somewhat simpler to work with.

Since the label interpretation is the remaining activity duration, we take the label set to be $\Omega_m = \{1, 2, \ldots\}$. For each m and $\ell_m \geq 1$, we define

$$g_m(\ell_m) = (1 - \rho_m)^{\ell_m - 1} \times \rho_m.$$

The completion probability function, relabel, and join pmf primitives are given by

$$\gamma_m^{(t)}(z, \ell_m) = 1(\ell_m = 1)$$

$$\pi_m^{(t),\text{rel}}(\ell_m' | z, \ell_m, z') = 1(\ell_m' = \ell_m - 1)$$

$$\pi_m^{(t),\text{join}}(\ell_m' | z, z') = g_m(\ell_m').$$

The specification of the three model primitives is standard and consistent with the remaining activity duration label interpretation; see Sect. 4.1. The routing pmf primitive $\pi^{(t)}(z'|z, d)$ is arbitrary.

One model that fits the above setup is the multiserver queueing model from Chap. 2. Proposition 18.3 below implies that the lifted QPLEX iterates are equal to the kernel iterates if the service duration distribution is geometric and there are initially no customers in the system. (Of course, it is not necessary to consider a QPLEX chain, as the process of the number of customers in the system constitutes a Markov chain. More generally, within the context of this section, it is possible to consider the dynamics of the counters without the histograms.)

Let $\mathcal{P}_{Z,H}^{(t),\text{geom}}$ denote the set of pmfs $\nu \in \mathcal{P}_{Z,H}^{(t),\text{mult}}$ that can be written as

$$\nu(z, h) = \nu(z) \times \prod_m \Pr\left[\text{mult}(x^{(t)}(z), g_m) = h_m\right].$$

The following proposition is our exactness result for this setting.

Proposition 18.3 *Consider the above setup. If the initial iterates satisfy* $\nu_{\text{ker}}^{(0)} = \mathcal{L}^{(0)}(\mu_Q^{(0)})$ *and* $\nu_{\text{ker}}^{(0)} \in \mathcal{P}_{Z,H}^{(0),\text{geom}}$, *then, for all* $t \geq 1$, *we have that* $\nu_{\text{ker}}^{(t)} = \mathcal{L}^{(t)}(\mu_Q^{(t)})$ *and that* $\nu_{\text{ker}}^{(t)} \in \mathcal{P}_{Z,H}^{(t),\text{geom}}$.

This proposition is a direct consequence of the following lemma.

Lemma 18.4 *Consider the above setup. We have, for all* $t \geq 0$, *that* $\mathcal{K}^{(t)}(\mathcal{P}_{Z,H}^{(t),\text{geom}}) \subseteq \mathcal{P}_{Z,H}^{(t+1),\text{geom}}$.

Indeed, suppose the initial iterates satisfy the two conditions given in the proposition. Since $\nu_{\text{ker}}^{(0)} \in \mathcal{P}_{Z,H}^{(0),\text{geom}}$, application of Lemma 18.4 yields that $\nu_{\text{ker}}^{(t)} \in \mathcal{P}_{Z,H}^{(t),\text{geom}}$ for all $t \geq 1$ by definition of the kernel iterates. The claim then follows from $\mathcal{P}_{Z,H}^{(t),\text{geom}} \subseteq \mathcal{P}_{Z,H}^{(t),\text{mult}}$ and Corollary 18.2.

The remainder of this section is devoted to proving Lemma 18.4. Fix some $\nu \in \mathcal{P}_{Z,H}^{(t),\text{geom}}$. We must show that $\mathcal{K}^{(t)}(\nu) \in \mathcal{P}_{Z,H}^{(t+1),\text{geom}}$.

First some notation. We define the histogram $\overleftarrow{h}_m$ in terms of the histogram h_m by $\overleftarrow{h}_m(\ell_m) = h_m(\ell_m + 1)$ for $\ell_m \geq 1$. In this setting, $\overleftarrow{h}_m$ is the histogram of remaining activity durations at time $t + 1$ for those entities represented in h_m that do not complete their activity during this period. We also let $h(1)$ denote the vector with m-th component equal to $h_m(1)$. Keep in mind that in this setting, $h_m(1)$ is the number of entities that complete their activity during this period.

Our starting point is the distributional program for the transition probabilities from Sect. 4.2 reproduced below.

```
1: function TRANSITIONPROBABILITY⁽ᵗ⁾(z, h)
2:     hˢᵗᵃʸ ~ STAYHISTOGRAMS⁽ᵗ⁾(z, h)
3:     for m = 1, ..., M do
4:         dₘ ← |hₘ| − |hₘˢᵗᵃʸ|
5:     z′ ~ π⁽ᵗ⁾(Z′ = ·|z, d)
6:     hʳᵉˡ ~ RELABELHISTOGRAMS⁽ᵗ⁾(z, hˢᵗᵃʸ, z′)
7:     hʲᵒⁱⁿ ~ JOINHISTOGRAMS⁽ᵗ⁾(z, d, z′)
8:     h′ ← hʳᵉˡ + hʲᵒⁱⁿ
9:     return (z′, h′)
```

In this setting, d_m now equals $h_m(1)$ and $h_m^{\text{stay}}(\ell_m) = h_m(\ell_m) \times 1(\ell_m > 1)$, so there is no need to sample h_m^{stay} and we can immediately substitute $d_m = h_m(1)$ and $h_m^{\text{rel}}(\ell_m) = \overleftarrow{h}_m(\ell_m)$. This results in the program below.

1: **function** TRANSITIONPROBABILITY$^{(t)}(z, h)$
2:    $z' \sim \pi^{(t)}(Z' = \cdot|z, D = h(1))$
3:    **for** $m = 1, \ldots, M$ **do**
4:        $h_m^{\text{join}} \sim \text{mult}\left(x_m^{(t+1)}(z') - (x_m^{(t)}(z) - d_m), g_m\right)$
5:        $h_m' \leftarrow \bar{h}_m + h_m^{\text{join}}$
6:    **return** (z', h')

Next, suppose that each h_m is independent and multinomial with parameters $x_m^{(t)}(z)$ and g_m. Note that the h_m appear only in Lines 2 and 5 of the above program. Using a standard property of multinomial distributions, also used in Sect. 4.3, each d_m is binomial with parameters $x_m^{(t)}(z)$ and $g_m(L_m = 1) = \rho_m$, and each $\bar{h}_m$ is conditionally multinomial given d_m with parameters $x_m^{(t)}(z) - d_m$ and pmf $\tilde{g}_m$ defined via $\tilde{g}_m(\ell_m) = g_m(L_m = \ell_m + 1|L_m > 1)$. By the memoryless property of g_m, the pmfs $\tilde{g}_m$ are g_m equal. The histogram h_m^{join} is thus multinomial with the same pmf parameter as $\bar{h}_m$, and we can sample h_m' directly from a multinomial distribution with parameters $x_m^{(t+1)}(z')$ and g_m. This leads to the following distributional program for the kernel map.

1: **function** KERNELMAP$^{(t)}(\nu)$
2:    $z \sim \nu(Z = \cdot)$
3:    **for** $m = 1, \ldots, M$ **do**
4:        $d_m \sim \text{bin}\left(x_m^{(t)}(z), \rho_m\right)$
5:    $z' \sim \pi^{(t)}(Z' = \cdot|z, d)$
6:    **for** $m = 1, \ldots, M$ **do**
7:        $h_m' \sim \text{mult}\left(x_m^{(t+1)}(z'), g_m\right)$
8:    **return** (z', h')

Lines 6 and 7 in this program immediately show that $\mathcal{K}^{(t)}(\nu) \in \mathcal{P}_{Z,H}^{(t+1),\text{geom}}$, as required.

18.3 Poisson Propagation

This section provides an exactness result for QPLEX chains with simple transition dynamics under the following assumptions:

- There is a single counter and a single label variable.
- The size function is given by, for $t \geq 0$,

$$x^{(t)}(z) = z.$$

- The completion probability function, relabeling pmfs, and join pmfs are constant in the (current and next) counters, and we denote them by $\gamma^{(t)}(\ell)$, $\pi^{(t),\text{rel}}(\ell'|\ell)$, and $\pi^{(t),\text{join}}(\ell')$, respectively. We denote the label set by Ω_L.
- The routing pmf is of the form, for $t \geq 0$,

$$\pi^{(t)}(z'|z, d) = \Pr[z - d + \text{pois}(\Gamma^{(t)}) = z'] \tag{18.1}$$

for some parameter $\Gamma^{(t)} \geq 0$. Thus, given (z, d), the number of entities $Z' - (z - d)$ that start their activity between times t and $t + 1$ is a Poisson random variable with parameter $\Gamma^{(t)}$.

One model that fits the above setup is the multiserver queueing model from Chap. 2 with $n = \infty$ servers. Note that the routing pmf primitive is given by $\pi^{(t)}(z'|z, d) = \alpha^{(t)}(z' - z + d)$, where $\alpha^{(t)}$ is a Poisson pmf, and therefore it satisfies (18.1).

First some notation. Let $\mathcal{P}^{\text{pois}}$ denote the set of pmfs $\tilde{\nu}$ on histograms on Ω_L satisfying

$$\tilde{\nu}(h) = \prod_{\ell \in \Omega_L} \Pr[\text{pois}(\kappa(\ell)) = h(\ell)]$$

for some nonnegative function κ on Ω_L with $\sum_\ell \kappa(\ell) < \infty$. We abbreviate the right-hand side by $\Pr[\text{pois}(\kappa) = h]$. A pmf in the set $\mathcal{P}^{\text{pois}}$ is characterized by such a function κ. Let the set $\mathcal{P}^{\text{pois}}_{Z,H}$ consist of pmfs ν of (Z, H) satisfying

$$\nu(z, h) = 1(z = |h|) \times \Pr[\text{pois}(\kappa) = h]$$

for some nonnegative function κ on Ω_L.

The following proposition is our exactness result for this setting.

Proposition 18.5 *Consider the above setup. If the initial iterates satisfy $\nu^{(0)}_{\text{ker}} = \mathcal{L}^{(0)}(\mu^{(0)}_Q)$ and $\nu^{(0)}_{\text{ker}} \in \mathcal{P}^{\text{pois}}_{Z,H}$, then, for all $t \geq 1$, we have that $\nu^{(t)}_{\text{ker}} = \mathcal{L}^{(t)}(\mu^{(t)}_Q)$ and that $\nu^{(t)}_{\text{ker}} \in \mathcal{P}^{\text{pois}}_{Z,H}$.*

This proposition is a direct consequence of the following lemma.

Lemma 18.6 *Consider the above setup. We have, for all $t \geq 0$, that $\mathcal{K}^{(t)}(\mathcal{P}^{\text{pois}}_{Z,H}) \subseteq \mathcal{P}^{\text{pois}}_{Z,H}$ and $\mathcal{P}^{\text{pois}}_{Z,H} \subseteq \mathcal{P}^{(t),\text{mult}}_{Z,H}$.*

Indeed, suppose the initial iterates satisfy the two conditions given in the proposition. Since $\nu^{(0)}_{\text{ker}} \in \mathcal{P}^{\text{pois}}_{Z,H}$, application of Lemma 18.6 yields that $\nu^{(t)}_{\text{ker}} \in \mathcal{P}^{\text{pois}}_{Z,H}$ for all $t \geq 1$ by definition of the kernel iterates. The claim then follows from $\mathcal{P}^{\text{pois}}_{Z,H} \subseteq \mathcal{P}^{(t),\text{mult}}_{Z,H}$ and Corollary 18.2.

The remainder of this section is devoted to proving Lemma 18.6. We begin by establishing that $\mathcal{P}^{\text{pois}}_{Z,H} \subseteq \mathcal{P}^{(t),\text{mult}}_{Z,H}$.

Consider some $\nu \in \mathcal{P}_{Z,H}^{\text{pois}}$ with underlying function κ on Ω_L. A well-known property of independent Poisson random variables (or direct verification) shows that

$$\nu(z, h) = \Pr[\text{pois}(|\kappa|) = z] \times \Pr\left[\text{mult}\left(z, \frac{\kappa}{|\kappa|}\right) = h\right],$$

where we write $|\kappa| = \sum_\ell \kappa(\ell)$. The right-hand side is $\mathcal{L}^{(t)}(\mu)$ with

$$\mu(z, \ell) = \Pr[\text{pois}(|\kappa|) = z] \times \frac{\kappa(\ell)}{|\kappa|}.$$

We have thus established that $\mathcal{P}_{Z,H}^{\text{pois}}$ is a subset of $\mathcal{P}_{Z,H}^{(t),\text{mult}}$ for all t.

It remains to establish that $\mathcal{K}^{(t)}(\mathcal{P}_{Z,H}^{\text{pois}}) \subseteq \mathcal{P}_{Z,H}^{\text{pois}}$. We use an argument reminiscent of multinomial propagation from Sects. 4.3 and 6.4, but now Poisson random variables propagate instead of multinomial random variables.

The histogram h^{stay} is constructed as in the following pseudocode fragment:

$h \sim \text{pois}(\kappa)$
for $\ell \in \Omega_L$ **do**
  $h^{\text{stay}}(\ell) \sim \text{bin}\left(h(\ell), 1 - \gamma^{(t)}(\ell)\right)$

The histogram h is no longer needed once h^{stay} is determined. Writing

$$\kappa^{(t),\text{stay}}(\ell) = \kappa(\ell) \times (1 - \gamma^{(t)}(\ell)),$$

by the well-known Poisson thinning property, this pseudocode fragment can be replaced with the following single sampling statement:

$h^{\text{stay}} \sim \text{pois}(\kappa^{(t),\text{stay}})$

Next, consider the relabeling step, specifically the following pseudocode fragment:

$h^{\text{stay}} \sim \text{pois}(\kappa^{(t),\text{stay}})$
$h^{\text{rel}} \leftarrow 0$
for $\ell \in \Omega_L$ **do**
  $h^{\text{rel}} +\sim \text{mult}\left(h^{\text{stay}}(\ell), \pi^{(t),\text{rel}}(L' = \cdot|\ell)\right)$

The histogram h^{stay} is no longer needed once h^{rel} is determined. Writing

$$\kappa^{(t),\text{rel}}(\ell') = \sum_\ell \kappa^{(t),\text{stay}}(\ell) \times \pi^{(t),\text{rel}}(\ell'|\ell),$$

by the well-known Poisson splitting property, this pseudocode fragment can similarly be replaced with the following single sampling statement:

$h^{\text{rel}} \sim \text{pois}(\kappa^{(t),\text{rel}})$

We next consider the aggregation step, specifically the following pseudocode fragment:

$$h^{\text{rel}} \sim \text{pois}(\kappa^{(t),\text{rel}})$$
$$z^{\text{join}} \sim \text{pois}(\Gamma^{(t)})$$
$$h^{\text{join}} \sim \text{mult}\left(z^{\text{join}}, \pi^{(t),\text{join}}(L' = \cdot)\right)$$
$$h' \leftarrow h^{\text{rel}} + h^{\text{join}}$$

The sum of independent Poisson random variables is another Poisson random variable. Using the Poisson splitting property again, this pseudocode fragment can be replaced with the following single sampling statement:

$$h' \sim \text{pois}(\kappa^{(t),\text{rel}} + \Gamma^{(t)} \times \pi^{(t),\text{join}})$$

We must have that $z' = |h'|$ in view of the size function, so the pmf of (z', h') lies in $\mathcal{P}_{Z,H}^{\text{pois}}$, as claimed.

18.4   Indicator Size Functions

This section provides an exactness result for *arbitrary* QPLEX chains with a single activity if the size functions $x^{(t)}(z)$ only take on the value 0 or 1. In particular, we do not assume simple transition dynamics. Similar results do not hold when there is more than one activity; the QPLEX iterates effectively impose additional independence between the label variables given the counter variables.

One model that fits the above setup is the multiserver queueing model from Chap. 2 with only one server, i.e., $n = 1$. Since $x^{(t)}(z) = \min(z, 1)$, the size function only takes the values 0 and 1 in that case.

The following proposition is our exactness result for this setting.

Proposition 18.7 *Consider the above setup. If the initial iterates satisfy $\nu_{\text{ker}}^{(0)} = \mathcal{L}^{(0)}(\mu_Q^{(0)})$, then, for all $t \geq 1$, we have that $\nu_{\text{ker}}^{(t)} = \mathcal{L}^{(t)}(\mu_Q^{(t)})$.*

This proposition is a direct consequence of the following lemma in view of Corollary 18.2.

Lemma 18.8 *Consider the above setup. We have, for $t \geq 0$, that $\mathcal{P}_{Z,H}^{(t),\text{mult}} = \mathcal{P}_{Z,H}^{(t)}$.*

The remainder of this section is devoted to proving this lemma. The inclusion $\mathcal{P}_{Z,H}^{(t),\text{mult}} \subseteq \mathcal{P}_{Z,H}^{(t)}$ is trivial, and so it remains to establish the reverse inclusion.

To this end, for $\ell \in \Omega_L$, define the histogram δ_ℓ via

$$\delta_\ell(\tilde{\ell}) = \begin{cases} 1 & \text{if } \tilde{\ell} = \ell \\ 0 & \text{otherwise.} \end{cases}$$

Fix some $\nu \in \mathcal{P}_{Z,H}^{(t)}$. Since the size function $x^{(t)}$ takes the values 0 or 1, we may write

$$v(z, h) = \begin{cases} v(z, \delta_\ell) & \text{if } x^{(t)}(z) = 1 \text{ and } h = \delta_\ell \\ v(z) \times 1(h = 0) & \text{if } x^{(t)}(z) = 0 \\ 0 & \text{otherwise.} \end{cases} \qquad (18.2)$$

Let $\mu = \mathcal{E}^{(t)}(v)$. Recall from Definition 16.1 of the compression map $\mathcal{E}^{(t)}$ that, for each $(z, \ell) \in \Omega_{Z,L}$,

$$\mu(z, \ell) = \sum_h \frac{h(\ell)}{|h|} \times v(z, h),$$

where the ratio should be understood to mean $\theta^0(\ell)$ if $h = 0$. In view of (18.2), it follows that

$$\mu(z, \ell) = \begin{cases} v(z, \delta_\ell) & \text{if } x^{(t)}(z) = 1 \\ v(z) \times \theta^0(\ell) & \text{if } x^{(t)}(z) = 0 \end{cases}$$

and therefore that $\mu(z) = v(z)$ and, if $\mu(z) > 0$, that

$$\mu(\ell|z) = \begin{cases} v(\delta_\ell|z) & x^{(t)}(z) = 1 \\ \theta^0(\ell) & \text{otherwise.} \end{cases}$$

As a result, the right-hand side of (18.2) can be written as

$$\mu(z) \times \Pr[\text{mult}(x^{(t)}(z), \mu(L = \cdot|z)) = h],$$

which is $[\mathcal{L}^{(t)}(\mu)](z, h)$. We have shown that $v \in \mathcal{P}_{Z,H}^{(t),\text{mult}}$, as claimed.

18.5   Age Versus Remaining Activity Duration

In this section, we consider single activity QPLEX chains with simple transition dynamics where the underlying model primitives are defined to be consistent with an activity duration that does not change over time. We consider two label interpretations: one based on age and the other based on remaining activity duration. We show that the marginal pmfs of the QPLEX iterates for the counters are unaffected by the choice of label interpretation.

Setup We assume that the activity duration of each entity does not depend on time and is given by the pmf $g(\ell)$ supported on $\{1, 2, \ldots\}$. For convenience, we assume that its support is infinite, which avoids conditional probabilities being undefined. It is straightforward to modify the arguments below when this assumption does not hold. We also let γ denote the corresponding hazard function, defined, for $k \geq 0$, via

$$\gamma(k) = g(L = k + 1 | L > k). \tag{18.3}$$

If the labels are interpreted as ages, then the completion probability function, relabeling pmf, and join pmf are given by, for $\ell \geq 0$,

$$\gamma^{(t)}(z, \ell) = \gamma(\ell)$$

$$\pi^{(t),\mathrm{rel}}(\ell' | z, \ell, z') = 1(\ell' = \ell + 1)$$

$$\pi^{(t),\mathrm{join}}(\ell' | z, z') = 1(\ell' = 0).$$

If the labels are interpreted as remaining activity durations, then the completion probability function, relabeling pmf, and join pmf are given by, for $\ell \geq 1$,

$$\gamma^{(t)}(z, \ell) = 1(\ell = 1)$$

$$\pi^{(t),\mathrm{rel}}(\ell' | z, \ell, z') = 1(\ell' = \ell - 1)$$

$$\pi^{(t),\mathrm{join}}(\ell' | z, z') = g(\ell').$$

The routing pmf primitives $\pi^{(t)}(z' | z, d)$ are arbitrary.

Notation In the presentation to follow, it is necessary to distinguish the labels that correspond to the age and remaining activity duration interpretations as well as to distinguish various sets and functions. Fix time t.

- Instead of using the symbol ℓ (L) to denote a label, the symbol a (A) denotes an entity's age, and the symbol r (R) denotes an entity's remaining activity duration. Note that $a \geq 0$ and $r \geq 1$.
- We write $\mathcal{P}_{Z,A}^{(t)}$ and $\mathcal{P}_{Z,R}^{(t)}$ instead of $\mathcal{P}_{Z,L}^{(t)}$ depending on the label interpretation. The symbols θ_A^0 and θ_R^0 denote, respectively, the "dummy" pmfs θ^0 for the two different settings. Throughout, we work under the assumption that, for $r \geq 1$,

$$\theta_R^0(r) = \sum_{a \geq 0} \theta_A^0(a) \times g(L = a + r | L > a). \tag{18.4}$$

- $\mathcal{Q}_A^{(t)}$ and $\mathcal{Q}_R^{(t)}$ denote the QPLEX maps for the age and remaining activity duration interpretations for the time period between t and $t + 1$, respectively.
- $\mu_A^{(t)}$ and $\mu_R^{(t)}$ denote the QPLEX iterates obtained using the age and remaining activity duration interpretations, respectively. Note that we no longer use the subscript "Q."

Let $\Psi^{(t)} : \mathcal{P}_{Z,A}^{(t)} \to \mathcal{P}_{Z,R}^{(t)}$ denote the map that converts a pmf of (Z, A) based on age to a pmf of (Z, R) based on remaining activity duration. It is defined, for

$\mu_A \in \mathcal{P}_{Z,A}^{(t)}$, via

$$[\Psi^{(t)}(\mu_A)](z,r) = \sum_{a \geq 0} \mu_A(z,a) \times g(L = a + r | L > a). \qquad (18.5)$$

This map depends on t only through its domain and codomain, which is not visible in (18.5).

We now argue that this map is well-defined. Fix some $\mu_A \in \mathcal{P}_{Z,A}^{(t)}$. We must show that $\Psi^{(t)}(\mu_A) \in \mathcal{P}_{Z,R}^{(t)}$. To this end, fix some z with $x^{(t)}(z) = 0$, so that $\mu_A(z,a) = \mu_A(z) \times \theta_A^0(a)$. From (18.4), we see that (18.5) becomes $[\Psi^{(t)}(\mu_A)](z,r) = \mu_A(z) \times \theta_R^0(r)$. Upon summing over r, we thus find that

$$[\Psi^{(t)}(\mu_A)](z) = \mu_A(z) \qquad (18.6)$$

and therefore $[\Psi^{(t)}(\mu_A)](z,r) = [\Psi^{(t)}(\mu_A)](z,r) \times \theta_R^0(r)$, as required.

Main Result The diagram below illustrates our main result. Fix time t and start from some $\mu_A \in \mathcal{P}_{Z,A}^{(t)}$ in the lower left corner. Under the above setup, the following proposition states that we can obtain μ_R' in the upper right corner by following the path defined by the black arrows or by following the path defined by the red arrows.

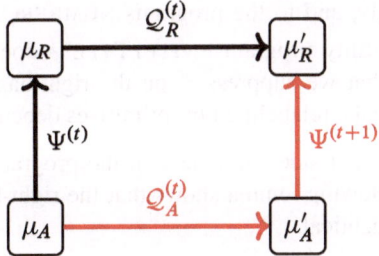

Proposition 18.9 *Let $\mu_A \in \mathcal{P}_{Z,A}^{(t)}$ be arbitrary, and set $\mu_A' = \mathcal{Q}_A^{(t)}(\mu_A)$, $\mu_R = \Psi^{(t)}(\mu_A)$, and $\mu_R' = \mathcal{Q}_R^{(t)}(\mu_R)$. Then, we have that $\mu_R' = \Psi^{(t+1)}(\mu_A')$.*

The proposition implies that if the initial pmfs $\mu_A^{(0)}$ and $\mu_R^{(0)}$ are appropriately aligned, then all iterates $\mu_A^{(t)}$ and $\mu_R^{(t)}$ are aligned as well.

Corollary 18.10 (Equivalence Between Label Interpretations) *If $\mu_R^{(0)} = \Psi^{(0)}(\mu_A^{(0)})$, then we have that $\mu_R^{(t)} = \Psi^{(t)}(\mu_A^{(t)})$ for all $t \geq 1$.*

The condition $\mu_R^{(0)} = \Psi^{(0)}(\mu_A^{(0)})$ is immediately satisfied when no entities initially undertake the activity. The conclusion $\mu_R^{(t)} = \Psi^{(t)}(\mu_A^{(t)})$ implies that $\mu_R^{(t)}(z) = \mu_A^{(t)}(z)$, so the marginal pmfs of counters generated over time are identical with the two label interpretations.

Proof Strategy We establish the proposition by comparing the two distributional programs shown below for $\mathcal{Q}_A^{(t)}$ and $\mathcal{Q}_R^{(t)}$ line by line under the assumptions that $\mu_A \in \mathcal{P}_{Z,A}^{(t)}$ and $\mu_R = \Psi^{(t)}(\mu_A)$. Note that this implies that $\mu_R \in \mathcal{P}_{Z,R}^{(t)}$.

1: **function** QPLEXMAPA$^{(t)}(\mu_A)$
2:     $z \sim \mu_A(\mathbf{Z} = \cdot)$
3:     $d \sim$ NUMBEROFCOMPLETIONSA$_{\mu_A}^{(t)}(z)$
4:     $z' \sim \pi^{(t)}(\mathbf{Z}' = \cdot | z, d)$
5:     $k' \sim$ ENTITYTYPE$^{(t)}(z, d, z')$
6:     $a' \sim$ NEXTLABELA$_{\mu_A}^{(t)}(z, k')$
7:     **return** (z', a')

1: **function** QPLEXMAPR$^{(t)}(\mu_R)$
2:     $z \sim \mu_R(\mathbf{Z} = \cdot)$
3:     $d \sim$ NUMBEROFCOMPLETIONSR$_{\mu_R}^{(t)}(z)$
4:     $z' \sim \pi^{(t)}(\mathbf{Z}' = \cdot | z, d)$
5:     $k' \sim$ ENTITYTYPE$^{(t)}(z, d, z')$
6:     $r' \sim$ NEXTLABELR$_{\mu_R}^{(t)}(z, k')$
7:     **return** (z', r')

To stress the differences between these two distributional programs pertaining to age and remaining activity duration, we have appended the symbol "A" or "R" to their names, respectively, and to the programs NUMBEROFCOMPLETIONS$_\mu^{(t)}$ and NEXTLABEL$_\mu^{(t)}$. (The utility program ENTITYTYPE$^{(t)}$ does not rely on the label interpretations.) Note that we suppress z' on the right-hand side in Line 6 since neither the join pmf nor the relabeling pmf primitives depend on it.

Derivation The right-hand sides in Line 2 in the programs are then identical in view of (18.6). The following lemma shows that the right-hand sides in Line 3 in the programs are also identical.

Lemma 18.11 *The distributional programs* NUMBEROFCOMPLETIONSA$_{\mu_A}^{(t)}$ *and* NUMBEROFCOMPLETIONSR$_{\mu_R}^{(t)}$ *are equivalent.*

Proof We begin by showing that the second parameter in the binomial sampling instruction is the same, i.e., that

$$\sum_a \mu_A(a|z) \times \gamma(a) = \mu_R(R = 1|z).$$

The definition of $\Psi^{(t)}$ implies that

$$\mu_R(z, R = 1) = [\Psi^{(t)}(\mu_A)](z, R = 1) = \sum_a \mu_A(z, a) \times g(L = a + 1 | L > a).$$

The right-hand side equals $\sum_a \mu_A(z, a) \times \gamma(a)$ by (18.3). The claim now follows upon combining this with the previously established fact that $\mu_R(z) = \mu_A(z)$ for all z. □

The right-hand sides in Lines 4 and 5 are identical in the two programs, so there is nothing to show. Having shown that Lines 2–5 jointly sample z, d, z', and k' from the same distribution in the two programs, to establish $\mu'_R = \Psi^{(t+1)}(\mu'_A)$, it remains to show that the pseudocode fragment

6:    $r' \sim \text{NEXTLABELR}_{\mu_R}^{(t)}(z, k')$
7:    **return** (z', r')

in the program for the remaining activity duration interpretation may be replaced with

$a' \sim \text{NEXTLABELA}_{\mu_A}^{(t)}(z, k')$
$r' \sim g(L = a' + \cdot | L > a')$
return (z', r')

to which we now turn.

The first line in this pseudocode fragment comes from the distributional program QPLEXMAPA$^{(t)}$ for $Q_A^{(t)}$, while the second line comes from the definition of $\Psi^{(t+1)}$ in (18.5). The equivalence of these pseudocode fragments follows from the lemma below.

Lemma 18.12 *Fix some z with $\mu_A(z) > 0$ and some k', and write θ'_A and θ'_R for the outputs of* $\text{NEXTLABELA}_{\mu_A}^{(t)}(z, k')$ *and* $\text{NEXTLABELR}_{\mu_R}^{(t)}(z, k')$, *respectively. We then have, for $r' \geq 1$, that*

$$\theta'_R(r') = \sum_{a' \geq 0} \theta'_A(a') \times g(L = a' + r' | L > a'). \qquad (18.7)$$

The remainder of this section is devoted to the proof of this lemma. There are three cases to consider, namely, $k' = \text{n/a}$, $k' = \text{new}$, and $k' = \text{old}$. We examine each case in turn.

Case $k' = \text{n/a}$ In this case, (18.7) is an immediate consequence of (18.4).

Case $k' = \text{new}$ In this case, $\text{NEXTLABELA}_{\mu_A}^{(t)}(z, k')$ is a pmf with mass 1 at 0 and $\text{NEXTLABELR}_{\mu_R}^{(t)}(z, k')$ equals g. Therefore (18.7) reads as $g(L = r') = g(L = r' | L > 0)$, which is evidently correct.

Case $k' = \text{old}$ From (4.4), we have that

$$\theta'_A(a') = \begin{cases} 0 & a' = 0 \\ \frac{\mu_A(A=a'-1|z) \times (1-\gamma(a'-1))}{\sum_{a \geq 0} \mu_A(a|z) \times (1-\gamma(a))} & a' > 0 \end{cases}$$

and, for $r' \geq 1$, that

$$\theta'_R(r') = \frac{\mu_R(R = r' + 1|z)}{\sum_{r>1} \mu_R(r|z)}.$$

We now argue that θ'_A and θ'_R satisfy (18.7). It is convenient to reason up to multiplicative proportionality constants depending on z, which can be done since z is fixed. The right-hand side of (18.7) is proportional to

$$\sum_{a'>0} \mu_A(z, A = a' - 1) \times (1 - \gamma(a' - 1)) \times g(L = a' + r'|L > a')$$

$$= \sum_{a \geq 0} \mu_A(z, a) \times (1 - \gamma(a)) \times g(L = a + r' + 1|L > a + 1). \qquad (18.8)$$

In view of

$$1 - \gamma(a) = g(L > a + 1|L > a),$$

the right-hand side of (18.8) can be written as

$$\sum_{a \geq 0} \mu_A(z, a) \times g(L > a + 1|L > a) \times g(L = a + r' + 1|L > a + 1)$$

$$= \sum_{a \geq 0} \mu_A(z, a) \times g(L = a + r' + 1|L > a),$$

which is $\mu_R(z, R = r' + 1)$ in view of $\mu_R = \Psi^{(t)}(\mu_A)$. This is proportional to $\theta'_R(r')$, as required.

Appendix A
Numerical Experiments

This chapter introduces a testbed of 180 experiments for the multiserver queueing model from Chap. 2, where the number of arrivals during each period is Poisson distributed. We report on the approximation quality of the QPLEX iterates with respect to two performance metrics over time: the number of customers in the system and the virtual waiting time. The experiments are designed to put the QPLEX methodology through a rigorous test. For each experiment, we compare the approximate distributions generated with the QPLEX iterates with estimated distributions generated by simulation with one million replications. Despite the simulation noise, we view the latter as the "exact." This appendix contains the details of the experiments summarized in Sect. 2.8.

A.1 Experimental Design

Each experiment starts with an empty system. The time horizon is $T = 10$ hours. The 180 experiments in the testbed consist of all possible combinations of six arrival rate functions, five service duration pmfs, and six values for the number of servers n (1, 2, 4, 8, 16, 32). The six arrival rate functions are obtained by scaling three arrival rate "scale" functions (time average set to 1) by two average utilization levels u_{avg} (0.8 and 0.95). We describe the choices for the arrival rate functions and the service duration distributions using continuous time and then describe in the next section how we obtain discrete-time analogs.

Service Duration Distributions We measure the service duration in hours and fix the mean service duration to 0.25 hours. In addition to the constant service duration distribution (labeled "CON"), we use four service-time distributions in the experiments:

- Uniform distribution ("UNI"). The squared coefficient of variation equals 1/3.

© The Author(s) 2025
A. B. Dieker, S. T. Hackman, *QPLEX: A Computational Modeling and Analysis Methodology for Stochastic Systems*, Springer Series in Operations Research and Financial Engineering, https://doi.org/10.1007/978-3-031-74870-7

- Erlang distribution or, equivalently, the Gamma distribution with first parameter equal to 2 ("ERh"). The squared coefficient of variation equals 1/2.
- Lognormal distribution with squared coefficient of variation equal to 2 ("LN2").
- Hyperexponential distribution with balanced means and squared coefficient of variation equal to 3 ("HE3").

The diagram below displays the chosen Erlang, lognormal, and hyperexponential probability densities.

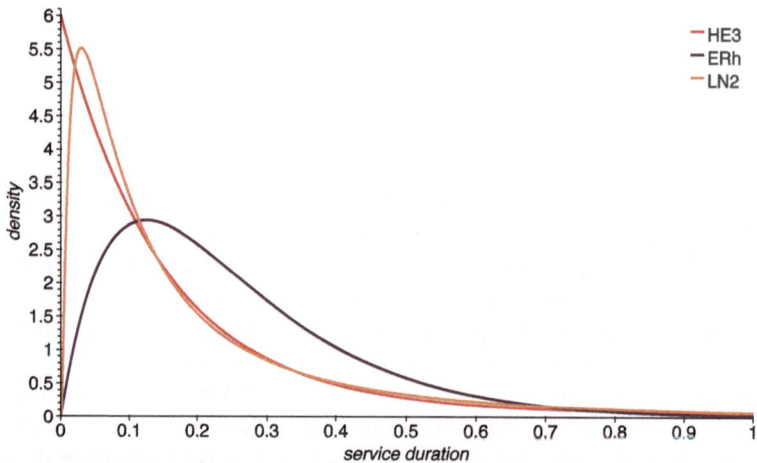

Arrival Rate Functions For each arrival rate function $\{\lambda(\tau) : 0 \leq \tau \leq T\}$, $\lambda(\tau)$ represents a rate per hour. (Here, τ represents continuous time, which is why we do not display it as a superscript.) Each arrival rate function is of the form

$$\lambda(\tau) = \lambda_{\text{avg}} \times s(\tau)$$

in which the scale function $\{s(\tau) : 0 \leq \tau \leq T\}$ has a time-average $\frac{1}{T} \int_0^T s(\tau) d\tau$ equal to 1, and so λ_{avg} is the average arrival rate per hour over the horizon. We use Little's Law to define the average utilization rate as

$$u_{\text{avg}} = \frac{\lambda_{\text{avg}} \times 0.25}{n}. \tag{A.1}$$

Given a target average utilization rate u_{avg}, we set the average arrival rate λ_{avg} via (A.1). We define the "instantaneous utilization" function $\{u(\tau) : 0 \leq \tau \leq T\}$ via

$$u(\tau) = \frac{\lambda(\tau) \times 0.25}{n} = \frac{\lambda_{\text{avg}} \times s(\tau) \times 0.25}{n} = u_{\text{avg}} \times s(\tau).$$

We use three scale functions $s(\tau)$ to reflect increasing levels of "stress" on the system:

- *Quadratic function* ("QUAD"). Its peak over average value is 1.5, and the length of time for which the instantaneous utilization exceeds 1 is 4.08 hours when $u_{avg} = 0.8$ and 5.46 hours when $u_{avg} = 0.95$.
- *Triangular function* ("TRIA"). Its peak over average value is 2.0, and the length of time for which the instantaneous utilization exceeds 1 is 3.75 hours when $u_{avg} = 0.8$ and 4.74 hours when $u_{avg} = 0.95$.
- *Burst function* ("BRST"). This is a step function. Its peak over average value is 5.5, and the length of time for which the instantaneous utilization exceeds 1 is 1.0 hours for both choices of the average utilization.

Our use of parametric service distributions and (piecewise) smooth scale functions is merely for convenience; our methods do not rely on parametric assumptions. Note that methods based on steady-state Markov-chain analysis generally cannot be used for systems where the instantaneous utilization exceeds 1. The diagram below displays the chosen scale functions.

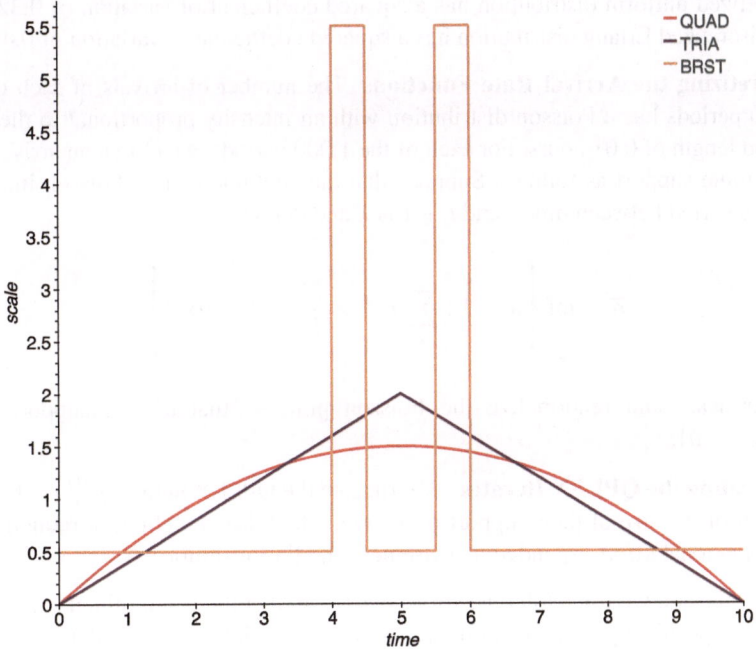

A.2   Discretization and Truncation

Each of the 10 hours is divided into 100 periods of equal length, so that there are 1,000 time periods, with the length of each time period being 0.01 hours. The time epoch t remains an integer as in the rest of this book, even though it typically corresponds to a fractional time epoch in the period of 10 hours. We thus calculate 1,000 (nontrivial) QPLEX iterates $\mu_Q^{(1)}, \ldots, \mu_Q^{(1000)}$ per experiment, and we calculate pmfs of the number of customers in the system and the virtual waiting time from each of these iterates.

Discretizing the Service Duration Distributions For each continuous service duration distribution, we obtain a corresponding service duration pmf g by moving the mass of each interval $(t, t + 1]$ to its right endpoint $t + 1$. For all distributions, we produce finite support pmfs by truncating from the right and renormalizing. We use a truncation tolerance of 10^{-4}.

We search over the parameter space of each service duration distribution with the objective of obtaining the (truncated and discretized) pmf g that has a squared coefficient of variation as close as possible to the desired squared coefficient of variation of the continuous-time distribution subject to the constraint that the mean service duration is 0.25 hours (which corresponds to 25 time periods). This target cannot be achieved exactly for the uniform and Erlang distributions: the discretized uniform distribution has a squared coefficient of variation of 0.32, and the discretized Erlang distribution has a squared coefficient of variation of 0.478.

Discretizing the Arrival Rate Functions The number of arrivals in each of the 1,000 periods has a Poisson distribution with an intensity proportional to the time period length of 0.01 hours. For each of the 1,000 periods, we obtain an arrival pmf with finite support as follows. Suppose that the continuous-time Poisson intensity for the period between time t and $t + 1$ is λ and define

$$\bar{a} = \inf \left\{ a \geq 1 : \sum_{k \leq a} e^{-\lambda} \times \frac{\lambda^k}{k!} \geq 1 - 10^{-8} \right\}.$$

We truncate and renormalize the Poisson pmf, so that $\alpha^{(t)}$ is supported on $\{0, 1, \ldots, \bar{a}\}$.

Truncating the QPLEX Iterates We truncate the QPLEX iterates $\mu_Q^{(t)}$ to slow the growth of the size of their support over time, which has the effect of reducing the computational effort expended on extremely small probabilities.

To this end, we use a tolerance of $\epsilon = 10^{-8}$ so that there is minimal sacrifice to quality. Specifically, after the pmf $\mu_Q^{(t)}$ has been calculated, we calculate lower and upper limits

$$\underline{z}^{(t)} = \inf \left\{ z : \mu_Q^{(t)}(Z \leq z) \geq \epsilon/2 \right\}$$

and

$$\bar{z}^{(t)} = \sup \left\{ z : \mu_Q^{(t)}(Z \geq z) \geq \epsilon/2 \right\},$$

and then replace $\mu_Q^{(t)}(z, \ell)$ with

$$\hat{\mu}_Q^{(t)}(z, \ell) = \frac{\mu_Q^{(t)}(z, \ell) \times 1(\underline{z}^{(t)} \le z \le \overline{z}^{(t)})}{\mu_Q^{(t)}(\underline{z}^{(t)} \le Z \le \overline{z}^{(t)})},$$

which has support included in $[\underline{z}^{(t)}, \overline{z}^{(t)}] \times \{1, 2, \ldots, \ell^{\max}\}$. Here, $\ell^{\max}$ stands for the largest possible remaining service duration, which is the largest integer in the support of the pmf g. Note that this operation amounts to applying a special case of the conditioning map introduced in Sect. 8.1.

A.3   Approximation Quality

The two performance metrics we consider are the number of customers in the system and the virtual waiting time. The pmf of the number of customers in the system can immediately be obtained from each QPLEX iterate by marginalization, but calculating the pmf of the virtual waiting time from a QPLEX iterate is more involved. We will discuss this topic and our definition of virtual waiting time in Appendix A.4.

Fix one of these two performance metrics. Let $F_{\text{QPLEX}}^{(t)}$ and $F_{1M}^{(t)}$ denote, respectively, the cumulative distribution functions of this performance metric at time t generated from the QPLEX iterate and by simulation using one million replications, the latter of which we treat as the true cumulative distribution function $F_\infty^{(t)}$. One measure of the difference between the two distributions of the performance metric is the exceptionally strong sup norm metric given by

$$\epsilon^{(t)} = \sup_{z \ge 0} \left| F_{\text{QPLEX}}^{(t)}(z) - F_{1M}^{(t)}(z) \right|.$$

We refer to this as the "sup norm error," even though $F_{1M}^{(t)}$ itself is a noisy approximation of $F_\infty^{(t)}$. For each experiment and each performance metric, we compute the *maximum* of the 1,000 sup norm errors generated for the horizon of 10 hours.

Sup Norm Errors   The average and maximum sup norm errors over time associated with each performance metric for each of the experiments are provided at the end of this section. Here, we provide a summary.

For the number of customers in the system metric:

- The histogram plot of all 180 average sup norm errors below shows that approximately 95% of all experiments result in an average sup norm error of less than 0.005.

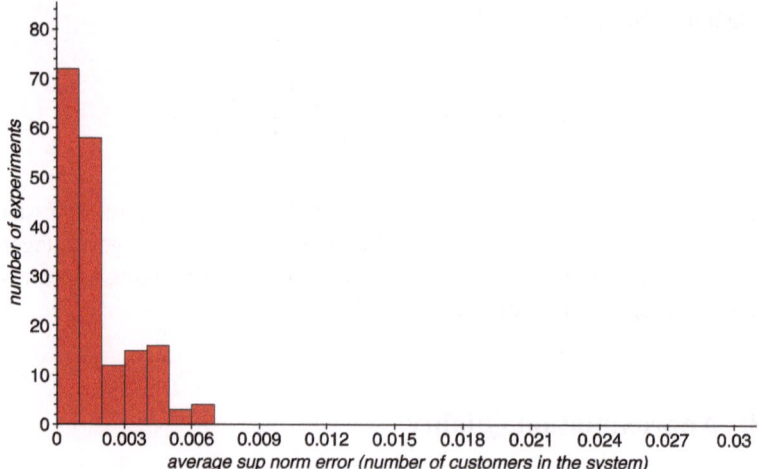

average sup norm error (number of customers in the system)

The largest average sup norm error is 0.00668. It corresponds to the experiment in which the number of servers is 4, the average utilization is 0.8, the service duration is constant, and the arrival rate scale function is quadratic.

- The histogram plot of all 180 maximum sup norm errors below shows that approximately 92% of all experiments result in a maximum sup norm error of less than 0.01.

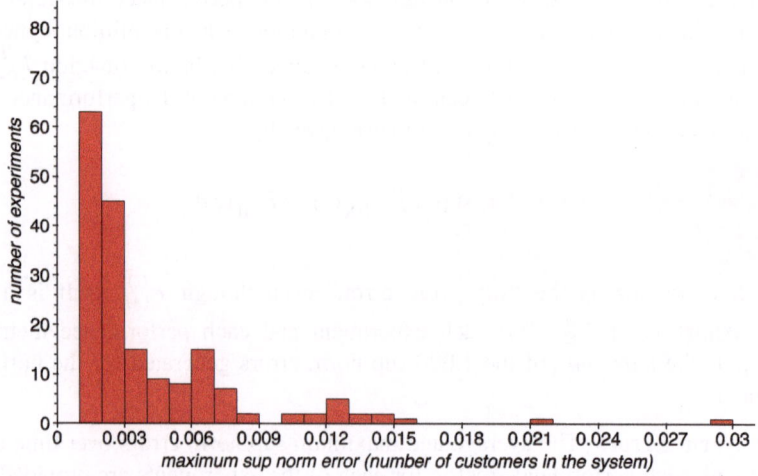

maximum sup norm error (number of customers in the system)

The largest maximum sup norm error is 0.02966. It corresponds to the experiment in which the number of servers is 32, the average utilization is 0.8, the service duration is constant, and the arrival rate scale function is triangular.

For the virtual waiting time metric:

- The histogram plot of all 180 average sup norm errors below shows that approximately 93% of all experiments result in an average sup norm error of less than 0.005.

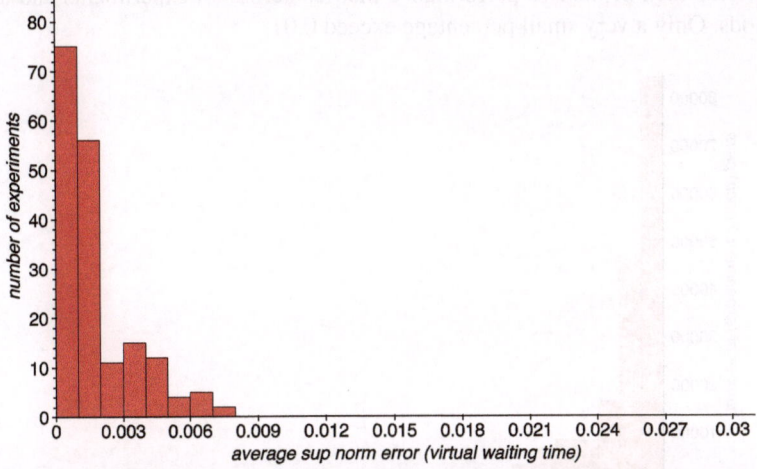

The largest average sup norm error is 0.00791. It corresponds to the experiment in which the number of servers is 4, the average utilization is 0.8, the service duration is constant, and the arrival rate scale function is quadratic.

- The histogram plot of all 180 maximum sup norm errors below shows that approximately 85% of all experiments result in a maximum sup norm error of less than 0.01.

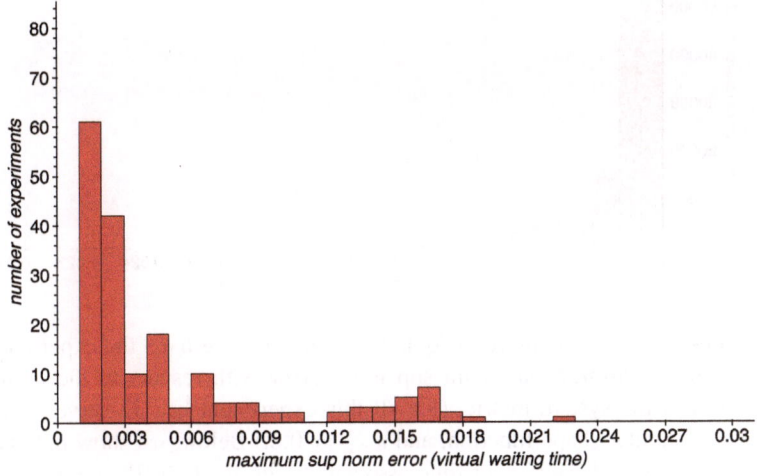

The largest maximum sup norm error is 0.02261. It corresponds to the experiment in which the number of servers is 32, the average utilization is 0.95, the service duration is constant, and the arrival rate scale function is the burst function.

Below are the histograms of all $180 \times 1{,}000 = 180{,}000$ maximum sup norm errors for each of the two performance metrics across all experiments and all time periods. Only a very small percentage exceed 0.01.

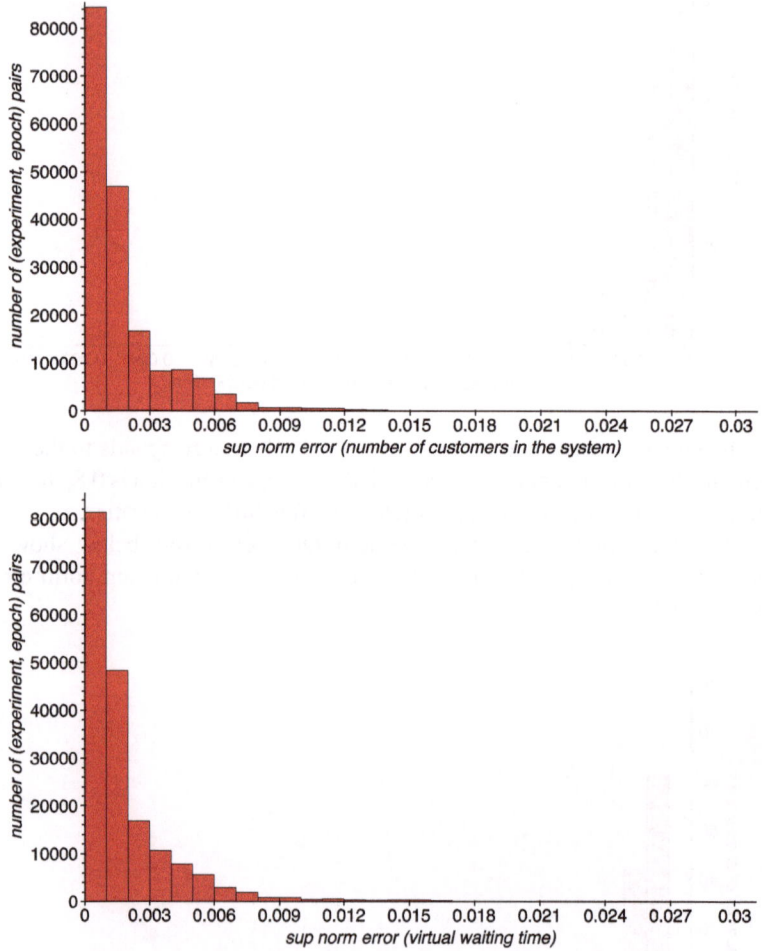

Quantiles To further explore the quality of QPLEX, we fixed the experiment that resulted in the *largest* maximum sup norm error with respect to the number of customers in the system metric. (Recall this experiment had 32 servers, constant service, triangular shape function, and $u_{avg} = 0.8$.) Below, we show the quantiles 0.5, 0.95, 0.99, and 0.995 over time associated with the QPLEX iterate marginals and corresponding simulation output for each performance metric. Recall that there are two curves for each quantile: a pink curve based on the QPLEX methodology and a purple curve based on simulation. The quantiles based on the QPLEX iterates are essentially spot-on. In fact, for each of the four quantiles, the maximum *absolute* deviation is only 1 for the number of customers in the system and 0.01 for the virtual waiting time.

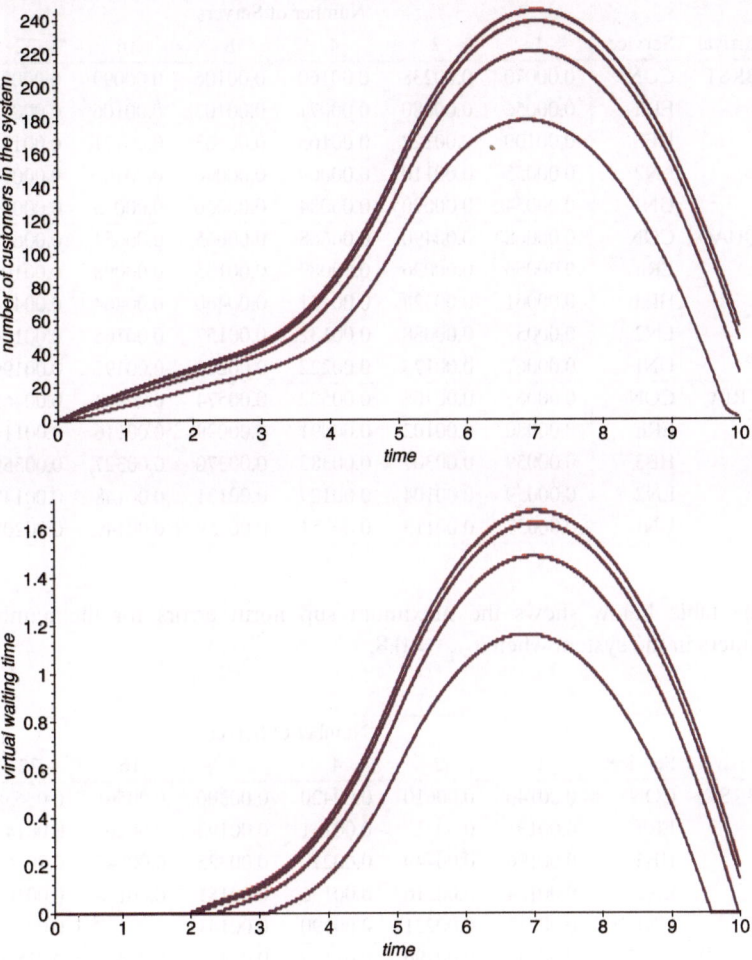

Approximation Quality for Each Experiment   The tables below give the average and maximal sup norm errors for each of the 180 experiments for both the number of customers in the system and the virtual waiting time performance metrics.

The table below shows the average sup norm errors for the number of customers in the system when $u_{avg} = 0.8$.

Arrival	Service	Number of Servers					
		1	2	4	8	16	32
BRST	CON	0.00070	0.00238	0.00160	0.00106	0.00099	0.00095
	ERh	0.00056	0.00080	0.00071	0.00103	0.00106	0.00073
	HE3	0.00100	0.00222	0.00165	0.00163	0.00171	0.00183
	LN2	0.00055	0.00110	0.00069	0.00090	0.00102	0.00096
	UNI	0.00054	0.00090	0.00084	0.00066	0.00075	0.00089
QUAD	CON	0.00068	0.00490	0.00668	0.00665	0.00654	0.00618
	ERh	0.00066	0.00096	0.00089	0.00135	0.00098	0.00104
	HE3	0.00061	0.00328	0.00431	0.00460	0.00464	0.00424
	LN2	0.00057	0.00088	0.00132	0.00157	0.00165	0.00160
	UNI	0.00067	0.00173	0.00222	0.00220	0.00197	0.00199
TRIA	CON	0.00067	0.00405	0.00522	0.00574	0.00494	0.00469
	ERh	0.00080	0.00102	0.00091	0.00099	0.00116	0.00114
	HE3	0.00059	0.00307	0.00382	0.00370	0.00327	0.00359
	LN2	0.00059	0.00104	0.00129	0.00131	0.00148	0.00147
	UNI	0.00057	0.00113	0.00151	0.00159	0.00149	0.00208

The table below shows the maximum sup norm errors for the number of customers in the system when $u_{avg} = 0.8$.

Arrival	Service	Number of Servers					
		1	2	4	8	16	32
BRST	CON	0.00146	0.00610	0.00420	0.00300	0.00561	0.00638
	ERh	0.00131	0.00211	0.00141	0.00193	0.00202	0.00147
	HE3	0.00196	0.00449	0.00279	0.00325	0.00347	0.00367
	LN2	0.00124	0.00216	0.00136	0.00181	0.00178	0.00184
	UNI	0.00212	0.00221	0.00190	0.00141	0.00147	0.00183
QUAD	CON	0.00145	0.00796	0.01179	0.01424	0.01422	0.01586
	ERh	0.00137	0.00207	0.00188	0.00231	0.00271	0.00210
	HE3	0.00159	0.00531	0.00711	0.00756	0.00791	0.00832
	LN2	0.00155	0.00175	0.00213	0.00267	0.00270	0.00265
	UNI	0.00133	0.00319	0.00400	0.00408	0.00505	0.00698
TRIA	CON	0.00142	0.00758	0.01101	0.01323	0.02174	0.02966
	ERh	0.00168	0.00176	0.00178	0.00292	0.00224	0.00230
	HE3	0.00111	0.00504	0.00647	0.00631	0.00606	0.00643
	LN2	0.00142	0.00204	0.00225	0.00203	0.00242	0.00242
	UNI	0.00137	0.00270	0.00396	0.00317	0.00606	0.00864

The table below shows the average sup norm errors for the number of customers in the system when $u_{avg} = 0.95$.

Arrival	Service	Number of Servers					
		1	2	4	8	16	32
BRST	CON	0.00046	0.00269	0.00225	0.00133	0.00097	0.00095
	ERh	0.00055	0.00101	0.00069	0.00063	0.00081	0.00080
	HE3	0.00053	0.00248	0.00210	0.00144	0.00160	0.00151
	LN2	0.00051	0.00097	0.00090	0.00072	0.00084	0.00099
	UNI	0.00067	0.00116	0.00099	0.00094	0.00073	0.00072
QUAD	CON	0.00062	0.00349	0.00498	0.00556	0.00442	0.00417
	ERh	0.00079	0.00079	0.00091	0.00114	0.00083	0.00121
	HE3	0.00056	0.00309	0.00403	0.00367	0.00417	0.00373
	LN2	0.00064	0.00090	0.00137	0.00150	0.00145	0.00164
	UNI	0.00075	0.00094	0.00173	0.00210	0.00174	0.00122
TRIA	CON	0.00052	0.00377	0.00415	0.00480	0.00423	0.00338
	ERh	0.00084	0.00091	0.00091	0.00107	0.00078	0.00101
	HE3	0.00053	0.00284	0.00295	0.00339	0.00350	0.00343
	LN2	0.00048	0.00097	0.00124	0.00134	0.00129	0.00151
	UNI	0.00060	0.00158	0.00154	0.00177	0.00139	0.00140

The table below shows the maximum sup norm errors for the number of customers in the system when $u_{avg} = 0.95$.

Arrival	Service	Number of Servers					
		1	2	4	8	16	32
BRST	CON	0.00139	0.00670	0.00518	0.00291	0.00393	0.00516
	ERh	0.00125	0.00179	0.00169	0.00151	0.00179	0.00160
	HE3	0.00109	0.00512	0.00326	0.00243	0.00312	0.00284
	LN2	0.00131	0.00207	0.00156	0.00166	0.00146	0.00174
	UNI	0.00170	0.00258	0.00231	0.00177	0.00188	0.00173
QUAD	CON	0.00177	0.00680	0.01046	0.01335	0.01232	0.01252
	ERh	0.00198	0.00248	0.00204	0.00232	0.00192	0.00192
	HE3	0.00106	0.00472	0.00641	0.00610	0.00716	0.00661
	LN2	0.00134	0.00143	0.00231	0.00226	0.00222	0.00268
	UNI	0.00161	0.00264	0.00323	0.00341	0.00389	0.00451
TRIA	CON	0.00160	0.00725	0.01038	0.01233	0.01207	0.01216
	ERh	0.00183	0.00176	0.00230	0.00269	0.00191	0.00195
	HE3	0.00124	0.00478	0.00483	0.00590	0.00618	0.00634
	LN2	0.00120	0.00185	0.00230	0.00229	0.00222	0.00278
	UNI	0.00168	0.00288	0.00352	0.00371	0.00395	0.00401

The table below shows the average sup norm errors for the virtual waiting time when $u_{avg} = 0.8$.

Arrival	Service	Number of Servers					
		1	2	4	8	16	32
BRST	CON	0.00085	0.00327	0.00186	0.00103	0.00097	0.00099
	ERh	0.00072	0.00088	0.00064	0.00099	0.00071	0.00047
	HE3	0.00102	0.00207	0.00143	0.00143	0.00151	0.00161
	LN2	0.00064	0.00133	0.00062	0.00070	0.00080	0.00067
	UNI	0.00070	0.00116	0.00090	0.00054	0.00051	0.00055
QUAD	CON	0.00086	0.00643	0.00791	0.00723	0.00650	0.00551
	ERh	0.00073	0.00106	0.00098	0.00131	0.00087	0.00089
	HE3	0.00078	0.00340	0.00447	0.00455	0.00455	0.00405
	LN2	0.00075	0.00087	0.00136	0.00156	0.00163	0.00146
	UNI	0.00086	0.00203	0.00242	0.00227	0.00185	0.00164
TRIA	CON	0.00083	0.00523	0.00612	0.00609	0.00474	0.00396
	ERh	0.00099	0.00114	0.00094	0.00097	0.00096	0.00095
	HE3	0.00076	0.00298	0.00365	0.00356	0.00307	0.00335
	LN2	0.00074	0.00109	0.00146	0.00131	0.00137	0.00131
	UNI	0.00074	0.00140	0.00166	0.00156	0.00134	0.00181

The table below shows the maximum sup norm errors for the virtual waiting time when $u_{avg} = 0.8$.

Arrival	Service	Number of Servers					
		1	2	4	8	16	32
BRST	CON	0.00146	0.00748	0.00821	0.01026	0.01578	0.01770
	ERh	0.00176	0.00157	0.00190	0.00202	0.00153	0.00121
	HE3	0.00177	0.00449	0.00243	0.00306	0.00314	0.00363
	LN2	0.00138	0.00222	0.00119	0.00168	0.00169	0.00149
	UNI	0.00212	0.00288	0.00167	0.00123	0.00131	0.00171
QUAD	CON	0.00156	0.01265	0.01687	0.01826	0.01678	0.01694
	ERh	0.00149	0.00230	0.00192	0.00248	0.00270	0.00239
	HE3	0.00176	0.00566	0.00724	0.00803	0.00827	0.00899
	LN2	0.00186	0.00175	0.00212	0.00257	0.00288	0.00272
	UNI	0.00168	0.00423	0.00454	0.00464	0.00461	0.00469
TRIA	CON	0.00149	0.01313	0.01611	0.01726	0.01601	0.01481
	ERh	0.00176	0.00188	0.00225	0.00292	0.00243	0.00219
	HE3	0.00121	0.00470	0.00666	0.00662	0.00669	0.00650
	LN2	0.00194	0.00176	0.00262	0.00237	0.00260	0.00246
	UNI	0.00137	0.00380	0.00448	0.00377	0.00375	0.00511

The table below shows the average sup norm errors for the virtual waiting time when $u_{avg} = 0.95$.

Arrival	Service	Number of Servers					
		1	2	4	8	16	32
BRST	CON	0.00061	0.00410	0.00279	0.00136	0.00097	0.00093
	ERh	0.00072	0.00096	0.00071	0.00060	0.00057	0.00051
	HE3	0.00069	0.00247	0.00189	0.00126	0.00127	0.00132
	LN2	0.00063	0.00103	0.00074	0.00059	0.00062	0.00081
	UNI	0.00081	0.00158	0.00109	0.00081	0.00045	0.00054
QUAD	CON	0.00079	0.00439	0.00583	0.00606	0.00467	0.00427
	ERh	0.00089	0.00096	0.00097	0.00112	0.00079	0.00105
	HE3	0.00070	0.00300	0.00380	0.00344	0.00383	0.00336
	LN2	0.00074	0.00088	0.00132	0.00150	0.00138	0.00142
	UNI	0.00083	0.00113	0.00187	0.00212	0.00166	0.00116
TRIA	CON	0.00066	0.00476	0.00497	0.00525	0.00446	0.00341
	ERh	0.00097	0.00096	0.00092	0.00091	0.00069	0.00090
	HE3	0.00067	0.00259	0.00270	0.00305	0.00310	0.00310
	LN2	0.00057	0.00105	0.00111	0.00127	0.00110	0.00141
	UNI	0.00074	0.00174	0.00162	0.00179	0.00127	0.00134

The table below shows the maximum sup norm errors for the virtual waiting time when $u_{avg} = 0.95$.

Arrival	Service	Number of Servers					
		1	2	4	8	16	32
BRST	CON	0.00163	0.00999	0.00923	0.01054	0.01359	0.02261
	ERh	0.00133	0.00197	0.00138	0.00109	0.00139	0.00174
	HE3	0.00125	0.00663	0.00368	0.00256	0.00262	0.00288
	LN2	0.00152	0.00236	0.00127	0.00157	0.00141	0.00179
	UNI	0.00140	0.00348	0.00252	0.00167	0.00155	0.00146
QUAD	CON	0.00191	0.01206	0.01546	0.01610	0.01539	0.01451
	ERh	0.00201	0.00250	0.00195	0.00240	0.00220	0.00192
	HE3	0.00122	0.00480	0.00625	0.00617	0.00752	0.00704
	LN2	0.00149	0.00170	0.00242	0.00232	0.00242	0.00282
	UNI	0.00168	0.00380	0.00401	0.00418	0.00468	0.00459
TRIA	CON	0.00194	0.01311	0.01545	0.01637	0.01548	0.01429
	ERh	0.00183	0.00220	0.00298	0.00264	0.00204	0.00226
	HE3	0.00127	0.00420	0.00534	0.00614	0.00655	0.00681
	LN2	0.00120	0.00162	0.00192	0.00249	0.00224	0.00258
	UNI	0.00179	0.00371	0.00430	0.00427	0.00470	0.00443

A.4   Virtual Waiting Time

Definitions We define the virtual waiting time associated with a pair (z, h) of a counter and a histogram as the time until there exists an idle server if the system evolves with arrivals turned off. This virtual waiting time can be deterministic (e.g., if $z < n$ the virtual waiting time is 0, and if $z = n$ it equals the infimum of the support of h) but is random in general. New arrivals can overtake customers represented in h. One can imagine inserting a $(z + 1)$-th "virtual" customer and defining the virtual waiting time as the time until the virtual customer enters service, explaining the terminology. This analogy critically relies on our assumption that the service discipline is First-In, First-Out (FIFO).

Given some pmf v of (Z, H), we define the virtual waiting time map $r^{(t)}$ by letting $r^{(t)}(v)$ be the pmf of the virtual waiting time associated with the random pair (Z, H). We then say that $r^{(t)}(v)$ is the *virtual waiting time pmf associated with* v. We use the symbol W for virtual waiting time, so $r^{(t)}(v)$ is a pmf of W. We similarly define the *virtual cycle time pmf associated with* v as the convolution of $r^{(t)}$ and the service duration pmf g.

Given kernel iterates $v_{\text{ker}}^{(t)}$ and QPLEX iterates $\mu_Q^{(t)}$, the QPLEX approximation of the pmf $r^{(t)}(v_{\text{ker}}^{(t)})$ of the virtual waiting time at time t is

$$(r^{(t)} \circ \mathcal{L}^{(t)})(\mu_Q^{(t)}).$$

If $\mu_Q^{(t)}$ is exact in the sense that $v_{\text{ker}}^{(t)} = \mathcal{L}^{(t)}(\mu_Q^{(t)})$, then the QPLEX approximation of the virtual waiting time pmfs is exact, too. Thus, the exactness results from Chap. 18 also apply to the virtual waiting (and cycle time) pmfs.

Efficient Calculation of Virtual Waiting Time Pmf We show that $(r^{(t)} \circ \mathcal{L}^{(t)})(\mu)$ can be calculated more efficiently than first calculating $\mathcal{L}^{(t)}(\mu)$ and then applying $r^{(t)}$. This more efficient calculation relies on the form of the multinomial lift map $\mathcal{L}^{(t)}$. Specifically, we can assume that $\mu(z)$ is the pmf of the number of entities in the system and that the remaining service durations of the $\min(z, n)$ entities in service are i.i.d. with pmf $\mu(\ell|z)$.

Our derivation uses a pmf p_θ defined via the following variation of the original model. There are n servers, and the service duration pmf is g. Suppose that initially all servers are busy serving customers and that there is an infinite supply of customers in the buffer waiting to be served. There are no external arrivals. The remaining service durations of the customers in service are i.i.d. with some pmf θ. The servers are now identified with a number between 1 and n. For $1 \le k \le n$ and $w \ge 1$, let N_w^k denote the number of service completions up to and including time w from servers 1 through k. For $d \ge 1$, let T_d denote the time of the d-th departure from server 1. We write p_θ for the (joint) pmf of these variables.

Fix some $\mu \in \mathcal{P}_{Z,L}^{(t)}$. The desired complementary cumulative distribution function of the virtual waiting time can be calculated by conditioning on the number of entities in the buffer, noting that the waiting time of an arriving customer who sees $i \geq 0$ entities waiting in the buffer coincides with the time it takes for the system to complete service for $i + 1$ entities. Consequently, we have that

$$[(r^{(t)} \circ \mathcal{L}^{(t)})(\mu)](W > w) = \sum_{i \geq 0} \mu(Z = n+i) \times p_{\mu(L=\cdot|Z=n+i)}(N_w^n \leq i). \quad (A.2)$$

Using this identity, the virtual waiting time pmf can be obtained from calculations of the pmf of N_w^n under p_θ for suitable choices of θ.

We now show how to efficiently calculate the pmf of N_w^n under p_θ. Fix $w \geq 1$. Given the pmf of N_w^1 under p_θ, the pmfs of $N_w^2, \ldots, N_w^n$ under p_θ can be sequentially calculated by noting that the pmf of N_w^k for $k \geq 2$ is a convolution of the pmf of N_w^{k-1} and N_w^1. Consequently, we have that

$$p_\theta(N_w^k = d) = \sum_{d'=0}^{d} p_\theta\left(N_w^1 = d'\right) \times p_\theta\left(N_w^{k-1} = d - d'\right). \quad (A.3)$$

(A more efficient approach reduces the number of k's to calculate by exploiting powers of two.) It remains to show how to calculate the pmf of N_w^1 under p_θ or, equivalently, its cumulative distribution function. For $d \geq 0$, we have that

$$p_\theta(N_w^1 \leq d) = p_\theta(T_{d+1} > w), \quad (A.4)$$

so it remains to show how to calculate the complementary cumulative distribution function of T_{d+1} or, equivalently, the pmf of T_{d+1}. The pmf of T_{d+1} is the convolution of the pmf of T_d and g, i.e., for $d \geq 1$, we have that

$$p_\theta(T_{d+1} = v) = \sum_{u=0}^{v} g(u) \times p_\theta(T_d = v - u), \quad (A.5)$$

whereas the pmf of the *first* departure time under p_θ is given by

$$p_\theta(T_1 = u) = \theta(L = u). \quad (A.6)$$

In sum, the pmf of N_w^n under p_θ can be calculated from Eqs. (A.2)–(A.6) in reverse order.

Extensions The pmfs of the number of arrivals between times t and $t + 1$ may depend on the number of customers in the system or even the QPLEX iterate $\mu_Q^{(t)}$. For example, each arriving customer may independently choose to balk and leave the system with a probability that depends on the number of customers in the system. This "thinning" probability could even be based on the (appropriately defined) conditional virtual waiting time pmf $[(r^{(t)} \circ \mathcal{L}^{(t)})(\mu_Q^{(t)})](W = \cdot|z)$. The actual value of this thinning probability is not prespecified initially. It is *endogenously* determined, as it depends on the effect of *past* thinning probabilities. See Sect. 4.5 for more discussion.

Time-Varying Service Duration Pmfs The above efficient calculation of virtual waiting (cycle) time pmfs critically relies on the service duration pmf not varying over time or the time epoch when service begins. In the case where service duration pmfs vary over time, a discrete-event simulation based on the QPLEX iterates can be used.

Fix $i \geq 0$. First, we obtain an estimate of the pmf of W_i, the virtual waiting time given that $Z = n + i$, to a desired degree of accuracy as follows. Draw n i.i.d. samples from the pmf $\mu(L = \cdot | Z = n+i)$ to obtain the remaining service durations of those entities currently in service, and sample the number of arrivals until the end of the time horizon. Calculate when the first customer(s) in the buffer will begin service, and calculate the number of customers in the system at that time. Sample remaining service duration(s) from the appropriate service duration pmf. Calculate when the next customer in the buffer will begin service, and calculate the number of customers in the system at that time. Draw a remaining service duration sample from the appropriate service duration pmf. Repeatedly continue in this manner to obtain the virtual waiting time along this one sample path. Sampling many such sample paths generates an estimate of the pmf of W_i to the desired degree of accuracy. The pmfs of the W_i so calculated are then weighted by the $\mu(Z = n + i)$ to obtain the pmf of W.

Bibliography

1. S. Allmeier, N. Gast, Mean field and refined mean field approximations for heterogeneous systems: it works! Proc. ACM Meas. Anal. Comput. Syst. **6**(1), 1–43 (2022)
2. D. Bertsimas, D. Gamarnik, *Queueing Theory: Classical and Modern Methods* (Dynamic Ideas, Waltham, 2022)
3. M. Bladt, B.F. Nielsen, *Matrix-Exponential Distributions in Applied Probability* (Springer, New York, 2017)
4. J.R.S. Blair, B. Peyton, An introduction to chordal graphs and clique trees, in *Graph Theory and Sparse Matrix Computation*, ed. by A. George, J.R. Gilbert, J.W.H. Liu (Springer, New York, 1993), pp. 1–29
5. M. Bramson, Y. Lu, B. Prabhakar, Asymptotic independence of queues under randomized load balancing. Queueing Syst. **71**(3), 247–292 (2012)
6. H. Chen, D.D. Yao, *Fundamentals of queueing networks: Performance, Asymptotics, and Optimization* (Springer, New York, 2001)
7. J.G. Dai, J.M. Harrison, *Processing Networks: Fluid Models and Stability* (Cambridge University Press, Cambridge, 2020)
8. J.G. Dai, V. Nguyen, M.I. Reiman, Sequential bottleneck decomposition: an approximation method for generalized Jackson networks. Oper. Res. **42**(1), 119–136 (1994)
9. Z. Feldman, A. Mandelbaum, W.A. Massey, W. Whitt, Staffing of time-varying queues to achieve time-stable performance. Manage. Sci. **54**(2), 324–338 (2008)
10. V. Gupta, M. Harchol-Balter, J.G. Dai, B. Zwart, On the inapproximability of $M/G/K$: why two moments of job size distribution are not enough. Queueing Syst. **64**(1), 5–48 (2010)
11. M. Harchol-Balter, *Performance Modeling and Design of Computer Systems: Queueing Theory in Action* (Cambridge University Press, Cambridge, 2013)
12. Q.M. He, *Fundamentals of Matrix-Analytic Methods* (Springer, New York, 2014)
13. F.P. Kelly, *Reversibility and Stochastic Networks* (Cambridge University Press, Cambridge, 2011)
14. D. Koller, N. Friedman, *Probabilistic Graphical Models: Principles and Techniques* (MIT Press, Cambridge, 2009)
15. P. Kuehn, Approximate analysis of general queuing networks by decomposition. IEEE Trans. Commun. **27**(1), 113–126 (1979)
16. S.L. Lauritzen, *Graphical Models* (Oxford University Press, Oxford, 1996)
17. B.N.M. Ma, J.W. Mark, Approximation of the mean queue length of an $M/G/c$ queueing system. Oper. Res. **43**(1), 158–165 (1995)

© The Author(s) 2025

351

A. B. Dieker, S. T. Hackman, *QPLEX: A Computational Modeling and Analysis Methodology for Stochastic Systems*, Springer Series in Operations Research and Financial Engineering, https://doi.org/10.1007/978-3-031-74870-7

18. H.P. McKean Jr., A class of Markov processes associated with nonlinear parabolic equations. Proc. Natl. Acad. Sci. USA **56**(6), 1907–1911 (1966)
19. K.P. Murphy, *Probabilistic Machine Learning: Advanced Topics* (MIT Press, Cambridge, 2023)
20. B.L. Nelson, *Foundations and Methods of Stochastic Simulation: A First Course* (Springer, New York, 2013)
21. B.L. Nelson, 'Some tactical problems in digital simulation' for the next 10 years. J. Simul. **10**(1), 2–11 (2016)
22. H. Ouyang, B.L. Nelson, Simulation-based predictive analytics for dynamic queueing systems, in *Proceedings of the 2017 Winter Simulation Conference (WSC)* (2017), pp. 1716–1727
23. R. Serfozo, *Introduction to Stochastic Networks* (Springer, New York, 1999)
24. J.S. Smith, B.L. Nelson, Estimating and interpreting the waiting time for customers arriving to a non-stationary queueing system, in *Proceedings of the 2015 Winter Simulation Conference (WSC)* (2015), pp. 2610–2621
25. R.E. Tarjan, M. Yannakakis, Simple linear-time algorithms to test chordality of graphs, test acyclicity of hypergraphs, and selectively reduce acyclic hypergraphs. SIAM J. Comput. **13**(3), 566–579 (1984)
26. H.C. Tijms, M.H. van Hoorn, A. Federgruen, Approximations for the steady-state probabilities in the $M/G/c$ queue. Adv. Appl. Probab. **13**(1), 186–206 (1981)
27. G. Weiss, *Scheduling and Control of Queueing Networks* (Cambridge University Press, Cambridge, 2021)
28. W. Whitt, The queueing network analyzer. Bell Syst. Tech. J. **62**(9), 2779–2815 (1983)
29. W. Whitt, Approximations for the $GI/G/m$ queue. Prod. Oper. Manage. **2**(2), 114–161 (1993)
30. W. Whitt, *Stochastic-Process Limits: An Introduction to Stochastic-Process Limits and Their Application to Queues* (Springer, New York, 2002)
31. W. Whitt, Time-varying queues. Queueing Models Serv. Manage. **1**(2), 79–164 (2018)
32. W. Whitt, W. You, A robust queueing network analyzer based on indices of dispersion. Naval Res. Logist. **69**(1), 36–56 (2022)
33. D.D. Yao, Refining the diffusion approximation for the $M/G/m$ queue. Oper. Res. **33**(6), 1266–1277 (1985)

Index

A
Activity, 42
Age, 43

B
Bayesian model selection, 126

C
Category, 49, 86
Chord, 180
Clique intersection property, 184
Counter, 42
  set, 42
  variable, 147
  vector, 42
Curse of dimensionality
  for counters, 147
  for histograms, 20, 46

D
Dead variable, 218
Decomposition approach, 149
Distributional program, 8
  efficient refactored, 212
  equivalence of, 9
  refactoring, 9
  structured, 235

E
Elapsed activity duration, 43

Entity
Entity, 42
Entity type, 62
  type of customer, 27

F
Factor, 235

G
G-information collection, 150
Global Markov property, 152
Graph, 12
  chordal (*see* Graph, triangulated)
  path, 178
  recursively simplicial, 181
  triangulated, 180
  weighted clique intersection, 182

H
Histogram, 5
  category assignment, 86
  current label category, 94
  entity-type flow, 87
  join, 56
  marginal, 94
  next label type, 94
  relabel, 55
  route, 97
  size of, 5
  stay, 55

© The Editor(s) (if applicable) and The Author(s) 2025
A. B. Dieker, S. T. Hackman, *QPLEX: A Computational Modeling and Analysis Methodology for Stochastic Systems*, Springer Series in Operations Research and Financial Engineering, https://doi.org/10.1007/978-3-031-74870-7